The Environmental Science of Drinking Water

The Environmental Science of Drinking Water

Patrick J. Sullivan
Franklin J. Agardy
James J. J. Clark

ELSEVIER
BUTTERWORTH
HEINEMANN

AMSTERDAM · BOSTON · HEIDELBERG · LONDON
NEW YORK · OXFORD · PARIS · SAN DIEGO
SAN FRANCISCO · SINGAPORE · SYDNEY · TOKYO

Elsevier Butterworth–Heinemann
30 Corporate Drive, Suite 400, Burlington, MA 01803, USA
Linacre House, Jordan Hill, Oxford OX2 8DP, UK

∞ Recognizing the importance of preserving what has been written, Elsevier
prints its books on acid-free paper whenever possible.

Library of Congress Cataloging-in-Publication Data

Sullivan, Patrick J., Ph.D.
  The environmental science of drinking water / Patrick J. Sullivan,
Franklin J. Agardy, James J.J. Clark.– 1st ed.
    p. cm.
  Includes bibliographical references and index.
  ISBN-13: 978-0-7506-7876-6 ISBN-10: 0-7506-7876-3 (alk. paper)
  1. Water-supply. 2. Drinking water. 3. Water–Pollution. 4. Water quality
management. I. Agardy, Franklin J. II. Clark, James J. J. III. Title.
  TD345.S77 2005
  628.1–dc22
ISBN-13: 978-0-7506-7876-6   2005009795
ISBN-10: 0-7506-7876-3

British Library Cataloguing-in-Publication Data
A catalogue record for this book is available from the British Library.

For information on all Elsevier Butterworth–Heinemann publications
visit our Web site at www.books.elsevier.com

Printed in the United States of America
06 07 08 09 10   10 9 8 7 6 5 4 3 2

Contents

Foreword

In an era of hyperbole and hubris, one hesitates to use current media terms like "sea-change" (despite its currency, not a new term—it was coined by Shakespeare almost 400 years ago) to describe the potential impact of any new book addressing important issues of broad public interest. But Patrick Sullivan and his coauthors, drawing on over 60 years of experience in the field of water quality science and engineering, have written such a book, an incisive and compelling manifesto on the present dangers inherent in the way our nation's drinking water supplies are being managed today to protect their quality.

Their thesis is that the current approach to water quality management, as codified through USEPA Water Quality Standards, is not sustainable, and cannot achieve its avowed goals because of an intractable complexity, not only of the cumbersome regulatory process through which drinking water standards are created for pollutant chemicals, but also the complexity that now attends the chemical nature of water pollution itself. As they point out, with at least 2000 new chemicals coming into production each year, with our nation already having 40% of its waters out of compliance with quality standards and facing water distribution infrastructure upgrades that will cost about 10% of its annual GDP, how can USEPA possibly assure the future availability of potable drinking water to everyone? And this dilemma does not even consider the many, many unregulated but potentially harmful chemicals now in use for which no water quality standards yet exist.

Everyone cares about the quality of drinking water, but not everyone has the educational or professional background to understand, without some guidance, why their thesis is a sound one. This book has been written to assist the motivated citizen, as well as the interested student, to achieve a solid working knowledge of how water quality is managed and why current approaches to water quality management are not sustainable, ending with a practical blueprint for how to improve drinking water quality, not by confronting head-on its overwhelming complexity as a technological problem, but instead by adapting to this complexity with sensible and feasible changes in both public policy and professional practice.

The first three chapters of this book offer a Cook's Tour of the science, technology, and policy underlying the chemical pollution and remediation of water supplies used for drinking. They can be read by environmental science and engineering students as a reliable guide to the technical issues involved. But, for the citizen-reader, I would suggest a different strategy: Begin with Chapters 4 and 5 instead. These two chapters offer a Grand Tour of the issues our nation faces in the water quality arena. They are both authoritative and wise, and they will leave the reader with the clear notion that the twenty-first century cannot be a simple technological and political tweaking of the previous one when it comes to protecting our drinking water. Something bolder—and more sustainable—than regulations and risk assessment is needed, and these chapters give a working outline of what that something might be. If an unfamiliar technical term, policy, or professional practice is mentioned, it can be looked up by consulting an appropriate page in the first three chapters. In the end, the reader will come to know that the solution to pollution is not dilution, but instead is a community-based plan of action rooted in that most American of civic values, personal responsibility.

Garrison Sposito

Department of Civil and Environmental Engineering and
Department of Environmental Science, Policy and Management,
University of California at Berkeley

Preface

When there is no boundary between wastewater and drinking water, biological pollution causes the spread of disease and death. This is a common condition within developing nations of the world where little or no clean water is readily available to the population. This situation is unfortunate, as the science and engineering necessary to provide clean water are well known. The boundary between wastewater and drinking water in the industrial nations of the world, for the most part, excludes biological threats. But as populations grow and chemical technology expands, the boundary between chemically polluted wastewater and drinking water is rapidly disappearing.

Because understanding chemical threats to humans is still more art than science, this book focuses on providing a basis for appreciating the threat posed by manufactured chemicals in water resources and the solutions available for minimizing the potential health risk associated with human abuse of this natural product. This focus is well justified given the history of global industrial development.

For example, before the 1900s, the dominant chemical pollutants were inorganic and coal-derived organic chemicals. As chemical manufacturing blossomed during the 1940s, a wide variety of synthetic petroleum-based chemicals were produced and released into the environment. Today in the United States, chemical manufacturers produce approximately 87,000 different chemicals. Current estimates predict the manufacture of at least 2,000 new chemical compounds each year. With so many chemicals being produced and used, it is no wonder that a significant portion of the nation's lakes, rivers, and groundwater contain a wide range of industrial inorganic and organic compounds, including pesticides and pharmaceuticals. These same water resources serve as the source for community drinking water systems and private groundwater wells. Today what is being evidenced in the United States is also recognized worldwide.

In 1925, the U.S. Public Health Service's Primary Drinking Water Standards required that the concentration of three trace metals not exceed certain limits in drinking water. This requirement can be considered to have been the first step in controlling chemical pollution. By 1962, nine trace metals and inorganic compounds were included in the drinking water standards. Today, in compliance with the 1974 Safe Drinking Water Act and its amendments, the Primary Drinking Water Standards still only limits the concentration of a very small number of inorganic chemicals, industrial organic compounds, and organic pesticides. These regulated compounds represent the major sources of pollution from chemical usage between the 1930s and the 1980s. Since then, the number of industrial chemicals, pesticides, and pharmaceuticals that have been produced and are now found in our waters has increased dramatically. Thus, many of the chemical pollutants found in our waters today are not regulated by current standards.

In 1998, the U.S. Environmental Protection Agency (USEPA) proposed adding 50 chemicals to the current list of Primary Drinking Water Standards. Yet, pharmaceuticals do not even appear on this list of 50. Given the number of chemicals being used in the United States today and the apparent inability of drinking water standards to rapidly adapt to these changes, it is clear that standards, as currently implemented, are incapable of keeping pace with the unregulated chemicals that are currently found in our drinking water. Furthermore, current methods of evaluating chemical health risks do not even address the known problem of consuming water containing chemical mixtures.

Approximately 50 years ago, the National Agricultural Chemicals Association argued that pesticides in low concentrations would have no effect on wildlife and that cancer, which had afflicted human beings for centuries, could not be shown to be caused by pesticides. After decades of research, these opinions have both been proven to be wrong. Given this history, why should consumers today believe that the ingestion of low levels of chemical mixtures are not a threat to human health, without sound scientific evidence to substantiate the premise?

Environmental science evaluates how humans use the earth's natural resources (in this case water resources), appraises the repercussions that occur as the result of this use, and evaluates how to mitigate these impacts. Therefore, it is the objective of this book, by focusing on the environmental science of drinking water, to (1) provide an introduction to the chemistry of the water we drink; (2) describe how water resources are polluted; (3) describe and evaluate how drinking water is protected; (4) assess the health risks associated with chemical pollution; and (5) evaluate and recommend approaches to managing chemically polluted drinking water.

Acknowledgments

We thank Mr. Jerome Gilbert, Past President of the American Water Works Association and former Executive Director of the California State Water Resources Control Board, for his constructive review of this text and Dr. Garrison Sposito, Department of Environmental Science, Policy and Management at UC Berkeley, for his support and guidance. We owe a sincere debt of gratitude to Paula Massoni for her time and energy in editing the text and Sterling Blanche for his contributions to data collection and analysis.

CHAPTER 1

The Water We Drink

"The boundaries between water and wastewater are already beginning to fade."
The American Water Works Association (2001)

In a landscape dominated and modified by human activity, it should not be surprising that the water we drink contains both chemical and biological pollutants. After all, the world is dependent on the ever expanding chemical wonders of our agricultural/industrial/pharmaceutical-based society that helps support a growing population. An increasing population in turn pollutes water resources with chemicals and biohazards from the production and use of consumer products, agricultural and animal production, and human waste disposal. This ever increasing spiral of population and pollution means that naturally pure sources of drinking water are almost extinct.

With our technological advances and increased population, there is a price we all pay. This price is drinking water that contains a mixture of manufactured chemicals or biohazards that are a potential threat to human health. This threat can be immediate when an individual is exposed to a water-borne disease, or long-term when drinking chemical pollutants. For those living in industrialized nations (e.g., United States, Canada, European Union), the threat from water-borne biohazards has been significantly reduced or eliminated. However, widespread exposure to specific chemicals or chemical mixtures in drinking water still remains. This condition illustrates the necessity to understand the environmental science of our drinking water if this expanding problem is to be managed or reduced.

Environmental science evaluates how humans use the earth's natural resources (in this case, water resources), appraises the repercussions that occur as the result of this use, and evaluates how to mitigate these impacts. Therefore, the objective of this book, by focusing on the environmental science of drinking water, is to provide a basis for understanding the threat posed by manufactured chemicals and biohazards in water resources and the solutions available for minimizing the potential health risk associated with our abuse of this natural product.

Natural Water[1]

Before human activity on this earth, the chemistry of water was initially influenced by the dissolution of minerals from soil, rock, biosynthesis, and biodegradation of organic matter. The chemical compounds that dissolve from minerals, biosynthesis, and biodegradation represent natural or background levels in the water we drink. In some cases natural water can contain elevated concentrations of trace elements (e.g., arsenic, copper, fluorine, lead, zinc) that are known to be detrimental to human health. Because many natural waters (i.e., advertised to contain little or no manufactured pollutants) from around the world are bottled and sold as a very expensive healthy alternative to tap water, it is important to at least understand the basic chemistry of natural water.

The hydrologic cycle is the process that has the greatest influence on the chemistry of natural water. For example, when precipitation falls on the land, it follows one of many paths that constitute what is known as the hydrologic cycle (Figure 1-1). As water runs off the earth's surface, it can infiltrate into the soil,[2] as well as form into bodies of surface water such as ponds, lakes, and rivers. Much of this surface water will evaporate back into the air, which can reform as clouds and subsequently result in precipitation, or it can continue to percolate into the soil. Once water moves into the soil, it can be removed by plants and evaporated back to the atmosphere (transpiration), or it can continue to move downward to form water-saturated zones in soil and fractured bedrock. These saturated zones are called aquifers. In aquifers, groundwater can find its way back onto the land as a spring, a natural artesian well, or by seeping into ponds, lakes, and rivers. Thus, water is naturally cycled from the earth's surface into groundwater and back again. This cycle may be only hours long or may occur over centuries. While groundwater remains in an aquifer (i.e., it is not being lost to the surface through a spring or being pumped from a well), it is essentially a stored water resource that can be tapped by a well. Based on this cycle, water resources are usually classified as being either surface water or groundwater. This is an important difference, as the chemistry of each resource is unique to its origin.

---

[1]Water that is free of manufactured pollutants.
[2]Unconsolidated porous media from the earth's surface to bedrock (e.g., earth that can be moved with a bulldozer without having to use explosives).

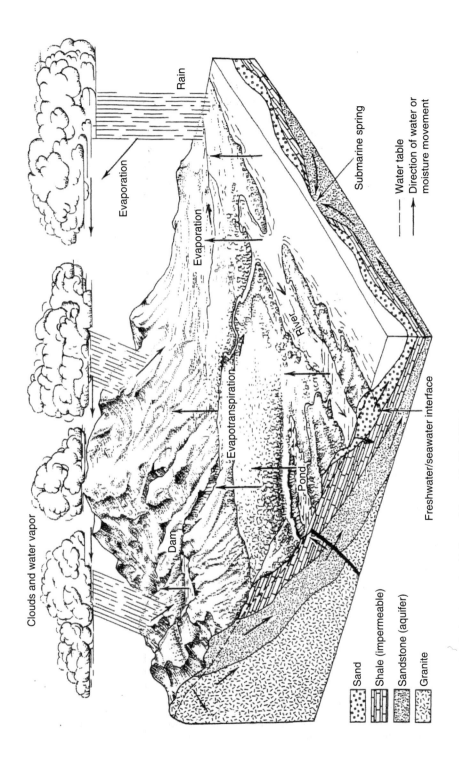

Clouds and water vapor

Rain

Evaporation

Evaporation

Evapotranspiration

Dam

Pond

River

Submarine spring

Water table

Direction of water or
moisture movement

Freshwater/seawater interface

Sand

Shale (impermeable)

Sandstone (aquifer)

Granite

FIGURE 1-1. Hydrologic cycle. (Adapted from Todd, 1980.)

Contact with Soil and Rock

Because water that exists at and below the earth's surface is in contact with soil and rock, some mineral or organic matter will be dissolved into the water. For the most part, the chemical elements that will be dissolved in water are generally preordained by their abundance. For example, the average abundance of the most common chemical elements in the earth's crust (in decreasing order of abundance, see Appendix 1-1) are oxygen (O), silica (Si), aluminum (Al), iron (Fe), calcium (Ca), sodium (Na), magnesium (Mg), and potassium (K). These first eight elements are the building blocks of the most common minerals that make up the earth's crust. The next 20 most common elements are titanium (Ti), hydrogen (H), phosphors (P), manganese (Mn), fluorine (F), barium (Ba), strontium (Sr), sulfur (S), carbon (C), zirconium (Zr), vanadium (V), and chlorine (Cl). All combined, these 28 elements make up 99.93 percent of the earth's crust.

Therefore, it would be anticipated that natural water would contain some or most of these dissolved elements. Based on the natural properties of water (H_2O) and the properties of each chemical element and mineral combination, the actual occurrence and concentration of an element in water can vary widely. When minerals dissolve in water, the chemical elements are usually ionized (i.e., they form a charged chemical species called an ion). These ions occur as either cations (a positively charged ion) or as anions (a negatively charged ion). The major cations and anions that occur in both surface water and groundwater are given in Table 1-1. These common ions usually occur in water at levels measured in the part-per-million[3] range.

Another crucial chemical property of water is its relative acidity or alkalinity. This chemical characteristic has a direct influence on the concentration of minor elements and trace elements[4] that can occur in

TABLE 1-1
Major Cations and Anions Found in Natural Water

Cations	Anions
Na+ (Sodium)	HCO_3^- (Bicarbonate)
K+ (Potassium)	SO_4^{2-} (Sulfate)
Ca^{2+} (Calcium)	Cl$^-$ (Chloride)
Mg^{2+} (Magnesium)	SiO_4^{4-} (Silicate and as aqueous SiO_2)

---

[3]One part-per-million (ppm) is equivalent to 1 milligram (mg) of any chemical element dissolved in 1,000 mg of water. Since 1 milligram of water has a volume of 1 milliliter and 1,000 milliliters equal a liter (L), the concentration of dissolved chemical elements in water is usually expressed as milligrams per liter or mg/L.
[4]Minor elements usually occur in natural water at concentrations less than 1 ppm but greater than 1 part-per-billion (ppb) or 1 microgram per liter (fg/L), whereas a trace element is usually less than 1 ppb.

natural water. For example, acid water will generally tend to have more dissolved trace elements at higher concentrations than alkaline water. Therefore, it is important that the relative acidity or alkalinity be measured by determining the pH. The pH of water is the negative logarithm of the hydrogen ion concentration. Although this is the exact definition, it is more important to understand its meaning in everyday use. Therefore, it is necessary to define the common terms associated with this measured value.

Since the 17th century, acids have been described as substances with a sour taste and the ability to dissolve many different substances. Recently, the simplest chemical definition of an acid was proposed by Arrhenius[5] to be a substance containing hydrogen, which, upon its dissolution in water, gives off hydrogen ions (H^+) into solution. This is illustrated by Equation 1-1, which represents sulfuric acid dissolved in water.

$$H_2SO_4 = 2H^+ + SO_4^{2-} \tag{1-1}$$

In this case, one molecule of sulfuric acid yields two hydrogen ions and one sulfate ion.

Bases have been described as substances that when dissolved in water feel soapy or slippery (that is because your skin is being dissolved), have a bitter taste, and can neutralize acids. According to Arrhenius, a base is a substance that gives free hydroxide ions (OH^-) when dissolved in water. For example, the ionization of calcium hydroxide would be represented by Equation 1-2.

$$Ca(OH)_2 = Ca^{2+} + 2OH^- \tag{1-2}$$

In this case, one molecule of calcium hydroxide yields one calcium ion and two hydroxide ions.

If an acid and base are mixed in equal proportion, the hydrogen ion and hydroxide ion will combine to form water so that there is no hydrogen or hydroxide ions dissolved in water. When such a reaction occurs the water is not acid or alkaline but neutral. For example, a neutralization reaction between sulfuric acid and calcium hydroxide is represented by Equation 1-3.

$$2H^+ + SO_4^{2-} + Ca^{2+} + 2OH^- = CaSO_4 + 2H_2O \tag{1-3}$$

In this reaction, the hydrogen and hydroxide ions form water, and the calcium and sulfate ions form a solid precipitate of calcium sulfate (i.e., a white mineral called gypsum). If the pH were measured after the

---

[5]Dr. Svante Arrhenius received the 1903 Nobel Prize in chemistry for the development of ionic theory, which he applied to the definition of acids and bases.

reaction was compete, the pH would be 7. If more hydrogen ions were added in excess of the available hydroxide ions for the neutralization reaction, the water would be acidic and the pH would measure less than 7 (i.e., the more acid the water the lower the pH). If the opposite were true and more hydroxide ions were added in excess of the available hydrogen ions for the neutralization reaction, the water would be alkaline and the pH would be greater than 7 (i.e., the more alkaline the water the higher the pH). These pH relationships are illustrated in Figure 1-2.

One of the most widespread and common mineral reactions that will influence the pH of natural water is the dissolution of carbonate rocks or sediment cemented by carbonates. For example, when limestone (primarily composed of calcium carbonate) and dolomite (dominated by a calcium-magnesium carbonate) contact water, some portion of these rocks will be dissolved. When calcite (a calcium carbonate mineral) comes in contact with water, some portion of that mineral will be dissolved, as shown in Equation 1-4.

$$CaCO_3 + H_2O = Ca^{2+} + HCO_3^- + OH^- \qquad (1\text{-}4)$$

In this reaction, one molecule of calcium carbonate dissolves in water to yield one calcium ion, one bicarbonate ion, and one hydroxide ion. Water in contact with carbonate rocks will usually have a pH around 8.5.

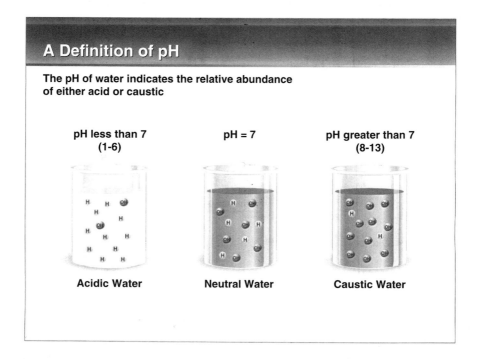

FIGURE 1-2. pH.

Other common silicate minerals that occur in soil and rock (e.g., albite, orthoclase, hornblende, augite, and biotite) will also react with water. This reaction will add metals such as Ca, Mg, Na, and K and hydroxide back into the water. An example of this reaction is shown in Equation 1-5.

$$\text{Metal}^+[\text{Silicate}]_{\text{SOLID}} + \text{H}^+_{\text{IN WATER}} = $$
$$\text{H}^+[\text{Silicate}]_{\text{SOLID}} + \text{Metal}^+_{\text{IN WATER}} \qquad (1\text{-}5)$$

The reaction of water in contact with both carbonate and silicate minerals in rock and soil tend to produce natural water that is alkaline. Because these minerals make up a large portion of the earth's crust, natural waters at and below the earth's surface tend to be alkaline. The one major exception occurs when iron sulfide minerals come into contact with water and oxygen. In some geologic environments, an abundance of iron sulfide minerals occurs in coal, oil shale, and economic ore deposits (e.g., gold, silver, zinc, copper, and lead deposits). When iron sulfide minerals in a coal or ore deposit react with water and oxygen, sulfuric acid will be produced. Therefore, water from coal and ore deposits can be acidic (i.e., have a pH less than 5). These waters also tend to contain elevated concentrations of aluminum, arsenic, cadmium, cobalt, copper, iron, lead, nickel, and zinc.

The pH of natural water is not only influenced by chemical reactions with minerals in soil and rock but also by the atmosphere. Water in contact with the atmosphere (i.e., surface water and shallow groundwater) will absorb carbon dioxide from the air. The resulting chemical reactions are given in Equations 1-6 and 1-7.

$$CO_2 \,[\text{carbon dioxide}] + H_2O = H_2CO_3 \,[\text{carbonic acid}] \qquad (1\text{-}6)$$

$$H_2CO_3 = H^+ + HCO_3^- \qquad (1\text{-}7)$$

The resulting acid-forming reaction causes natural rainwater to be slightly acidic (approximate pH of 5.7). However, as previously discussed, when this slightly acid water contacts most soil and rock the acidity can be neutralized. This means that because of all the potential chemical reactions that can occur, the pH of natural water usually ranges from 6 to 9.

As a consequence of elemental abundances, mineral variations and reactions of water within the environment, the actual chemistry of natural water can be highly variable. In the United States, one of the first summaries of this variability in regional water quality for both surface water and groundwater was compiled by Clarke, in 1908, from water samples taken from approximately 1880 through 1905. The data taken from Clark (1908) and summarized in Table 1-2 can be considered to generally represent the natural mineral content of river water before the explosive growth of the chemical industry in the early 1900s. The data in Table 1-2 are representative of river waters in the United States and around the world. These data show that, for the most part, river water is dominated by calcium and bicarbonate. This result is consistent with rock-weathering trends (i.e., Fe, Al, and Si tend to occur in less water-soluble minerals), the influence of carbon dioxide in the atmosphere on water chemistry, and the common occurrence of calcium carbonate minerals in rock and soil. These data generally show that metal cation concentrations have the sequence Ca > Na > Mg > K > Fe and Al, while the anion concentration sequence is HCO_3 > SO_4 > Cl.

Clark (1908) also provided data on the chemistry of groundwater, but focused on mineral water from springs and wells, as they have "the greatest commercial importance." According to Clark (1908), "all springs are mineral springs, for all contain mineral impurities; but in a popular sense the term is restricted to waters of abnormal or unusual composition." The analyses provided by Clark (1908) are grouped by the dominant anion (i.e., chloride, sulfate, bicarbonate). Representative mineral waters are given in Table 1-3, Chloride Dominated, Table 1-4, Sulfate Dominated, Table 1-5, Bicarbonate Dominated, and Table 1-6, Mixed Anion Dominated. These data demonstrate that there can be fairly significant variations in the number and concentration of chemical constituents in water around the world. Yet the chemical similarities of water should not be surprising given their contact with the earth's crust.

The chemical data reported by Langmuir (1997) (Table 1-7) also show that surface water and groundwater are dominated by bicarbonate and calcium. In other words, with all their variability, natural water can be anticipated to contain the major chemical constituents shown in Table 1-7. The only unknowns are their actual concentrations. As shown by Clark (1908), some water can contain elevated concentrations of dissolved minerals. In those special cases, where chemical constituents dissolved in *groundwater* exceed 250 mg/L, this groundwater is defined as a *mineral water*.

Rocks and Radioactivity

Some naturally occurring chemical elements are also radioactive and occur throughout the environment. Those natural radioactive elements of concern that are known to affect water quality are radium, radon, and uranium. All three of these elements emit alpha particles. Alpha particle decay radioactivity occurs when a parent nucleus (e.g., radium) expels an alpha particle[6]. The amount of alpha particle decay is defined in terms of the number of decays per unit of time. In water this decay process is measured in units of alpha radiation or in picocuries per liter (pCi/L). A curie is the number of alpha particles emitted by 1 gram of radium in one second.

---

[6]An alpha particle is equivalent to a helium nucleus (i.e., two protons and two neutrons).

TABLE 1-2
National and International Data on the Mineral Content of River Water

River	ppm									
	HCO_3	SO_4	Cl	Ca	Mg	Na	K	SiO_2	Fe_2O_3	Al_2O_3
St. Lawrence, Montreal	44.43	11.17	2.41	20.67	6.44	4.87	nr	10.01	nr	nr
Genesse, Rochester, NY	37.94	25.29	1.47	24.48	5.29	2.59	1.35	0.82	0.83*	nr
Merrimac, Concord, NH	28.15	12.78	8.78	17.14	4.18	6.16	trace	18.14	3.33	1.34
St. James, Richmond, VA	42.52	5.26	1.51	18.49	5.44	3.52	3.58	14.74	0.96	0.58
Mississippi, Minneapolis, MN	47.04	9.61	0.85	20.59	7.67	5.33*		8.01	0.05	nr
Mississippi, New Orleans, LA	34.74	14.90	6.23	20.42	5.21	4.92	4.65	6.77	0.15	0.44
Kentucky, Frankfort, KY	39.06	7.32	1.42	22.62	4.06	5.71*		15.32	0.94	nr
Missouri, Great Falls, MT	27.10	25.45	7.62	8.93	5.48	17.13*		8.29	nr	nr
Kansas, Lawrence, KS	23.83	18.15	18.80	14.76	3.51	15.45	nr	4.83	0.67*	nr
Arkansas, Little Rock, AR	10.80	12.61	38.55	7.60	1.67	25.92	0.74	1.81	0.06	0.24
Brazos, Waco, TX	8.83	20.74	33.71	9.54	2.04	23.27*		1.69	0.03	nr
Salt, Mesa, AZ	9.61	8.29	41.56	7.15	2.69	26.38	1.38	2.94	nr	nr
Sacramento, Sacramento, CA	27.36	16.24	7.18	12.34	5.83	10.05	1.26	16.69	3.05*	nr
Yukon, Eagle, AL	46.16	10.75	0.41	22.21	4.71	6.14	trace	7.78	nr	1.48
Elbow, Calgary, Canada	44.66	18.80	0.56	24.39	6.55	2.77	0.42	1.85	trace	nr
Plata, Buenos Aires, Brazil	11.59	17.97	18.11	3.71	1.42	24.89	nr	10.82	4.81*	nr
Rio de Arias, Salto, Argentina	39.13	13.24	2.77	19.63	5.20	1.82	5.75	11.57	0.89	nr
Seine, Bercy, France	39.78	8.57	2.95	29.13	0.63	2.87	0.86	9.59	0.99	0.19
Rhone, Geneva, Switzerland	27.92	23.18	0.55	24.89	1.48	2.75	0.88	13.08	nr	2.14
Rhine, Strassburg, France	36.69	8.38	0.52	25.30	0.61	2.17	0.66	21.07	2.51	1.09
Moldau, Prague, Czech Rep	32.86	11.95	10.69	13.52	4.88	10.22	5.19	8.96	1.26*	nr
Elbe, Celakowitz, Czech Rep	45.87	8.95	3.27	26.41	3.21	3.93	2.46	4.09	0.91*	nr
Danube, Regensburg Germany	51.70	8.54	1.31	27.40	6.00	1.12	0.72	2.42	0.06	0.42
Lago di Garda, northern Italy	53.29	4.17	3.13	24.56	6.66	2.49	2.01	2.33	0.15	1.21
Klarelf, Sweden	38.68	7.63	2.24	11.67	0.51	8.42	3.78	19.17	7.44*	nr
Om, Omsk, Russia	43.73	2.15	12.81	11.24	9.68	9.64	2.28	6.51	1.42	nr
White Nile, Khartoum, Sudan	42.97	0.25	4.58	9.78	3.00	17.66	6.79	14.72	nr	nr

Adapted from Clark, 1908.
*Elemental data for Na and K are combined and Fe and Al are combined.
nr = not reported.

TABLE 1-3
National and International Data on Chloride Dominated Mineral Waters

Mineral Water	ppm														
	HCO_3	SO_4	Cl	Br	I	AsO_4	Ca	Mg	Ba	Sr	Na	K	SiO_2	Fe_2O_3	Al_2O_3
Artesian well, Abilene, KN	nr	0.07	61.59	0.29	nr	nr	4.85	1.52	nr	nr	31.57	trace	trace	nr	nr
Montesano Springs, MI	nr	nr	57.38	0.31	nr	nr	6.15	2.09	nr	nr	28.17	0.15	0.17	nr	nr
Utah Hot Springs, Ogden, UT	0.61	0.94	58.79	trace	nr	nr	4.90	0.40	nr	nr	30.38	3.76	0.20	nr	nr
Spring at Pahua, New Zealand	0.17	0.15	60.78	trace	0.11	nr	3.14	0.60	nr	nr	34.81	0.02	0.12	trace	nr
Water of Salsomaggiore, Italy	nr	0.18	61.09	0.15	0.03	nr	3.21	0.82	nr	0.24	34.04	nr	0.01	0.03	0.01
Harrogate Spa, England	2.12	nr	58.81	0.19	0.01	trace	2.65	1.37	0.42	trace	36.18	0.48	0.07	nr	trace

Adapted from Clark, 1908.
nr = not reported.

TABLE 1-4
National and International Data on Sulfate Dominated Mineral Waters

Mineral Water	CO$_3$	SO$_4$	Cl	H$_3$AsO4	Ca	Mg	Na	K	PO$_4$	Mn	Fe	Cu	SiO$_2$	Zn	Cd
							ppm								
Spring Joplin, MO	8	53	.48	nr	11.32	.71	.67	.46	nr	.42	.11	.04	2.54	22	.1
Mine Water Mo Zn Reg.	nr	63.3	.03	nr	3.55	.26	.5	tr	nr	.02	4.88	.04	1.11	25	.09
Cottage Well England	6.5	57	6.56	nr	8.33	7.03	13.5	.48	nr	tr	tr	nr	.16	nr	nr
Spring, Bosnia	nr	65	.48	nr	3 1	48	.31	.32	tr	.12	25	.36	1.7	.3	nr
So. Tyrol	nr	71	.03	1.93	7	1	1.25	.23	.23	.78	.03	.15	1.61	.06	nr

Adapted from Clark, 1908.
nr = not reported.

TABLE 1-5
National and International Data on Carbonate Dominated Mineral Waters

Mineral Water	ppm												
	CO_3	HCO_3	SO_4	Cl	Ca	Mg	Na	K	PO_4	Mn	Fe	SiO_2	AL_2O_3
Private well Missouri	nr	1,287	88	94	4	2	581	nr	nr	nr	nr	12	nr
Artesian Lajunta, CO	791	nr	71	67	4.4	2.5	669	669	nr	tr	2.4	51	3.4
Spring Water Brit. Col.	nr	6,339	60	1.5	117	1,152	52	12	tr	nr	6.7	83	6.5
Silesia Austria	nr	405	6.5	1	66	19	5.2	1.8	.54	.05	47	69	.3

Adapted from Clark, 1908.
nr = not reported.

TABLE 1-6
National and International Data on Mixed Anion Mineral Waters

Mineral Water	ppm													
	CO_3	SO_4	Cl	Ca	Mg	Na	K	PO_4	Fe	SiO_2	AL_2O_3	NH_4	NO_3	B_4O_7
Hot Spring Clear Lk., CA	22	tr	16.5	tr	tr	25	tr	nr	nr	2.6	.40	7.9	nr	25.6
Phosphatic Water France	19.5	7.7	5.1	30.4	1.2	3.3	tr	22.4	.04	4	nr	nr	6.3	nr
Holy Well Mecca, Arabia	12.8	14	16	48	72.7	12.7	6.7	nr	nr	1.4	nr	nr	24.6	nr

Adapted from Clark, 1908.
nr = not reported.

TABLE 1-7
Water Chemistry of Natural Water

	Median	*Median*
Cations	**Surface water (mg/L)**	**Groundwater (mg/L)**
Calcium	15	50
Sodium	6.3	30
Magnesium	4.1	7
Potassium	2.3	3
Anions		
Bicarbonate	58	200
Silica (aqueous)	14	7.4
Chloride	7.8	20
Sulfate	3.7	30

Minor Elements (less than 1 ppm but greater than 1 ppb)

Aluminum	Lead
Arsenic	Lithium
Antimony	Manganese
Barium	Molybdenum
Boron	Phosphorus
Beryllium	Rubidium
Bromine	Strontium
Chromium	Titanium
Fluoride	Vanadium
Iodine	Zinc
Iron	

Trace Elements (less than 1 ppb)
Cadmium
Cobalt
Cesium
Gold
Mercury
Nickel
Selenium
Silver
Tin
Thallium
Tungsten
Uranium

Adapted from Langmuir, 1997.

Furthermore, both radium and uranium emit beta particles. Beta-particle decay radioactivity is created when neutron/proton conversions occur. The amount of beta-particle decay is defined in terms of REMs. A REM (Roentgen Equivalent Man) is the dosage of ionizing radiation that will cause the same amount of injury to human tissue as 1 roentgen of x-rays. In water, a measurable unit of beta-particle emissions is millirems/year (mrem/yr). Regardless of the form of radioactivity, the occurrence of radioactivity in water tends to be associated with specific rock types. As a result, in most developed regions of the world, water resources that may contain measurable quantities of radioactive elements have been characterized. For example, radon exposure in surface water and groundwater resources was mapped by the United States Geological Survey (Schumann, 1993) and is illustrated in Figure 1-3.

Organic Matter

Given the biodiversity of our water resources and the occurrence of organic matter in our soil resources, natural water will contain organic compounds (i.e., those chemical compounds that are composed primarily of carbon). Many of the compounds found in natural water are the result of biosynthesis and biodegradation. As a result, natural water can contain a wide range of organic compounds. Examples of the types of organic compounds that can occur in natural water are listed in Table 1-8. Although these types of organic compounds are known to occur in natural water, the actual organic chemistry of various aquatic environments (e.g., lakes, rivers, estuaries) remains largely uncharacterized, as most water quality studies only determine the total amount of dissolved organic carbon in water and not the concentrations of individual natural organic compounds. According to Stumm and Morgan (1981), natural freshwater bodies of water will contain a few milligrams of carbon per liter. However, some freshwater bodies of water that are low in calcium

TABLE 1-8
Organic Compounds Found in Natural Water

Biological Substance	Intermediate Decomposition Product	Decomposition Products Found in Natural Water
Proteins	Amino Acids	Ammonia (NH_4^+), Nitrate (NO_3^-), Methane (CH_4), Hydrogen Sulfide, Phosphates (PO_4^{3-}), Phenols, Mercaptans, Fatty Acids
Fats, Waxes, and Oils	Fatty Acids and Glycerol	Methane, Acids (acetic, lactic, citric, glycolic, malic, palmitic, stearic, oleic), Carbohydrates
Carbohydrates	Mono and Poly Saccharides & Alcohols	Methane, Phosphates, Glucose, Fructose, Galactose
Complex Organic Substances		Humic and Fulvic Acids

Adapted from Stumm and Morgan, 1981.

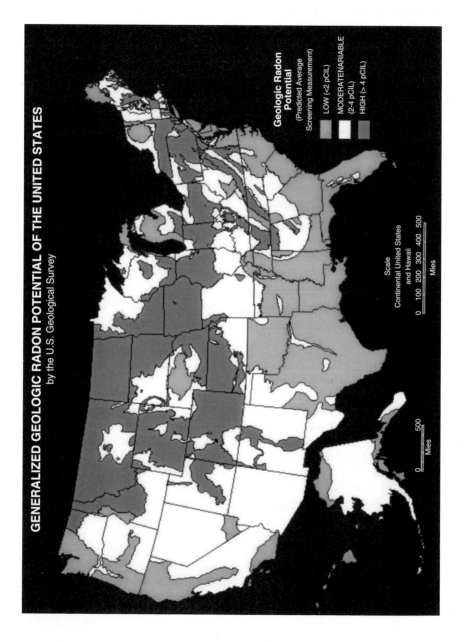

FIGURE 1-3. Generalized geologic radon potential of the United States. (From Schumann, R.R., ed., 1993.)

and magnesium can support higher concentrations of humic substances,[7] which can result in natural waters having carbon concentrations that reach 50 milligrams per liter.

Humic substances, which are also a key chemical component of soil, owe their origin to the composition and decomposition of polysaccharides that are common in plants and fruits. For example, polysaccharides contain fructose, xylose, and hemicelluloses, which decompose into butenoic acids that polymerize into humic acids. These polymers contain a heterogenous mixture of organic molecules (e.g., carboxyls, ketones, ethers, anhydrides, lactones, methylenes). The existence of natural humic substances in water is very important. When water containing humic substances is disinfected with halogens (i.e., chlorine and bromine), toxic halogenated organics are formed.

Water and the Public Health

The chemistry of natural water is highly dependent on its geologic and hydrologic origins as well as its biological contributions. The imparted chemical characteristics to a natural water can be highly valued or a potential threat to public health. Historically, mineral springs in countries around the world have been sought after for their therapeutic effects. For example, mineral waters known to contain arsenic were used for medical purposes, whereas mineral waters with high levels of magnesium and sulfates provided a natural laxative. Too much arsenic was also known to be toxic. Just because water comes from a natural source does not guarantee that it is necessarily safe to drink. This can be true for some mineral waters that can have high levels of sodium, trace metals, and total dissolved solids with a very alkaline pH.

Given a choice, most individuals won't drink water that is dirty or has a noticeable taste or odor. Our reliance on these criteria (i.e., clarity, taste, and odor) have served us well over the centuries, yet with increasing population and pollution there has been an increasing need to understand the impact of water quality on public health and how to manage water quality to ensure abundant sources of safe drinking water. To manage the quality of the water we drink, it is necessary to first understand how pollution is managed by public health agencies.

Managing Pollution

Since the 1700s, rivers have been the natural extension of municipal and industrial wastewater discharge systems (i.e., wastes were discharged

---

[7]Calcium and magnesium tend to precipitate out humic substances (i.e., these substances are no longer dissolved in water).

directly into lakes and rivers). With increasing population, pollution increased to the point that by the late 1800s, vocal public concern forced the development of state regulations to control pollution. These early regulations controlled the discharge of untreated raw sewage into our water resources to prevent the spread of water-borne disease. By 1904, laws in the United States prohibited water pollution in general, with specific attention to biological pollutants and selected industries (Goodell, 1904). However, it was not until 1907 that information concerning the toxic effect of waste discharges on aquatic organisms was widely publicized (Parker, 1907). The first real effort to describe and record the specific chemical concentrations that were lethal to various aquatic organisms was compiled by Ellis in 1937. In essence, this was the first publication of what could be considered water quality standards in the United States.[8]

The first extensive compilation of water quality criteria was published by the State of California by McKee in 1952. In this document, a "potential pollutant" was defined as: "Any substance that may enter or be contained in the waters of the State is . . . a potential pollutant, if concentrated sufficiently, it can adversely and unreasonably affect such waters for one or more beneficial uses."[9] The chemical compounds listed in 1952 with identified water quality criteria are given in Appendix 1-2. By 1976, the USEPA published Quality Criteria for Water. The current National Recommended Water Quality Criteria (USEPA, 2002) from freshwater ecosystems are given in Appendix 1-3.

By setting specific chemical water quality criteria with defined concentrations that are assumed to be safe, the federal government established a policy of permissible pollution. In other words, the federal government essentially issued a license to allow water resources to be polluted up to an acceptable level. As a natural consequence of setting water quality criteria, specific definitions for the terms *contamination* and *pollution* were developed. These definitions are presented in Exhibit 1-1. Thus, chemicals and biological pollutants are allowed to be in water resources as long as water quality criteria are not exceeded. This condition is not unique to the United States and generally reflects how the world manages water pollution.

With prohibitions against biological pollution and the creation of water quality standards, agencies with the mandate and authority to protect public health usually regulate (1) the concentration of specific pollutants that can be discharged from an industrial/commercial source, sewage treatment plant or a specific land use and (2) the fundamental

---

[8]Earlier standards were published in various European Countries. For example, in England, water quality recommendations for mine sites were proposed by A.E. Hunt and G. H. Clapp in The Impurities of Water, *Transactions of the American Institute of Mining Engineers*, Vol. XVII (1889).

[9]The beneficial uses for water are drinking water, aquatic habitat (fresh and saltwater), recreational, agricultural, and industrial.

Exhibit 1-1. Definitions

Because water quality standards set acceptable levels of pollution based on their potential for health effects, some states have defined pollution based on the knowledge of health hazards. For example, the State of California has used the following definitions since the 1960s:

- Contamination is defined as any impairment of the quality of the water of the State by sewage or industrial waste to a degree which creates an actual hazard to public health through poisoning or through spread of disease.
- Pollution is defined as an impairment of the quality of the water of the State by sewage or industrial waste to a degree that does not create an actual hazard to public health but does adversely and unreasonably affect such water for domestic, industrial, agricultural, navigational, recreational, or other beneficial use.
- Nuisance is defined as damage to any community by odors or unsightliness resulting from unreasonable practices in the disposal of sewage or industrial waste.

Source: McKee, J. E. and H. W. Wolf, 1963, "Water Quality Criteria, California State Water Quality Control Board," Publication No. 3-A.

---

quality of a body of water that receives any waste discharge. Ultimately, the way in which pollution is managed is directly related to the nature of the pollution source. These sources of pollution are usually classified as either point or nonpoint sources. Examples of both point and nonpoint sources of pollution are given in Exhibit 1-2. As a result of these sources of pollution, a vast array of chemicals and biological organisms are distributed throughout surface water and groundwater resources.

Exhibit 1-2. Point and Nonpoint Sources of Pollution

When attempting to evaluate the magnitude and extent of pollution on water resources, it is critical to distinguish between point and nonpoint sources of pollution. Point source pollution is typified by the discharge of sewage or chemicals (i.e., in either gas, liquid, or solid form) from a facility that has an identified point of release. For example:

(continued)

Exhibit 1-2. (continued)

- Chemicals are released into the atmosphere in the form of aerosols, gases and/or particulates from stacks, flares, conveyors, and towers.
- Chemicals or biological wastes are discharged into surface water (rivers, lakes, bays, and oceans) from industrial wastewater and sewage outfalls.
- Chemicals or biological wastes are released onto the land in the form of wastewater irrigation or wastewater injected into deep groundwater aquifers.

Pollutants released into the atmosphere have the potential to travel extremely long distances and can subsequently be deposited in a surface water resource. Wastewater, and any solids contained therein, is an obvious source of pollution to surface water. However, it should not be forgotten that surface water, in all but the most rare circumstances, will recharge groundwater. In a similar fashion, chemicals placed into or on land may leach into both surface water and groundwater.

Nonpoint source pollution is typified by processes that result in the release of chemicals across a fairly broad or diffuse geographic area. For example:

- The extraction of coal and metal sulfides (i.e., gold, silver, copper, lead, zinc, etc.) from the earth can disturb thousands of acres of land. Because of this disturbance, trace metals and acid can be released from areas of exposed coal and minerals (e.g., high walls, overburden, shafts, adits, waste rock piles, tailings, and mineral processing waste piles).
- The widespread application of pesticides to agricultural lands can pollute irrigation water, surface water runoff from rainfall and ultimately creeks, rivers, ponds, lakes, oceans, and groundwater.
- Rural and agricultural areas used to feed livestock are a source of nitrogen, phosphors, hormones, and antibiotics to both surface water and groundwater.
- Urban environments are subject to the accumulation of chemicals on buildings, streets, and impervious surfaces, as well as in storm channels and sewers and areas of waste storage or disposal. Common pollutants include pesticides, nitrogen, phosphorus, metals, solvents, and petroleum products. During rainfall events, these chemicals will be washed off and carried in the runoff to local receiving waters.

- Nitrogen and phosphorus can be found in the runoff from agricultural lands. Residential lawns and parks can also contribute these same pollutants to surface waters.
- Certain rocks and ore bodies are sources of nitrogen, perchlorate, fluoride, and radioactive elements that can pollute groundwater.
- Pollutants that have been deposited in creek, river, lake, and ocean sediments in the past may continuously leach into water or be released when the sediment is disturbed (i.e., by erosion or dredging).

Another category of nonpoint pollution sources are chemical releases from industrial facilities, commercial businesses, and waste disposal sites that either manufactured, handled, stored, or disposed of hazardous chemicals, solid wastes, and sewage sludge. Although pollution from such facilities may have a general point source location, the chemical pollution to either surface water and/or groundwater can be spread over a wide geographic area.

The distinction between point and nonpoint sources of pollution is important because our ability to prevent a point source of pollution from entering a water resource is much easier than preventing pollution from nonpoint sources because of the diffuse nature of the pollution. Although the USEPA has begun to address runoff from urban and agricultural land uses and mining, there are no quick economical or easy solutions on the horizon. Regardless of the level of point source control, widespread pollution of water resources from nonpoint sources will continue to exist for the foreseeable future.

Protecting the Public Health

Given the fact that pollution of our water resources is a fact of life, how can the public be sure that drinking water resources are safe to drink? The answer, once again, is the application and use of standards to protect the public health. For example, if a drinking water meets all of the criteria set forth in the USEPA Primary and Secondary Drinking Water Standards (Table 1-9), then that water is deemed safe for human consumption. The USEPA drinking water standards address both chemical and biological pollutants. Interestingly, when the drinking water standards are compared to the US ambient water quality criteria for beneficial uses (i.e., Appendix 1-2 and 1-3), a large number of compounds listed in the ambient water quality criteria are not regulated in drinking water. These differences beg the question: if the chemicals listed in Appendix 1-3 are sufficiently harmful to be listed in the ambient water quality criteria, why aren't they also listed in the drinking water standards?

TABLE 1-9
USEPA National Primary and Secondary Drinking Water Standards (June 2003)

Type of Constituent	Maximum Contaminant Level (µg/L) or Treatment		Standard
Organic	Acrylamide	Does not exceed 0.05% of dose used to treat water	Primary
Organic-p	Alachlor	2	Primary
Radioactive	Alpha particles	15 picocuries/L (pCi/L)	Primary
Inorganic	Aluminum	50 to 200	Secondary
Inorganic	Antimony	6	Primary
Inorganic	Arsenic	10	Primary
Inorganic	Asbestos (fibers>10µm)	7 million fibers/L	Primary
Organic-p	Atrazine	3	Primary
Inorganic	Barium	2,000	Primary
Organic	Benzene	5	Primary
Organic	Benzo(a)pyrene	0.2	Primary
Inorganic	Beryllium	4	Primary
Radioactive	Beta particles	4 millirems/year	Primary
Organic*	Bromate	10	Primary
Inorganic	Cadmium	5	Primary
Organic-p	Carbofuran	40	Primary
Organic	Carbon tetrachloride	5	Primary
Disinfectant	Chloramines (as Cl_2)	4,000	Primary
Organic-p	Chlordane	2	Primary
Inorganic	Chloride	250,000	Secondary
Disinfectant	Chlorine (as Cl_2)	4,000	Primary
Disinfectant	Chlorine dioxide (as ClO_2)	800	Primary
Inorganic*	Chlorite	1,000	Primary
Organic	Chlorobenzene	100	Primary
Inorganic	Chromium (total)	100	Primary
	Color	15 color units	Secondary
Inorganic	Copper	1,300	Primary
	Corrosivity	Noncorrosive	Secondary
Microbial	Cryptosporidium	99% removal	Primary
Inorganic	Cyanide (as free cyanide)	200	Primary
Organic-p	2,4-D	70	Primary
Organic-p	Dalapon	200	Primary
Organic-p	1,2-Dibromo-3-chloropropane	0.2	Primary
Organic	o-Dichlorobenzene	600	Primary
Organic	p-Dichlorobenzene	75	Primary
Organic	1,2-Dichloroethane	5	Primary
Organic	1,1-Dichloroethylene	7	Primary
Organic	cis-1,2-Dichloro-ethylene	70	Primary

TABLE 1-9 *(continued)*

Type of Constituent	Maximum Contaminant Level (μg/L) or Treatment		Standard
Organic	trans-1,2-Dichloro-ethylene	100	Primary
Organic	Dichloromethane	5	Primary
Organic-p	1,2-Dichloropropane	5	Primary
Organic	Di(2-ethylhexyl) adipate	400	Primary
Organic	Di(2-ethylhexyl) phthalate	6	Primary
Organic-p	Dinoseb	7	Primary
Organic	Dioxin (2,3,7,8-TCDD)	0.00003	Primary
Organic-p	Diaquat	20	Primary
Organic-p	Endothall	100	Primary
Organic-p	Endrin	2	Primary
Organic	Epichlorohydrin	Does not exceed 0.01% of dose used to treat water	Primary
Organic	Ethylbenzene	700	Primary
Organic	Ethylene dibromide	0.05	Primary
Inorganic	Fluoride	4,000	Primary
	Foaming Agents	500	Secondary
Microbial	Giardia lamblia	99.9% removal/ inactivation	Primary
Organic-p	Glyphosate	700	Primary
Organic*	Haloacetic acids (HAA5)	60	Primary
Organic-p	Heptachlor	0.4	Primary
Organic-p	Heptachlor epoxide	0.2	Primary
Microbial	Heterotropic plate	No more than 500 bacterial count colonies per million	Primary
Organic	Hexachlorobenzene	1	Primary
Organic	Hexachlorocyclo-pentadiene	50	Primary
Inorganic	Iron	300	Secondary
Inorganic	Lead	15	Primary
Microbial	Legionella	If *Giardia* is controlled so is *Legionella*	Primary
Organic-p	Lindane	0.2	Primary
Inorganic	Mercury	2	Primary
Organic-p	Methoxychlor	40	Primary
Inorganic	Manganese	50	Secondary
Inorganic	Nitrate (as nitrogen)	10,000	Primary
Inorganic	Nitrile (as nitrogen)	1,000	Primary
	Odor	3 threshold odor numbers	Secondary
Organic-p	Oxamyl (Vydate)	200	Primary

(continued)

TABLE 1-9 *(continued)*

Type of Constituent	Maximum Contaminant Level (µg/L) or Treatment		Standard
Organic	Pentachlorophenol	1	Primary
Organic-p	Picloram	500	Primary
Organic	Polychlorinated biphenyls (PCB)	0.5	Primary
	pH	6.5-8.5 pH units	Secondary
Radioactive	Radium 226 and 228	5 (pCi/L)	Primary
Inorganic	Selenium	50	Primary
Inorganic	Silver	100	Secondary
Organic-p	Simazine	4	Primary
Inorganic	Sulfate	250,000	Secondary
Organic	Styrene	100	Primary
Organic	Tetrachloroethylene	5	Primary
Inorganic	Thallium	2	Primary
Organic	Toluene	1,000	Primary
Microbial	Total Coliforms	No more than 5.0%	Primary
Organic*	Total Trihalo-methanes	80	Primary
Inorganic	Total dissolved solids	500,000	Secondary
Organic-p	Toxaphen	3	Primary
Organic-p	2,4,5-TP (Silvex)	50	Primary
Organic	1,2,4-Trichloro-benzene	70	Primary
Organic	1,1,1-Trichloroethane	200	Primary
Organic	1,1,2-Trichloroethane	5	Primary
Organic	Trichloroethylene	5	Primary
Microbial	Turbidity	Cannot exceed 5 NTU	Primary
Radioactive	Uranium	30	Primary
Organic	Vinyl chloride	2	Primary
Microbial	Total viruses (enteric)	99.9% removal/inactivation	Primary
Organic	Xylenes (total)	10,000	Primary
Inorganic	Zinc	5,000	Secondary

Organic-p: Pesticides.
Organic*: Compounds that result from disinfection using chlorine or bromine.
Secondary Drinking Water Standards are nonenforceable guidelines.
The trihalomethanes chloroform, bromoform, bromodichloromethane, and dibromochloromethane.
The Haloacetic acids (HAA5) include monochloroacetic acid, dichloroacetic acid, trichloroacetic acid, monobromacetic acid, and dibromoacetic acid.

Given this dichotomy, it is important to assess the scientific process of selecting chemical pollutants to be regulated in drinking water. This can be accomplished by comparing the chemical drinking water guidelines for the European Union (EU) and World Health Organization (WHO), which are given in Table 1-10, with the USEPA Primary and Secondary Drinking Water Standards.

TABLE 1-10
European Union (EU) and World Health Organization (WHO) Drinking Water
Guidelines for Chemical Compounds (2003)

Constituent	EU (µg/L)	WHO (µg/L)
Acrylamide	0.10	0.50
Alachlor*	20	
Aldicarb*		10
Aldrin and dieldrin*		0.03
Aluminum	200	
Ammonium	500	
Antimony	5	20
Arsenic	10	10
Atrazine*	2	
Barium		700
Benzene	1	10
Benzo(a)pyrene	0.01	0.7
Boron	1	500
Bromate	10	10
Bromodichloromethane	60	
Bromoform	100	
Cadmium	5	3
Carbofuran*		7
Carbon tetrachloride		4
Chlorate		700
Chlordane*		0.2
Chloride	250,000	
Chlorine		5,000
Chlorite		700
Chloroform		200
Chlorotoluron*	30	
Chlorpyrifos*		30
Chromium	50	50
Copper	2	2,000
Cyanazine	0.6	
Cyanide	50	70
Cyanogen chloride		70
2,4-D*	30	
2,4-DB*	90	
DDT and metabolites*	1	
Di(2-ethylhexyl)phthalate	8	
Dibromoacetonitrile	70	
Dibromochloromethane	100	
1,2-Dibromo-3-chloropropane*		1
1,2-Dibromoethane		0.4
Dichloroacetate		50
Dichloroacetonitrile	20	
1,2-Dichlorobenzene		1,000
1,4-Dichlorobenzene		300
1,2-Dichloroethane	3	30

(continued)

TABLE 1-10 *(continued)*

Constituent	EU (µg/L)	WHO (µg/L)
1,1-Dichloroethene		30
1,2-Dichloroethene		50
Dichloromethane		20
1,2-Dichloropropane		40
1,3-Dichloropropene*		20
Dichlorprop*	100	
Dimethoate*		6
Edetic acid (EDTA)		600
Endrin*		0.6
Epichlorohydrin	0.1	0.4
Ethylbenzene		300
Fenoprop*	9	
Fluoride	1.5	1.5
Formaldehyde		900
Hexachlorobutadiene	0.6	
Iron	200	
Isoproturon*	9	
Lead	10	10
Lindane*		2
Manganese	50	400
MCPA*		2
Mecoprop*	10	
Mercury	1	1
Methoxychlor*		20
Metolachlor*	10	
Microcystin-LR	1	
Molinate*	6	
Molybdenum		70
Monochloramine		3,000
Monochloroacetate	20	
Nickel	20	20
Nitrate	50	50
Nitrilotriacetic acid		200
Nitrite	0.5	200
Pendimethalin*	20	
Pentachlorophenol		9
Pyriproxyfen*	300	
Pesticides (see note 1)	0.1	
Pesticides-total (see note 2)	0.5	
Polycyclic aromatic hydrocarbons	0.1 (see note 3)	
Selenium	10	10
Simazine*		2
Sodium	200,000	
Styrene		20
Sulfate	250,000	
2,4,5-T*		9
Terbuthylazine*	7	

TABLE 1-10 *(continued)*

Constituent	EU (µg/L)	WHO (µg/L)
Tetrachloroethylene	10	40
Toluene		700
Total organic carbon	No abnormal change	
Trichloroacetate		200
Trichloroethylene	10	70
2,4,6-Trichlorophenol		200
Tritium (radioactive)	100 Bq/L	
Trifluralin*	20	
Trihalomethanes-total (see note 4)	10	1
Uranium		15
Vinyl chloride	0.5	0.3
Xylenes		500

1. Pesticides means all insecticides, herbicides, fungicides, nematocides, acaricides, algicides, rodenticides, slimicides and metabolites, and decomposition products. In the case of aldrin, dieldrin, heptachlor, and heptachlor epoxide the criteria value is 0.03.
2. Total means the sum of all individual pesticides detected and quantified in the monitoring procedure.
3. This is the sum of benzo(b) fluoranthene, benzo(k) fluranthene, benzo(g,h,i) perylene, and indeno(1,2,3-cd)pyrene.
4. For the EU, the sum of the following compounds should be calculated: chloroform, bromoform, dibromochloromethane, bromodichloromethane; for WHO, the sum of the ratio of each compound concentration should not exceed 1.
* = Pesticide.
Bq = Becquerel, which is a measure of radioactivity (i.e., one disintegration/sec).

When comparing all of these chemical standards, there are several obvious differences between the chemicals being regulated and their concentrations at which they are allowed in drinking water. The most significant difference pertains to pesticides. For example, the United States lists 21 pesticides (4 of which are banned), WHO lists 31 pesticides (with 11 common to the United States), and the EU essentially requires their drinking water to be almost free of all pesticides. Another striking difference is the EU's significant lack of regulated industrial petroleum hydrocarbons, both halogenated (i.e., chlorine and bromine) and nonhalogenated, in water. Yet, the EU regulates a suite of polycyclic aromatic hydrocarbons, but the United States and WHO do not.

Given these significant differences, how can the consumer of this public product be assured that the water resources are really safe to drink? In attempting to understand this dichotomy and the environmental science of drinking water, this text provides a detailed description of how and why our water resources are polluted (Chapter 2), how our current drinking water programs strive to protect the public health (Chapter 3), why the protection of our drinking water resources is scientifically questionable (Chapter 4), and options available to us as individuals and as a community to ensure that water is actually safe to drink (Chapter 5).

References

Clarke, Frank W., 1908, "The Data of Geochemistry," U. S. Geological Survey, Bulletin No. 330.

Ellis, M.M., 1937, "Detection and Measurement of Stream Pollution," U.S. Bureau of Fisheries, Bulletin 48, pp 365–437.

Goodell, Edwin B., 1904, "A Review of the Laws Forbidding Pollution of Inland Waters in the United States," U. S. Geological Survey, Water-Supply and Irrigation Paper No. 103.

Krauskopf, Konrad B., 1967, *Introduction to Geochemistry*, McGraw-Hill, New York.

Langmuir, Donald, 1997, *Aqueous Environmental Geochemistry*, Prentice Hall, Upper Saddle River, New Jersey.

McKee, Jack, 1952, *Water Quality Criteria*, California State Water Pollution Control Board, Sacramento, California.

Parker, Horatio N., 1907, "Stream Pollution, Occurrence of Typhoid Fever, and Character of Surface Waters in Potomac Basin," U. S. Geological Survey Water-Supply and Irrigation Paper No. 192, Washington, D.C.

Schumann, R.R., ed., 1993, *Geological Radon Potential of the United States*, U. S. Geological Survey Open-File Report 93-292, Parts A-J.

Stumm, Werner and James J. Morgan, 1981, *Aquatic Chemistry*, John Wiley & Sons, New York.

Todd, David K., 1980, *Groundwater Hydrology*, John Wiley & Sons, New York.

USEPA, 2002, National Recommended Water Quality Criteria, EPA-822-R-02-047.

CHAPTER 2

Water Pollution

"The most alarming of all man's assaults upon the environment is the contamination of air, earth, rivers, and sea with dangerous and even lethal materials."

Rachel Carson, Silent Spring, 1962

The increasing complexity of chemical pollution is evolutionary. As society has become more technologically advanced, pollution has evolved from being primarily biohazards in our water to containing an ever expanding mixture of dissolved manufactured chemicals. In addition to the increasing complexity of pollution, the sources of pollution are now evident throughout every region of the world. In response to the increases and complexity of both human and industrial pollution, methods for treating pollution have also evolved. Primarily because of cost, however, advanced treatment technologies have been implemented only selectively and nonpoint sources of pollution remain virtually uncontrolled. As a result, a wide range of pollutants from a variety of sources are being discharged into our water resources. Because of the biological and chemical diversity and complexity of today's pollution, the environment simply cannot assimilate all of these potentially harmful discharges. Furthermore, pollution regulations around the world allow chemical pollution of water resources as long as ambient water quality criteria are not exceeded (for example, see Appendix 1-3). Thus, pollution of our water resources has become unavoidable.

Human Waste and Pollution

The first serious waste disposal problem encountered by humans was the disposal of their own waste products. Although these wastes were usually biodegradable,[1] they did contain organisms (bacterial and viral)

29

that spread disease and thus posed the greatest threat. As long as the population remained small and widely distributed, however, the disposal of human waste usually created little or no problem. With increasing population and population density, waste disposal became a greater nuisance and posed a major health problem.

In rural areas worldwide, privies, septic tanks, and cesspools continue to represent the "state of practice" with regard to the disposal of human wastes. It has been reported that septic tanks have their origins in ancient India. In any case, these methods of human waste disposal continue today even in highly industrialized nations. As recently as 1990, the Census Bureau (American Housing Survey) estimated that almost 10 percent of the U.S. population was served by either septic tanks or cesspools.

In an urban environment, the solution to this dilemma was to discharge waste into sewers that then went out of the city itself. In most cases, these wastes ultimately discharged to nearby rivers and carried to downstream communities. The impact on rivers, which were unable to effectively dilute or biodegrade an ever increasing waste load, was described in an 1885 report to the Boston Health Department (Clark, 1885) as leaving surrounding communities "enveloped in an atmosphere of stench so strong as to arouse the sleeping, terrify the weak, and nauseate and exasperate everybody." This serious problem was addressed much earlier in European nations. For example, England was one of the first countries to attempt to address stream pollution by adopting a comprehensive plan, backed by passing legislation (the Rivers Pollution Commission of 1855) for county-wide abatement of pollution (Maxcy, 1955). These laws tried to address concerns about waste discharges by regulating the amount and type of wastewater that could be released to surface bodies of water.

In the United States, it rapidly became clear to many communities that the "solution to pollution" was not "dilution." Although sewage discharged to a body of water is normally affected (decomposed) by naturally occurring microorganisms as well as being further diluted, the capacity of a body of water to handle sewage flows was soon found to be "less than infinite." One answer to this problem was to put the sewage back onto the land, thereby reducing discharges to surface bodies of water. An 1896 water supply text (Mason, 1896) reported that the soil could be used to "purify" sewage, but the soil would be "overtaxed" if too much waste was applied and the groundwater polluted.[2] Thus, at the turn of the 20th century, sewage was managed by either dilution and biodegradation in rivers, lakes, and oceans or purified by the action of soil microorganisms. Scientists and engineers at the time referred to nature's

---

[1]Organic wastes could be degraded by microorganisms into simpler compounds (e.g., carbon dioxide, water, and less complex organic materials).

[2]Thus, transferring the pollution from surface water to groundwater.

ability to handle such waste as its "assimilative capacity." In other words, biological wastes could be assimilated with little or no nuisance as long as the nature's capacity to purify waste was not exceeded.

To ensure the environment would not be overtaxed, individual states passed laws to control pollution from the 1890s to early 1900s (Goodell, 1904). For example, a 1902 New Jersey statute contains the following language: "It shall be unlawful for any person, corporation, or municipality to build any sewer or drain or sewerage system from which it is designed that any sewage or other harmful and deleterious matter, solid or liquid, shall flow into any of the waters of this State so as to pollute or render impure said waters, except under such conditions as shall be approved by the State sewerage commission."

Although a concern in the United States with increasing population and industrial development, pollution began to migrate across state lines and forced basin-wide approaches toward pollution control. For example, a 1907 report on stream pollution described pollution in the Potomac Valley (Parker, 1907):

> The prosperity of the industries of the Potomac Valley, with its attendant increase of population, is justly a cause of congratulation to the several States within which the basin lies. Yet this success brings responsibilities that cannot be shirked, . . . Acts which may be viewed with indifference in a sparsely settled country become crimes in densely populated communities. No resource will be more seriously affected by changed conditions than water. . . . For, one by one, the sources of pure water which are not too expensive to utilize will be preempted, and then will come the time when the supplies that have been ruthlessly damaged must be purified. . . . The silver river threads are direct lines of communication between each individual and every other below him on the stream. The offenses that he commits against the water are paid for by his fellow countrymen in the basin, and the bill is larger or smaller according to the gravity of the transgressions.

Industrial Pollution

As population centers grew during the 19th century, the obvious "stench and sights" of waste in major cities such as London, Boston, New York, and Chicago demanded better methods of pollution control. In the late 1800s, the commonly expressed environmental phrase that the "solution to pollution is dilution" was already being taxed in many metropolitan environments (Mason, 1896). Thus, the need to maintain clean water resources led to the development of the "sanitary engineering" discipline as an outgrowth of civil engineering. This development ultimately led to the creation of technologies and methodologies designed to treat both liquid and solid waste before its discharge to the environment.

Historically, industrial development and the pollutants associated with such development have evolved over time. In the 1850s, manufactured gas plants, which used coal as a feedstock, released organics such as coal tars to our water resources. It was also recognized that chemical residuals stored on site could subsequently contaminate water supplies (Peckston, 1841; Hairns, 1884). By the late 1800s, mining and the waste generated from the extraction and processing of mineral ores/coal was recognized as a major source of metals and sulfuric acid to both surface water and groundwater (Hunt, 1889; Hillebrand, 1893; Jones, 1897).

Near the turn of the 19th century, a wide range of liquid and solid wastes also were being discharged by major industries that did not manufacture or use synthetic materials. Examples include canning, cement manufacturing/products, coal chemicals, dairies and feedlots, distilling, fertilizer production, foundries, glass manufacturing, iron and steel, inorganic chemicals, metal plating, nonferrous metals, petroleum refining, pulp and paper, paint formulation, meat packing and rendering, rubber production, tanning, textiles, and timber products/wood treating (Mason, 1896).

All of these industries disposed of their wastes to adjacent bodies of water, groundwater, and onto the land. As a general rule, industry continued these practices until (1) water resources were no longer available for their use, (2) their pollution created a public relations problem, (3) pollution led to a regulatory response, and/or (4) they were sued. Litigation associated with water resources damage has a long history. For example, in 1861 the Pennsylvania Supreme Court ruled that the Pottstown Gas Works (Pottstown, 1861) was subject to the existing nuisance laws when their waste disposal practices contaminated a neighbor's water well. By 1905, Johnson documents numerous lawsuits from the pollution of groundwater resources. According to Johnson (1905):

> The fact that a man has absolute right to the underground waters within his territory, and may abstract those waters entirely, even to the point of draining his neighbor's land, does not give him the right to poison or foul those waters and allow them to pass into his neighbor's land in such condition. Such an act is illegal, and he who causes the damage is generally held liable even if he is not guilty of negligence.

In addition to lawsuits pertaining to groundwater pollution, Goodell (1904) reports on surface water pollution-related lawsuits in Alabama, Arkansas, California, Colorado, Connecticut, Georgia, Indiana, Iowa, Kentucky, Maryland, Massachusetts, Minnesota, Mississippi, Missouri, New Hampshire, New Jersey, New York, Ohio, Pennsylvania, Rhode Island, South Carolina, Vermont, Wisconsin, and Wyoming. While many of these early lawsuits were associated with coal mine pollution, a 1922 petroleum industry textbook also advises engineers to consider "whether drainage effluent (from the site) may later form the basis of a pollution suit" (Day, 1922).

The industries responsible for the pollution in the Potomac River Valley at the turn of the 20th century included mining, leather tanning, manufacture of textiles, manufactured gas from coal (commonly referred to as illuminating gas for gas lights), and the manufacture of whisky

(Parker, 1907). Wastes discharged from these industries, with the exception of the manufacture of whisky, were significantly different than domestic sewage. These wastes contained nonbiological pollutants, which could not be "purified."

For example, mining and metallurgy, tanning, and textile processes released toxic metals (e.g., arsenic, cadmium, chromium, copper, lead, mercury, silver, zinc), whereas manufactured gas and textile industries released toxic organic coal tars and dyes to the environment. Chemicals from manufactured gas waste tars were characterized in *The Handbook of American Gas-Engineering Practice* (Nisbet-Latta, 1907) to contain water-soluble concentrations of benzene, toluol, paraffin, xylol, phenol, naphtha, creosote, and naphthalene. These compounds were not readily biodegradable, as sewage treatment works could not accept manufactured gas waste if it exceeded more than 9 percent of the total sewage volume (Frankland and Silvester, 1907). Both the trace metals and organic compounds released by these industries could not be substantially biodegraded. As a result, these wastes accumulated within the environment and became a continuous source of toxic materials that could leach into a water resource for many years. When nonbiodegradable chemicals (i.e., persistent chemicals) are discharged, one cannot rely on nature's assimilative capacity to repair this insult to the environment and, as a result, past chemical discharges continue to pollute our water resources even today.

By the 1930s, technological advances not only resulted in the expansion of existing industries, but created major new industries, such as battery manufacturing, electroplating and metal finishing, electronics, fibers (natural and synthetic), pesticides, pharmaceuticals, plastics, as well as a variety of other synthetic chemicals. These industries created new sources of chemical pollution and resulted in numerous cases of pollution-related litigation from the 1930s through the 1950s.

In the early years of industrial development, persistent wastes were generally discharged without any regard for the environment. As population centers grew, industrial pollution became more egregious. As the nation's water resources became more polluted, scientists and engineers reported on the hazard, prevention, treatment, and regulations related to chemical pollution (for example, Vilbrandt, 1934; Dickey, 1949; Anonymous, 1952; Rudolfs, 1953). Examples of surface and groundwater pollution from common sources of industrial discharges were published by the State of California, Los Angeles Regional Water Quality Control Board in 1952. These common sources of pollution are applicable to any industrial environment and are illustrated in Figure 2-1. Given these conditions, pollution of our water resources is understandably a serious worldwide problem.

Even today, in the 21st century, industrial pollution bedevils even the most industrialized nations. Although one tends to categorize wastes as to their sources, such as agricultural pollution, industrial pollution, and municipal pollution, there is no lack of information as to the pollution loads contributed to the world's water resources. On an international scale, agriculture contributes more pollution than either

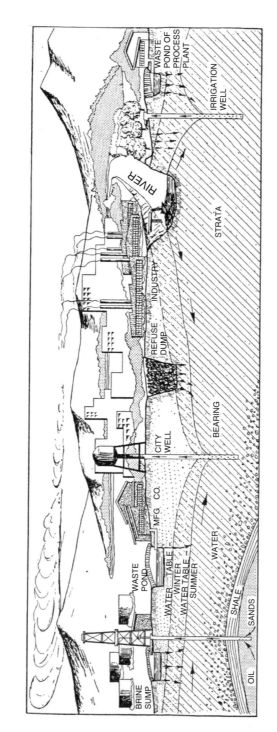

COMMON SOURCES OF POLLUTION

FIGURE 2-1. Common Sources of Pollution. (From: Anonymous, 1952, *Underground Water Pollution Resulting from Waste Disposal*, State of California, Los Angeles and Santa Ana Regional Water Quality Control Boards, June 1952.)

industry or municipalities (Population Reports, 1998). Among the industrialized countries of Europe several examples are worth citing. In Poland three fourths of the country's river water is too polluted even for industrial use. Of Hungary's 1,600 well fields tapping groundwater, 600 are already contaminated. In the Czech Republic 70 percent of all surface waters are heavily polluted. Turning to examples within developing countries, all of India's 14 major rivers are badly polluted, and three fourths of China's 50,000 kilometers of major rivers are so polluted that they can no longer sustain fish life. In Sao Paulo, Brazil, the Tiete River contains high concentrations of lead, cadmium, and other heavy metals, introduced as the river passes through the city.

To better understand the amount of pollution that is being allowed into our environment, both the United States and the European Union have embarked on programs whereby pollutant emissions are reported in terms of both chemicals and amounts. Although somewhat different in reporting format, the intent is to make available, in a comprehensive manner, data allowing one to note progress (or regression) in the control of pollutants to the environment. Current information on emissions is found in Exhibit 2-1. In other words, pollutant emissions are a reality of

---

Exhibit 2-1. U.S. and European Toxics Release Inventory

Since 1987, the U.S. EPA has collected information on the disposal or other releases, as well as waste management practices, of more than 650 chemicals from industrial sources.[3] This Toxics Release Inventory (TRI) program includes information from each of the 50 states and compiles the information in multiple databases.

The 2002 data (released in June 2004) reports a total release of 4.79 billion pounds (yes, that number is billions not millions) from 24,379 U.S. facilities. EPA estimates that of this total approximately 38.1 percent represents air emissions. Of the remainder, releases are summarized as follows:

U.S. Toxic Release Inventory

On-site disposal (total)	2.65 billion pounds
Class I underground injection and RCRA Subtitle C landfills	597 million pounds
Waste piles, spills, leaks	982 million pounds
Off-site disposal (total)	514 million pounds
Injection & landfills	273 million pounds
Solidification and/or stabilization	127 million pounds

(continued)

---

[3]U.S. EPA Toxics Release Inventory—2002 Data Release—Summary of Key Findings, USEPA, 2004 (epa.gov/triexplorer/chemical).

Exhibit 2-1. *(continued)*

Of the total amount released, it was estimated that approximately 104.1 million pounds was released to U.S. surface waters.[4] Although this appears to represent a rather small contribution of the total reported toxic releases, when one factors in the methods of disposal (underground injection, landfills, waste piles, spills, and leaks), it appears probable that the actual release (over time) to surface waters is underestimated.

Not to be left out, federal facilities (including 315 facilities operated by both federal agencies and contractors) discharged 85 million pounds to the environment.

On a comparative basis, the overall release of toxics (as reported to EPA) decreased by 15 percent (819 million pounds) over the comparable data for 2001. According to the EPA, however, surface water discharges increased by more than 400,000 pounds, and therein lies the rub. Any and all toxic releases to the water environment represent an increased challenge to the goal of "pure drinking water." Since the beginning of the survey period (1998), there has been a substantial decrease (37 percent) of releases of TRI chemicals. This equates to a reduction of 2.50 billion pounds.

European Pollutant Emissions[5]

The European Union (EU) pollutant release inventory began in 2001. This inventory is similar in scope and content to the U.S. Toxics Release Inventory. Data are presented for 50 chemicals and are broken down by nation and sources (including about 10,000 industrial facilities). Emissions categories include discharges to air, direct to water and indirect to water. By way of reference, direct discharge to air and water are included. The discharge to air is relevant simply because there is a significant relationship between pollutants released to air traveling considerable distances and subsequently identified in bodies of water.

By way of illustration, the breakdown of biocides and explosives includes carbon dioxide, total organic carbon, and chlorides. Emission examples include:

Pollutant	To Air (kg/year)	Direct to Water (kg/year)
Nonmethane volatile organic compounds	3,589,000	
Nitrogen oxides	1,021,000	

---

[4]*Water Environment and Technology*, "Volume of Toxic releases in U.S. Remained Steady in 2002," p. 17, August 2004.
[5]European Pollutant Emission Register (www.eper.cec.eu.int.).

Sulfur oxides	695,000	
Dichloromethane	43,150	1,170
Trichloroethylene	62,600	
Phosphorus, total		12,930
Nitrogen		840,000
Examples of pharmaceutical products include[6]:		
Dichloroethane-1,2	20,870	1,705
Dichloromethane	1,192,130	11,170
BTEX		1,282
PAH		7,100
Cyanides		108
Fluorides		12,700
Phosphorus		548,190

current industrial, municipal, and agricultural practices. This means that wastewater treatment control is even more important.

Wastewater Control and Treatment

The combination of municipal sewage and industrial waste discharges coupled with an expanding economy required that pollutant waste loads be reduced. Steps were taken to ensure that solid wastes were not to be dumped into surface waters and that wastewater was treated to remove some fraction of its chemical and biological constituents before its discharge to surface waters. These actions gave rise to the development of centralized sewage and industrial treatment facilities.

From the early 1900s to the 1940s, more and more states imposed stricter pollution control requirements on the discharge of both sewage and industrial wastes. These state regulations often limited the amount of chemicals and bacteria in sewage and industrial waste discharges so as to reduce the pollution load on receiving waters. Greater emphasis was also placed on "engineered" waste treatment prior to waste discharge to the environment. On a community-wide or region-wide basis, this emphasis led to the development of publicly owned treatment works (POTW) that contained the following primary treatment processes:

- Physical methods (e.g., screens and racks) were used to remove debris from wastewater.

---

[6]The European emissions are reported as specific chemicals rather than pharmaceutical compounds.

- Treatment of the liquid portion of the waste stream was accomplished by coagulation and settling to remove suspended solids (sludge).
- The final effluent would then be chlorinated to destroy microorganisms such as coliform bacteria.
- The sludge resulting from this treatment would be placed in an anaerobic[7] digester where microbial decomposition and the resulting elevated temperatures served to disinfect sludge. The sludge (commonly referred to as "biosolids") was then either land farmed (tilled into soil), placed in a landfill, dumped into the ocean, or incinerated.
- The treated wastewater could, depending on the geographic location, be either evaporated, discharged into an adjacent body of water, used for irrigation, or injected into the groundwater.

A POTW receives primarily sanitary waste from homes and commercial business. It also accepts commercial and industrial waste, if these wastes do not contain levels of toxic chemicals sufficient to kill the microorganism in the plant's liquid treatment systems and digesters (i.e., causing an "upset"). The actual chemical makeup of industrial discharges was often unknown to the operators of a POTW because state wastewater permits during this period rarely required monitoring for any chemical pollutants; however, operators had other ways to determine the presence of chemical pollutants in a waste discharge. For example, the business or industry could disclose the actual composition of their wastewater, but they rarely did. More frequently, an unknown pollutant would be identified only when an industry discharged enough of that chemical to a POTW to cause a treatment upset. As a result, the real composition of wastewater discharges was unknown as long as no upsets occurred.

Furthermore, wastewater most often required treatment beyond the primary water treatment process described previously. An overview of wastewater treatment systems that can be used are diagramed in Figure 2-2. By showing the treatment process in stages, one can parallel the implementation that is based on local needs and requirements. Primary treatment, ending with sedimentation, is the accepted treatment standard in many nations and is still commonly found in industrialized nations. The inclusion of the next step, secondary treatment, is common in industrialized nations but has seen limited application in developing nations. The third stage, commonly referred to as tertiary treatment, is more common in the industrialized nations as the complexity of waste constituents has become the driving force behind the necessity to treat to higher effluent standards. As for the disinfection step, this is often found at the end of each stage of treatment.

To assist the reader, a brief discussion of each step in the treatment scheme merits inclusion. Bar screens and grinders are the initial step in

---

[7]This is an oxygen-deficient environment.

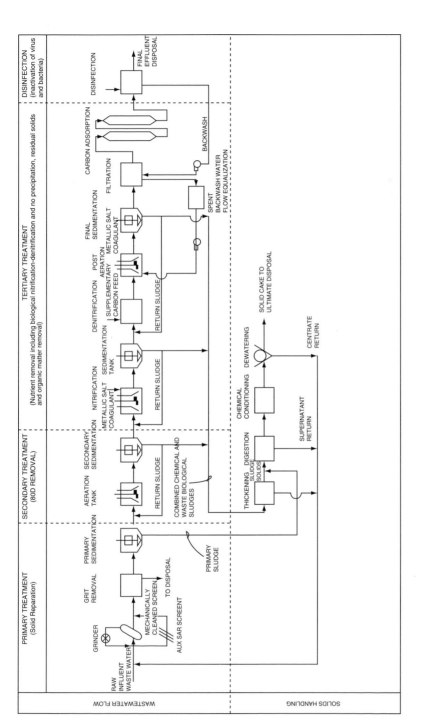

FIGURE 2-2. Typical schematic flow and process diagram of a sewage treatment plant. (From Anonymous, 1976, *Operation of Wastewater Treatment Plants*, Water Pollution Control Federation, Washington, D.C.)

most treatment plants in order to deal with large materials such as cloth, wood, metal (e.g., cans), glass, paper, and other items that might end up in a sewer. Sedimentation is the step whereby remaining suspended solids are removed from the wastewater. These solids are commonly directed to a digester where the organic fraction is decomposed anaerobically (in the absence of oxygen) and subsequently disposed of to the land (commonly used as a soil supplement or fertilizer). Secondary treatment involves biological processes to decompose the remaining organic fractions found in solution. The activated sludge process is usually used to accomplish this. Tertiary treatment involves a combination of biological and chemical treatment to remove whatever is remaining in the waste stream after secondary treatment. Constituents that may be viewed as detrimental to the receiving environment are treated and removed. Examples include nitrification and denitrification to control the release of nitrates that might encourage algal aftergrowths in the receiving water. Coagulation is used to address the removal of any remaining suspended, colloidal, and dissolved solids, whereas filtration acts as a polishing step to free the effluent from any remaining solids that might bind the final step, carbon adsorption. Carbon adsorption removes a significant portion of whatever chemical remains in the waste stream. As stated earlier, the disinfection step is used to address biological or viral constituents that might have passed through the system. All in all, complete treatment, including all of the steps shown in Figure 2-2, does result in a rather high-quality effluent.

Like industries, residential dischargers also disposed of an equally complex array of organic chemicals down their sinks, drains, and toilets. Because of these sources, most POTWs received an extremely diverse mixture of waste that contained a wide variety of chemicals including:

- Household cleaning chemicals, personal care products, medicines, petroleum products, solvents, and poisons
- Human excrement that contained pharmaceuticals and their by-products
- Chemicals from commercial/light industry that included process chemicals (e.g., degreasing solvents, ink, dyes, petroleum, paints and paint solvents, dry-cleaning solvents, poisons, metal plating chemicals, acids) used in everyday business operations
- Industrial discharges containing a wide spectrum of both organic and inorganic chemicals

The standard methods of wastewater treatment described previously have the ability to remove many of the organic and inorganic compounds found in raw effluent. However, POTWs must rely on dischargers to limit the release of chemical pollutants.

In addition to POTW discharges, during the early 1900s to the 1940s, many industries also discharged their waste directly to water resources.

In virtually all cases, industries would discharge chemically polluted wastewater into a water resource during periods of high flow when dilution would mask any observable damage. However, some industries, as part of their discharge permit requirements, did test for chemical and biological indicators of pollution in the receiving body of water after dilution had occurred. The standard test methods used are discussed in Exhibit 2-2. In the summer months when river flow was low (i.e., less dilution) or when a chemical pollutant even at low concentrations would create observable changes in receiving water quality, the responsible industry would have to either (1) build holding ponds so that wastewater could subsequently be released during periods of high flow or (2) provide treatment before discharge. A good example of how industry dealt with wastewater treatment during this time is illustrated in articles published by the Dow Chemical Company (Harlow, 1938; Powers, 1945).

Exhibit 2-2. Monitoring Pollution

Before World War II and through the early 1960s, the major constituents found in domestic sewage were from households as opposed to industrial sources. Because these wastes were found to be relatively biodegradable, methods of monitoring pollution measured the effect of these biodegradable wastes on water quality. The standard methods used to monitor the effects of biodegradable wastes were the biochemical oxygen demand (BOD), suspended solids, floating solids, pH, coliform organisms, and dissolved oxygen (DO). Because these methods were used to characterize the degree of pollution in receiving waters, these parameters were also used to address and establish effluent discharge standards. In other words, the receiving waters were assumed to be protected if the specified limits were not exceeded. For example, the following parameter limits were in general use:

BOD: The BOD is a measure of the amount of oxygen consumed as biodegradable material is aerobically decomposed. Thus, the greater the BOD, the less oxygen available in the water and the greater potential harm to fish and aquatic organisms. BOD limits are usually set at 10 to 20 ppm to ensure that overload conditions, which would cause excessive oxygen depletion in the receiving body of water, do not occur.

(continued)

Exhibit 2-2. (continued)

Suspended Solids:	Suspended solids were usually limited to 20 to 100 ppm. This limit was necessary to prevent either excessive sludge buildup in the receiving water and to ensure that sunlight is not prevented from penetrating the water environment (i.e., turbidity associated with suspended solids could affect the growth of marine flora).
Floating Solids:	This limit was an aesthetic standard that required the full removal of all floating solids.
pH:	To maintain a natural acid/base balance, pH was usually set between 5.5 and 8.5.
Coliforms:	Coliform limits were determined by "frequency of tests" with a statistical average and allowable maximums over time. Actual values were usually set as follows: the most probable number of coliforms should not exceed more than 1 per milliliter in 50 percent of the 1 milliliter samples.
Dissolved Oxygen:	Dissolved oxygen was usually set at 5 ppm to maintain this level in the receiving waters.
Chemical Oxygen Demand (COD):	This test is an analog to the BOD test in that it measures the oxidation of organic and oxidizable inorganic materials in water chemically, whereas the BOD test measures constituents biologically. An advantage of the test (over the BOD test) is rapid determination of the oxygen demand. However, a more significant advantage of this analytical tool is the ability to indicate the presence of industrial chemicals, which may not exert a BOD. Thus, the ratio of COD to BOD is an excellent indicator of industrial pollution, as well as the presence of toxics in wastewaters.

In addition to these general standards, each region of the country often had additional requirements based on local conditions. For example, discharges in the Ohio River Basin usually had restrictions on chlorides. Other parameters such as metals (e.g., zinc, copper, arsenic, cadmium), sulfides, nitrates, phosphates, and phenols were occasionally incorporated into state discharge permits. The concentration limits were usually determined by the state and local governments, as there were no national standards.

These same general pollution test methods are still used today along with monitoring for a number of specific organic compounds (e.g., the Total Toxic Organics list of compounds). Just like past monitoring programs, however, testing for the vast majority of chemical pollutants in our water resources is never required.

By the early 1940s, the Dow Chemical Company at its Midland, Michigan facility manufactured various industrial chemicals, organic solvents, pharmaceuticals, aromatic organic compounds, insecticides, and dyes.[8] Dow identified and focused on only one class of chemicals (phenols) that was recognized as causing a major problem in its wastewater discharge. According to Dow (1938), "phenol is used as a standard in testing disinfectants and its germicidal action is used in many ways, for example, in the manufacture of germicidal and disinfecting paints and germicidal soaps, and is a preservative in leather, glue, and adhesives industries." Thus, phenol is clearly toxic.[9] The 1946 United States Public Health Service Drinking Water Standards (USPHS, 1946) state that "phenolic compounds should not exceed 0.001 ppm."[10]

According to Dow (Harlow, 1938), the manufacture of phenol from benzol and the production of salicylates added chemical wastes that could be objectionable in water in very "minute amounts." Because phenol concentrations in their waste streams were too large to release even during the periods of dilution (i.e., high flow conditions in the Tittabawassee River), Dow built a treatment plant designed to remove phenols. Before this plant went online, the concentration of phenol in Dow's wastewater ranged from 305 to 458 ppm. After treatment operations began, phenol concentrations in the treated wastewater dropped to 0.01 to 1.7 ppm. This reduction is a good example of just what can be accomplished through good engineering practice and when the incentive exists to address a problem, even in the early 1940s.

As a result, Dow was able to discharge the treated effluent to the Tittabawassee River and rely on a dilution factor of anywhere from

---

[8]For a list chemicals manufactured by the Dow Chemical Company in 1938, see Appendix 2-1. This list also provides an example of the types of chemicals that were potential sources of pollution more than 60 years ago.

[9]The 1946 Manufacturing Chemists Association, Chemical Safety Data Sheet for Phenol establishes that chronic poisoning by phenol may be fatal.

[10]Parts-per-million.

10,000 to 100,000 times to further reduce the concentration of its phenolic discharges. Under these conditions, although phenol would still be a pollutant in the river, the resulting concentration would be so low as to be nearly undetectable. What was more significant, however, was the fact that no other chemicals were specifically treated and removed from this wastewater. Thus, the release of this wastewater would have polluted the Tittabawassee River with an unknown suite of chemicals that were being manufactured by Dow. The chemicals that were being manufactured by Dow in the 1930s are listed in Appendix 2-1. Given that this approach was typical of how industry[11] dealt with its wastewater, it is not surprising that water resources became grossly polluted.

Individual states continued to regulate municipal and industrial discharges up until the passage of the Clean Water Act in 1972, when national standards for regulating wastewater discharges were incorporated into the National Pollutant Discharge Elimination System (NPDES) permit program.[12] Sadly, this program does not "eliminate" pollution. The regulation "limits" pollution by establishing levels of permissible pollution.

Basic Components of the NPDES Program[13]

The Clean Water Act provides the regulatory basis for controlling the quality of wastewater discharges of both toxic and nontoxic pollutants to surface bodies of water through the NPDES permit. This permit covers discharges from municipal and industrial point sources as well as non-point sources of pollution. In setting up the national program, the Clean Water Act required or allowed the following:

- Each state had to establish ambient water quality standards (i.e., receiving water quality standards).
- The (U.S. Environmental Protection Agency) USEPA had to identify and set discharge controls for toxic compounds that are classified as priority pollutants. This list is given in Appendix 2-2.
- The USEPA had to define effluent limitation guidelines for municipal and industrial point sources of water pollution. In other words, the USEPA had to establish which chemicals could be discharged from a specific industry and the chemical concentrations at which they could be discharged.

---

[11]This finding is based on the authors' review of the historical records for hundreds of industrial companies in the United States.
[12]Although the NPDES program was a federal regulation, it was implemented at the state level.
[13]NPDES regulations are extremely complex. This discussion is a simplified introduction to the NPDES program and is not intended to be comprehensive.

- Industrial sources of pollution were allowed to be discharged under permit to a municipal POTW.
- All industrial sources under the NPDES permit process had to either adhere to specified effluent discharge limits or pretreat their wastes so as to be in compliance.
- All municipal POTWs were required to have an NPDES permit.
- Each state had to establish programs aimed at controlling nonpoint pollution (e.g., storm water runoff controls).

These basic features of the NPDES permit program have been in place for approximately 30 years. As of today, all facilities that discharge wastewater to waters of a state must have an NPDES permit. Point source permits regulate the amount of pollution that is allowed to be discharged into a receiving body of water. This approach is usually referred to as a *standards*-based permit system. In such a system, the discharger cannot exceed the set discharge standard for each chemical parameter listed in the NPDES permit. The discharger must also monitor their effluent to demonstrate performance with the permit requirements.

The NPDES program for point sources of pollution as currently implemented allows specific concentrations of regulated chemical pollutants into our water resources as long as these levels are below acceptable criteria. This program also allows for the discharge of an unknown number of unregulated chemicals. For example, when the USEPA began the development of effluent limitations for the pharmaceutical manufacturing industry, it requested chemical discharge information in a questionnaire sent to pharmaceutical manufacturing facilities. Based on the results of this questionnaire, the chemicals listed in Table 2-1 were reported in wastewater discharges. The vast majority of these chemicals are nonpriority pollutants, and are thus unregulated. In developing the effluent limitations for the pharmaceutical industry, the USEPA (1998) chose not to limit the discharge of most of these chemicals because these chemical pollutants were ". . . discharged on an industry-wide basis at less than 3,000 pounds per year." What is telling about the development of this discharge effluent limitation is that the actual pharmaceutical products, which should obviously be small, are not even measured or reported. Furthermore, the process followed for this industry category is typical of the effluent limitations that have been developed for other major industries (e.g., pulp and paper industry, nonferrous metals industry, textile industry, etc.). In other words, there is a vast array of both regulated and nonregulated chemicals (i.e., no effluent limitation) that are discharged to our water resources.

Finally, the NPDES program allows permitted facilities to self-monitor the quality of their discharges. As a result, the point source NPDES program does not really protect sources of drinking water from pollution. These permits allow a defined level of pollution. In fact, the use of the name National Pollutant Discharge Elimination System misrepresents the actual goals of this program. Pollutants are not "eliminated" from

TABLE 2-1
Reported Chemical Pollutants in Pharmaceutical Manufacturing Industry Wastewater

Compound	Pollutant Type[14]	Selected by USEPA for Regulation[15]
Acetaldehyde	NPP	NO
Acetic acid	NPP	NO
Acetone	NPP	NO
Acetonitrile	NPP	YES
Allyl chloride	NPP	NO
Ammonium hydroxide	NPP	NO
n-Amyl acetate	NPP	YES
Amyl alcohol	NPP	NO
Aniline	NPP	NO
Benzaldehyde	NPP	NO
Benzene	PP	YES
Benzyl alcohol	NPP	NO
Benzyl chloride	NPP	NO
Bis(chloromethyl)ether	NPP	NO
2-Butanone (MEK)	NPP	NO
N-Butyl acetate	NPP	NO
N-Butyl alcohol	NPP	NO
Tert-butyl alcohol	NPP	NO
Carbon disulfide	NPP	NO
Chloroacetic acid	NPP	NO
Chlorobenzene	PP	YES
Chloroform	PP	YES
Chloromethane	PP	NO
Cyclohexane	NPP	NO
Cyclohexanone	NPP	NO
Cyclohexylamine	NPP	NO
Cyclopentanone	NPP	NO
Dichlorobenzene	PP	YES
1,2-Dichloroethane	PP	YES
Diethylamine	NPP	NO
Diethylaniline	NPP	NO
Diethyl ether	NPP	NO
N,N-Dimethylacetamide	NPP	NO
Dimethylamine	NPP	NO
N,N-Dimethylaniline	NPP	NO
Dimethylcarbamyl chloride	NPP	NO
N,N-Dimethylformamide	NPP	NO
Dimethyl sulfoxide	NPP	NO
1,4-Dioxane	NPP	NO
Ethanol	NPP	NO
Ethyl acetate	NPP	YES

[14]NPP = Nonpriority pollutant; PP = priority pollutant (see Appendix 2-2).
[15]Indicates compounds not selected for removal and monitoring for all discharge classes.

TABLE 2-1 *(continued)*

Compound	Pollutant Type[14]	Selected by USEPA for Regulation[15]
Ethylamine	NPP	NO
Ethylbenzene	PP	NO
Ethyl bromide	NPP	NO
Ethyl cellosolve	NPP	NO
Ethylene glycol	NPP	NO
Ethylene oxide	NPP	NO
Formaldehyde	NPP	NO
Formamide	NPP	NO
Formic acid	NPP	NO
Furfural	NPP	NO
Glycol ethers	NPP	NO
n-Heptane	NPP	NO
n-Hexane	NPP	NO
Hydrazine	NPP	NO
Iodomethane	NPP	NO
Isopropanol	NPP	NO
Isobutyl alcohol	NPP	NO
Isobutyraldehyde	NPP	NO
Isopropyl acetate	NPP	YES
Isopropyl ether	NPP	YES
Methanol	NPP	NO
Methylal	NPP	NO
Methylamine	NPP	NO
Methyl cellosolve	NPP	NO
Methylene chloride	PP	YES
Methyl formate	NPP	NO
Methyl isobutyl ketone	NPP	NO
2-Methylpyridine	NPP	NO
Methyl-t-butyl-ether (MTBE)	NPP	NO
Naphtha	NPP	NO
n-Octane	NPP	NO
Phenol	PP	YES
Polyethylene glycol	NPP	NO
n-Propanol	NPP	NO
Propylene oxide	NPP	NO
Pyridine	NPP	NO
1,1,2,2-Tetrachloroethane	PP	NO
Tetrachloromethane	PP	NO
Tetrahydrofuran	NPP	NO
Toluene	PP	YES
1,1,1-Trichloroethane	PP	NO
1,1,2-Trichloroethane	PP	NO
Trichlorofluoromethane	NPP	NO
Triethylamine	NPP	NO
Xylenes	NPP	NO

From USEPA, 1998.

our water resources under an NPDES permit; they are only "limited." The real name for this program should be the National Pollutant Discharge Limitation System. Furthermore, this program still relies on dilution and nature's assimilative capacity to mask the discharge of pollutants.

Point Source Water Pollution and the NPDES Program

The point sources that are regulated by the NPDES program are (1) POTWs that discharge treated wastewater or that use wastewater for spraying or land irrigation and (2) industrial facilities that either discharge to a POTW, discharge directly to a receiving water, or use the industrial wastewater for spraying or land irrigation. In each case, the amount of pollution that can be discharged by a given facility is based on the site-specific and region-specific characteristics of the receiving water. For example, the amount of metals that can be discharged depends on local water quality criteria, the use of the river (e.g., used only for recreation, aquatic habitat, or as a drinking water source), the ability of the river to dilute the metal load, and the amount of metals added by other facilities both up- and down-gradient of the facility in question. Because of these characteristics, it is necessary to discuss the NPDES program with respect to each type of discharge it is designed to regulate.

POTW Discharges

A POTW that receives wastewater from industrial sources has two compliance problems. First, the POTW must ensure that the industrial sources meet their pretreatment requirements before releasing their wastewater to the POTW. For example, an industrial facility that has an electroplating operation must remove cyanide and metals (such as copper, nickel, chromium, and zinc) to levels specified in their permit in order to discharge their wastewater to the POTW. In addition, the electroplating operation must not exceed the total toxic organics (TTO) limit.[16] The TTO is determined by summing of all chemical concentrations greater than 0.01 ppm for the list of organic priority pollutants presented in Appendix 2-2. A facility may avoid this requirement by certifying that "no dumping of concentrated toxic organic chemicals" into the wastewater will occur (e.g., see Code of Federal Regulations, Title 40, Part 413, Electroplating Point Source Category). This issue aside, industrial discharges may have effluent limitations for both toxic organic and inorganic compounds to ensure that these discharges do not "upset" the POTW. To assure the POTW of their compliance with these limitations, each industry that discharges to the POTW must monitor their effluents for the chemicals identified in their pretreatment permit and submit these monitoring results to the POTW.

---

[16]The TTO requirement may be applied to any industry that discharges any of the compounds listed in Appendix 2-2.

Based on the monitoring data, a POTW can assess civil penalties against industries that do not meet their effluent limitations. Such penalties ensure that industries install and maintain wastewater treatment systems that meet their permit requirements. By using this type of program, the POTWs minimize the concentration of toxic substances discharged by industrials. Some municipalities also set up household hazardous waste collection facilities to reduce the amount of hazardous chemicals that households and small businesses dispose via the sewers. Some amount of hazardous inorganic and organic compounds, however, still make their way to POTWs. As a result, POTWs also have to meet effluent limitations. The second compliance problem faced by a POTW is this need to meet their own NPDES permit requirements. Because most primary wastes treated by a POTW are of biological origin, POTW systems are based on the treatment and destruction of biological hazards.

At a minimum, most POTWs use basic treatment methods to reduce the concentration of biological pollutants in their wastewater discharge. These methods remove approximately 85 percent of oxygen-demanding pollutants as measured by biochemical oxygen demand (BOD) from wastewater by using sedimentation and biological treatment (e.g., activated sludge). Depending on the NPDES permit requirements, a POTW might use (1) advanced treatment methods that remove up to 95 percent of oxygen-demanding pollutants (i.e., advanced activated sludge methods) and/or (2) tertiary treatment methods to further remove nitrogen and phosphorus compounds from the wastewater. These treatment methods are not designed to remove toxic metals and synthetic organic compounds. Their inability to do so is a significant problem, as not only industrial discharges, but also households and commercial businesses can and do dispose of hazardous materials (e.g., cleaning products, solvents, petroleum products, medications) into sewers.

Although treatment to remove biological pollutants will result in some portion of toxic metals and synthetic organic compounds partitioning into the sludge,[17] biological treatment methods are not designed to remove metals and toxic organics from wastewater. As a result, some portion of soluble metal and synthetic organic chemicals will pass through a POTW and be discharged to a receiving body of water or sprayed on the land. Such discharges can and do pollute groundwater resources (Bouwer et al., 1998). Because of this problem, POTWs also have effluent limitations.[18] These effluent limitations are usually facility specific. An example of typical effluent limitations for a POTW that receives wastewater from both urban and rural areas is illustrated by the NPDES permit (CA0037648) for the Central Contra Costa Sanitary

---

[17]As a result, toxic metals and organics will be disposed of with the sludge by land farming, land filling, or incineration.
[18]These limitations are primarily focused on regulated compounds.

District in Martinez, California. These limitations are primarily focused on regulated compounds.

This treatment plant consists of screening facilities, primary sedimentation, an activated sludge biological treatment process, secondary clarification, and ultraviolet disinfection. The plant discharges 45 million gallons of treated effluent per day to Pacheco Slough, which flows to Walnut Creek, and then into Suisun Bay. The chemical effluent limitations for this POTW are:

Constituent	Daily Maximum	Annual Average
Oil and grease (ppm)	20	
Ammonia (ppm)	0.16	
Dissolved sulfide (ppm)	0.1	
Copper (ppb)	19.5	
Lead (ppb)	8.2	
Mercury (ppb)	0.16	
Acrylonitrile	7	
Bis(2-ethylhezyl)phthalate (ppb)	190	
4,4-DDE (ppb)	0.05	
Dieldrin (ppb)	0.01	
Tributylin (ppb)	0.06	
Dioxin compounds (mg/year)		9.45

Based on these effluent limitations, the Central Contra Costa Sanitary District Wastewater Treatment Plant must also have a self-monitoring program to determine whether it is in compliance with its permit requirements. Under that program, the following chemical parameters must be monitored:

- pH and dissolved sulfides are monitored daily.
- An effluent sample is analyzed monthly for oil and grease, ammonia, cadmium, copper, cyanide, lead, mercury, nickel, and tributyltin.
- Quarterly samples are analyzed for arsenic, trivalent, and hexavalent chromium, selenium, silver, zinc, 4,4-DDE, and dieldrin.
- Samples are analyzed twice a year for dioxin compounds, diazinon, and the remaining organic compounds on the TTO list not previously analyzed.

Based on the frequency of this monitoring, there is ample opportunity for toxic pollutants to be discharged from the POTW without detection. In addition to these monitoring requirements, industrial dischargers must also provide monitoring data for nonpriority pollutants (i.e., unregulated toxic pollutants) if the industry "believes" that one of these compounds may pass through the POTW and into the receiving body of water. For example, the Central Contra Costa Sanitary District

Wastewater Treatment Plant monitors for diazinon twice a year and tributylin once a month. However, these two compounds represent only the tip of the "unregulated chemical" iceberg. Although these self-monitoring programs provide some indication of the regulated pollutants that are discharged with the treated wastewater effluent, the pass-through of unregulated organic chemicals in POTW effluents has only begun to be significantly addressed within the last several years.

For example, researchers at the University of West of England (Bristol) have developed a very sensitive method of testing for low levels of estrogen in wastewater effluent[19] and river water (Anonymous, 2001). Although monitoring for estrogen and hormones is important to define the extent of any problem, it is unlikely that such monitoring will result in the installation of additional wastewater treatment. Unfortunately, a vast array of pharmaceuticals, chemicals from personal care products (Daughton and Ternes, 1999), and disinfection by-products,[20] other than estrogen, are present in treated wastewater for which neither monitoring requirements nor simple methods of monitoring are available. Because most pharmaceuticals occur in wastewater as a result of the public's use of medications, there is no easy way for regulatory agencies to impose discharge limits. The extensive degree of pharmaceutical pollution in the water resources in the United States was further characterized by Hunt (1998) and is illustrated in Exhibit 2-3. A study by Ongerth and Khan (2004) compared several common pharmaceutical compounds that pass through secondary sewage treatment plants in Australia, Germany, and California. This comparison is given in Table 2-2. Although these studies give us a hint as to the problem, they do not provide a true sense of the magnitude of pharmaceuticals in our water resources, as monitoring for these compounds is not widespread. However, it can be assumed that many pharmaceuticals, and their decomposition products, will be passed through the human body and directly into sewers, which ultimately discharge to water resources (Exhibit 2-4).

Another example is provided by a study on trace organic compounds in wastewater from Lake Arrowhead, California (Levine et al., 2000), which reported a wide range of both regulated (i.e, Appendix 2-2 compounds) and unregulated organic chemicals in their discharge effluent. The compounds found in the Lake Arrowhead study are listed in Table 2-3. This list differs considerably from the compounds that are regulated under the NPDES program. Studies conducted at Clemson University (Magbanua et al., 2000) analyzed effluents from a mixed activated sludge

---

[19]The most important application of this method will be for the monitoring of estrogen in drinking water.

[20]For example, N-nitrosodimethylamine (NDMA), which has a California MCL of 0.002 ppb and is commonly found in treated wastewater and groundwater where treated wastewater was used to recharge aquifers (see California Department of Health Services web site on California Drinking Water and NDMA-Related Activities).

Exhibit 2-3. *Pharmaceuticals in Water Resources*

According to the article by Tara Hunt in the July 1998 issue of *Water Environment and Technology*, the following pharmaceuticals have been found in the environment:

Type of Drug	Human Drug	Veterinary Drug
Analgesic	Aspirin	
	Dextropropoxyphene	
	Diclofenac	
	Ibuprofen	
	Indomethacin	
Antibiotic	Erythromycin	Oxytetracycline
	Penicilloyl groups	
	Sulfamethoxazole	
	Tetracycline	
Antiparasitic		Ivermactin
Anxiolytic	Diazepam	
Cancer treatment	Bleomycin	
	Cyclophosphamide	
	Ifosfamide	
	Methotrexate	
Hormone	Estrogen	Estrogen
	Estradiol	Testosterone
	Estrone	
	Ethinylestradiol	
	Norethisterone	
	Oral contraceptives	
	Testosterone	
Cholesterol-lowering	Clofibrate	
	Clofibric acid	
Narcotic	Morphinan-structure	
Psychomotor	Caffeine	
	Theophylline	

Most researchers and government officials interviewed have said that more research is required to determine the extent of any pharmaceutical pollution, but that finding the necessary financial support to do so would be difficult. In addition, the ability to control such pollution requires that the source be identified. Currently, there are three potential sources: the chemical manufacture, medical facilities that distribute pharmaceuticals, and the patients who receive and use them. Ultimately, some portion of waste generated by all of these sources may end up at a public sewage treatment plant. As a result, municipal wastewater treatment plants may become responsible for the removal of pharmaceuticals from wastewater discharges. If this occurs, the cost of treating wastewater to remove pharmaceuticals could increase treatment costs by at least 10 percent. Given this cost, wastewater treatment engineers argued that it would be more cost effective to remove pharmaceuticals from drinking water instead. Regardless of where the treatment is installed, "until now we've only seen the top of a large iceberg . . . many compounds, especially some metabolites, cannot be detected or extracted in water using our current methods."

TABLE 2-2
A Comparison of Pharmaceutical Concentrations in Discharged Effluent (ppb)

Compound	Use	Australia	Germany	California
Salicylic acid	Analgesic	0.38	0.23	0.58
Ibuprofen	Analgesic	0.22	0.16	0.11
Acetaminophen	Analgesic	0.39	0.79	0.67
Gemfibrozil	Cholesterol reducer	0.25	0.11	0.65
Naproxen	Anti-inflammatory	0.35	0.35	0.40

Exhibit 2-4. Top 10 Pharmaceuticals

According to IMS Health, a global pharmaceutical marketing company, the top 10 pharmaceuticals sold in the "13 leading countries worldwide" in 2000 were:

Pharmaceutical	Active Ingredient	Metabolites	Excretion
Losec gastric acid reducer	Omeprazole	Hydroxyomeprazole	79% of dose found in urine
		Omeprazole acid	
Lipitor cholesterol reducer	Atorvastatin	Ortho- & para-hydroxylated derivatives	<2% of dose found in urine
Zocor cholesterol reducer	Simvastatin	β-hydroxyacid metabolites	13% of dose found in urine and 60% in feces

(continued)

Exhibit 2-4. *(continued)*

Norvasc relieves angina	Amlodipine Besylate	90% converted to inactive metabolites	60% of metabolites found in urine
Prevacid gastric acid reducer	Lansoprazole	Hydroxylated sulfinyl & sulfone	33% of dose found in urine and 66% in feces
Prozac antidepressant	Fluoxetine Hydrochloride	Norfluoxetine	Excreted to urine
Celebrex inflammation reducer	Celecoxib	Carboxylic acids	<3% of dose found in urine and feces; 73% of metabolites found in urine
Paxil antidepressant	Paroxetine	Methylation Hydrochloride products	2% of dose found in urine with 62% of metabolites found in urine; 1% was found in feces
Claritin antihistamine	Loratadine	Descarboethoxy-loratadine-distributed in both urine and feces	80% of dose equally
Zyprexa psychotropic	Olanzapine	10-N-glucuronide 4'-N-desmethyl-olanzapine	7% of dose found in urine; 57% and 30% of metabolites found in urine and feces

Based on sales of these pharmaceuticals and their chemical characteristics, it is clear that urine and feces will contain a significant amount of both the original pharmaceutical active ingredients and their metabolites. Therefore, it is not surprising that pharmaceuticals and their metabolites pollute our water resources. For example, the BBC News reported on August 8, 2004, the discovery of Prozac in England's drinking water. If statistics were available for the top 10 pharmaceuticals in 2004, it is certain that this list would still contain many of these same chemicals but would most likely also include compounds such as Viagra.

TABLE 2-3
Trace Organics in Lake Arrowhead, California Wastewater

Benzaldehyde
1,2-Benzene dicarboxylic acid
Benzene, 1-methyl-4-2(methyl propyl)
Benzophenone
Benzothiazole, 2,2-(methylthio)
Bromodichloromethane
Butyl 2-methylpropyl ester
Cholestanol
3-Chloro-2-butanol
1,3,5-Cycloheptatriene
Cyclohexanone
Cyclohexanone, 4-(1,1-dimethyethyl)
Cyclopentanol 1,2-dimethyl-3-(methylethenyl)
Decahydro naphthalene
Decanal
Diacetate, 1,2-ethanediol
Dibromochloromethane
N,N-Diethyl-3-methyl benzamide
2,5-Dimethyl 3-hexanol
2,2-Dimethyl 3-pentanol
Ethanol, 2-butoxy-phosphate
Ethanone 1-(2-naphthalenyl)
Ethyl citrate
Fluoranthene*
Fluorene*
Heptanal
2-Heptanone,3-hydroxy-3-methyl
Hexadecanoic acid
Hexanol
3 Hexanol
3-Methoxy-3-methyl-hexane
2-Methoxy-1-propanol
Nonanal

(continued)

TABLE 2-3 *(continued)*

Octandecanoic acid
Octadiene,4,5-dimethyl-3,6-dimethyl
Octanal
Phenol 2,4(bis(1,1-dimethylethyl))
Phenol 4,4(1,2-diethyl-1,2-ethanediyl)bisphenol nonyl
1-Phenyl ethanone
Propanic acid 2-methyl-2,2-dimethyl-1-(2 hr . . .)
1-Propanol, 2-(2-hydroxypropoxy)
2-Propanone, 1-(1-cyclohexen-1-yl)
Tetradecanal
Tetradecanoic acid
Tribromomethane
Trichloromethane

*Denotes priority pollutants.

unit for a small suite of organic compounds. Of the chemicals analyzed, those identified in the effluent included acrylonitrile, acrylamide, 4-chlorophenol, dichloromethane, 1,2-dichloropentane, isophorone, methyl ethyl ketone, 2-nitrophenol, 4-nitrophenol, phenol, m-toluate, and m-xylene. Yet another study by the U.S. Geological Survey (USGS) in Nevada (Roefer et al., 2000) found that the following chemicals passed through the wastewater treatment systems: estradiol, nonylphenol ethoxylates, and octylphenol.

In a study on pharmaceuticals and personal care product (PPCP) chemicals in wastewater in both the United States and Germany (Daughton and Ternes, 1999), it was concluded that, "As opposed to the conventional, persistent priority pollutants, PPCPs need not be persistent if they are continually introduced to surface waters, even at low parts-per-trillion/parts-per-billion concentrations (ng-μg/L). Even though some PPCPs are extremely persistent and introduced to the environment in very high quantities and perhaps have already gained ubiquity worldwide . . . ," some of the compounds of concern are composed of a wide range of complex aromatic amines, phenols, and halogenated aromatics.

Given this extensive nonregulated pollution problem, the only method available for protecting the environment from these discharges is to rely on wastewater treatment to remove these pollutants. Doing so is not a simple task. For example, researchers at the University of California at Berkeley (Sedlak et al., 2000), predict that 90 percent of hormones can be removed by a municipal wastewater treatment plant, but some compounds such as N-nitrosodimethylamine (NDMA) are not removed at all. Because there are so many unregulated chemicals discharged to POTWs and so few actual studies on the pass-through of specific organic chemicals, the extent of the problem is truly unknown. In other words, we have just begun to define the problem of micropollutants in wastewater discharges. As a result, scientists recommend that additional treatment methods be evaluated and developed for treatment of such pollutants. More specifically, they recommend the use of ultraviolet light or ozone in place of chlorine/bromine disinfection to eliminate disinfection by-products such as NDMA, and reverse osmosis to remove additional and unspecified chemical pollutants. Without using these treatment methods, most of these unmonitored chemicals will continue to be discharged into the environment unnoticed. It is clear from these studies that a wide range of organic pollutants pass through POTWs and pollute the water. The only questions that remain are (1) how many different manufactured organic chemicals are discharged in wastewater effluents and (2) what are their concentrations.

Another important aspect of this problem occurs when treated effluent from human waste or raw waste is disposed into groundwater resources. For example, a U.S. Geological Survey investigation in Nevada (Seiler et al., 1999) showed the soluble pass-through of pharmaceuticals from septic systems into groundwater. These compounds include chlorpropamide (used for the treatment of diabetes), phensuximide, and carbamazepine, which are used to treat seizures.

Given the potential for the pass-through of chemical pollutants in POTW effluents, injection of these effluents into groundwater aquifers as advocated by various wastewater associations constitutes a serious potential threat to the public health. This practice is usually termed *wastewater reuse*. The WateReuse Association contends that (Saunders, 2001) "the Colorado River receives the treated effluent from Las Vegas, and the Sacramento/San Joaquin Delta is downstream of the discharge of dozens of Central Valley communities. Several Southern California projects recycle more than 170,000 acre-feet of highly treated effluent every year into underground water supplies used by three million to four million people. Some of these projects have operated safely and reliably for nearly 40 years." In general, wastewater treatment focuses on reducing biological pollutant as opposed to chemical pollution. Yet in Los Angeles, it was reported that reclaimed wastewater that was injected into groundwater was polluted with EDTA (ethylenediaminetetraacetic acid) (Barber et al., 1997). To say that this practice is safe, when only biological issues are typically evaluated, is totally misleading. In a groundwater environment, the fate of chemically resistant compounds is not necessarily predictable (see Exhibit 2-5 on the fate of chemicals in groundwater).

Scientists have also expressed concern over micropollutants in recycled wastewater. According to researchers at the University of California at Berkeley (Sedlak et al., 2000), "the effluent-derived contaminants that are currently being discussed in the scientific community account for only a small fraction of the organic compounds in recycled water. As analytical techniques and our understanding of aquatic and human toxicology improve, it is likely that other chemical contaminants and disinfection by-products will be detected in recycled water."

Water reuse projects also rely on the basic chemical nature of unconsolidated sediments in an aquifer, particularly clays and organic

Exhibit 2-5. The Fate of Chemicals in Groundwater

Inorganic and organic chemicals introduced into a groundwater environment may either (1) stay dissolved and move along unimpeded with the groundwater flow, (2) interact with the porous media to a point where the chemicals will be retarded relative to the rate of groundwater flow, or (3) be substantially removed from the groundwater (i.e., sorption to solids, precipitated as an insoluble solid phase, lost as a volatile gas into the unsaturated soil above the groundwater, decomposed by microbial activity, or undergo chemical alteration). Therefore, the ability to predict the fate (i.e., will a specific chemical stay dissolved in groundwater or be substantially altered or removed from the groundwater?) of a specific chemical compound in groundwater is very complex.

Given this complexity, it is important to distinguish between the fate of inorganic versus organic compounds. The majority of inorganic compounds introduced into groundwater, if not immediately precipitated as the result of the chemical characteristics of the groundwater and aquifer materials, will generally exist as dissolved cations (e.g., positively charged metals such as sodium, calcium, iron) or anions (e.g., negatively charged nonmetals such as chloride and sulfate). Because the aquifer media contain natural organic matter and inorganic solids that may exhibit negatively charged surfaces, cations can sorb (i.e., be removed from the groundwater) and desorb (i.e., be introduced back into the groundwater) as groundwater moves through an aquifer. Therefore, in general, inorganic anions can be more mobile (i.e., retarded less) in groundwater than metal cations. For example, a common form of arsenic in groundwater is the arsenate ion (AsO_4^{3-}), which is both soluble and mobile. Other than this very general statement, the ability to predict the fate of inorganic compounds in groundwater requires an extensive amount of chemical data on both the nature of the pollution, the groundwater chemistry, and the chemistry of the aquifer (i.e., mineral, soil, rock, natural organic makeup, and characteristics).

Predicting the fate of an organic chemical is even more complex since the fate of an organic compound in groundwater can be greatly affected by microbial decomposition and volatilization. For those organic compounds that are nonvolatile and not amenable to microbial attack, however, they can persist in the groundwater environment for a very long time. Given this problem, it is important to understand the potential fate of organic chemicals before they are released into our water resources. In general, the fate and transport of organic chemicals in both the soil environment and aquifers have been shown to be generally correlated to their retention (or nonretention) by natural geochemical sorbents (e.g., minerals, soil, rock, and organic matter). In other words, dissolved organic chemicals being transported in soil water or groundwater can adsorb to soil, rock, mineral, and organic matter surfaces or be absorbed into the sorbent matrix. The interaction of dissolved organic compounds relative to surface adsorption is especially important when the dissolved organic compounds are either polar or ionized (i.e, soil media tend to adsorb cations or anions depending on the pH).

Sorption of organic compounds is usually estimated using the sorption coefficient (K_d in units milliliters/gram) that represents the ratio of the sorbed-organic compound concentration (micrograms/gram) divided by the solution-organic compound concentration in micrograms/milliliter. Generally, it has been determined that the sorption coefficient (K_d) is related to the determination of how much an organic chemical is retarded in the groundwater flow by the following equation:

$$R_d = B \, K_d / n_e \qquad (2\text{-}1)$$

where R_d = retardation coefficient, B = average soil bulk density (g/cm^3), and n_e = effective porosity (unitless).

In other words, the larger the K_d value, the greater the chemical is retarded by the soil or aquifer media. Since the average soil bulk density and effective porosity can be measured directly, the only value that needs to be determined is the sorption coefficient. The sorption coefficient can be measured in the field (i.e., if soil is sampled and analyzed for sorbed organic chemicals) or estimated using the following relationship:

$$K_d = K_{oc} \times f_{oc} \qquad (2\text{-}2)$$

where f_{oc} is the fraction of naturally occurring organic carbon in soil and K_{oc} can be correlated to an organic chemical's tendencies to bioaccumulate. The value of K_{oc} has been shown to be related to the water solubility of organic compounds and has also be determined using high performance liquid chromatography. For example, a list of K_{oc} values for pesticides can be found in Wauchope et al., 1991).

The mobility of pesticides in soil water or groundwater can be simply illustrated using the following K_{oc} data and assuming that f_{oc} is constant:

Compound	K_{oc}
Dicamba	2
Atrazine	100
Malathion	1,900

(continued)

Exhibit 2-5. *(continued)*

Based on this information, malathion would be retarded approximately 20 times more than atrazine and 100 times more than dicamba. Clearly, the percentage of organic carbon in a soil or aquifer is a critical number when considering the fate and transport of organic chemicals in the environment.

When experimentally determined K_{oc} values are not available, then K_d values can be approximated by using n-octanol/water partition coefficient (K_{ow}). For example, Schwarzenbach and Westfall (1981) observed that for soil water and aquifer systems with low-organic-carbon content, the following relationship could be used to predict K_d:

$$\log K_d = \log f_{oc} + 0.72 \log K_{ow} + 0.49 \qquad (2\text{-}3)$$

where K_{ow} values are available from handbooks or publications like the *Groundwater Chemicals Desk Reference* (Montgomery and Welkom, 1990).

Generally, the ability to predict the fate and transport of dissolved organic compounds can only be accomplished by completing the necessary data collection, geochemical modeling, and hydrological studies before any discharge. Unfortunately, comprehensive technical studies to assess chemical (regulated and unregulated) fate and transport are not part of any permit system. This means that the true extent and magnitude of mixing polluted water resources with sources of drinking water are, for the most part, unknown and unmonitored. With this degree of uncertainty, the intentional injection of treated wastewater into groundwater aquifers must be carefully evaluated.

---

matter, to adsorb chemical pollutants that may be present in wastewater. Although this process may in fact remove some pollutants from wastewater, these pollutants may be released back into groundwater as a result of natural variations in water chemistry. Furthermore, no studies have evaluated the ability of the sediments in aquifers to attenuate the majority of organic chemicals that can occur in municipal wastewater. As a result, some of these compounds may pass through an aquifer with little or no attenuation, polluting down-gradient groundwater extraction wells used for drinking water. A good example of this problem has occurred in Los Angeles, California, where groundwater basins were polluted with NDMA from wastewater recharge (see California Department of Health Services web site on California Drinking Water and NDMA-Related Activities).

Although water reuse facilities use advanced treatment technologies to remove potential pollutants, both regulated and unregulated, without extensive chemical monitoring it is not certain that pollutants are not being discharged to the environment. As of January 2002 (Crook and Vernon, 2002), a total of 26 water reuse facilities using membrane-based technologies were operating in the United States. The degree of pollutant removal achievable using these technologies was not reported so that the overall safety of these treated waters could be evaluated. One thing is clear, however, the use of membrane-based technologies is becoming more and more widespread.

Similar types of pass-through problems occur at industrial facilities that are regulated under permits. In these cases, the problem is much more egregious because many of these industries know the exact chemical specificity of their discharges or have the resources and technical ability to fully characterize their wastewater. They also have the technical knowledge required to design and operate state-of-the-art systems to treat their effluents to virtual purity. Yet, industries continue to discharge regulated organic pollutants at or below discharge standards and unregulated organic compounds in their waste streams.

Industrial Discharges

Industrial facilities that have NPDES permits for discharge to either a POTW or directly into the environment must meet, at a minimum, the effluent limitations published in the Code of Federal Regulations. Although more advanced methods for the treatment of pollutants[21] have been available since the 1950s, the NPDES program still allows the discharge of regulated pollutants to permitted levels, does not monitor unregulated pollutants except in very general consolidated tests, and assumes that the environment will assimilate or mask the discharge of both regulated and unregulated chemicals in the receiving body of water.

It cannot be assumed, however, that any given environment will assimilate these chemicals. Given the complexity of the organic compounds discharged today, scientists and engineers cannot just rely on dilution and biodegradation to predict the ability of the environment to assimilate a specific compound or compounds. Prediction of the fate of organic and inorganic chemicals in the environment usually requires the use of complex models that are based on either general or site-specific assumptions. An example of the factors that must be considered in predicting the fate of environmental pollutants discharged into surface water resources is given in Exhibit 2-6.

Based on pollutant loading data for a watershed and predictions of chemical behavior, an amount of allowable pollution is incorporated into the NPDES permit. For example, the toxic organics that must be

---

[21]See Chapter 3 for a discussion of water treatment methods.

Exhibit 2-6. *The Fate of Pollutants in Surface Water*

When a chemical pollutant is discharged into a surface water, the potential exists for specific chemical, physical, and biological changes that can influence the fate of that chemical. In other words, what happens to a chemical once it is released into a river? The first thing that happens is that the chemical becomes diluted. For example, in some rivers a wastewater that has 500 ppm arsenic can be diluted over 100,000 times to 0.005 ppm. After dilution, the most important factors that influence the concentration of a pollutant in water are those that cause its concentration to decrease. For example, organic compounds that are biodegradable can be converted to carbon dioxide, water, and other intermediate chemicals. Furthermore, both organic and inorganic compounds can be removed from surface water by (1) volatilization to the atmosphere, (2) sorption to sediments, and (3) chemical reactions that cause new insoluble compounds to form. Finally, some organic chemicals can be broken down by sunlight.

Ultimately, what must be remembered is that each chemical species, whether organic or inorganic, will most likely be affected by one or more of these factors. Thus, our ability to predict the fate (i.e., concentration in water) of any chemical species requires site-specific environmental and chemical information. Without this information, chemical predictions cannot be reliable.

Chemicals that do not readily biodegrade in the environment (i.e., they tend to persist in the environment) can be a significant hazard if they are toxic and accumulate in aquatic organisms. These environmentally persistent compounds have been identified by the USEPA and are to be either prohibited or significantly reduced in wastewater discharges over the next 10 years. These compounds are alkyl-lead, octachlorostyrene, aldrin/dieldrin, DDT, DDD, DDE, mirex, toxaphene, hexachlorobenzene, chlordane, benzo(a)pyrene, mercury, polychlorinated biphenyls, dioxin, and furans. Seven of these compounds, octachlorostyrene, aldrin/dieldrin, DDT, DDD, DDE, mirex, and furans, do not currently have drinking water standards.

---

removed to "acceptable" levels are usually composed of the compounds listed in Appendix 2-2 and those compounds listed as toxic pollutants in the Code of Federal Regulation, Title 40, Part 129. The Part 129 compounds are aldrin/dieldrin, DDT, DDD, DDE, endrin, toxaphene, benzidine, and polychlorinated biphenyls.[22] With effluent limitations

---

[22]All of these compounds except DDT, DDD, and DDE are included in Appendix 2-2.

established, a given facility must then use an appropriate waste treatment technology to meet its permit conditions on the removal of toxic compounds. Once again, such limitations are facility-specific.

An example of effluent limitations for an industrial facility is illustrated using the NPDES permit (CA0005134) for the Chevron refinery in Richmond California. This facility discharges an average of 4.0 million gallons per day of wastewater from petroleum refining, petrochemical manufacturing and research, storm water runoff, construction dewatering at the refinery and pipeline facilities, groundwater monitoring and remediation activities, tank wash water, ship ballast water discharges, and wastewater from storage tanks. The discharge is through a deepwater outfall into San Pablo Bay. The chemical effluent limitations for this facility are:

Constituent	Monthly Average	Daily Maximum
Oil and grease (lb/day)	1,728	
Ammonia (lb/day)	2,052	
Phenolic compounds (lb/day)	20.66	
Dissolved sulfide (lb/day)	30	
Total chromium (lb/day)	24	
Hexavalent Chromium (lb/day)	1.98	
Lead (ppb)	39.9	
Zinc (ppb)	358.6	
Benzo(a)Pyrene (ppb)	0.94	
Chrysene (ppb)	0.91	
Dibenzo(a,h)Anthracene (ppb)	0.87	
Indeno(1,2,3-cd)Pyrene (ppb)	0.91	
Heptchlor Epoxide (ppb)	0.00132	
Copper (ppb)		14.11
Mercury (ppb)		0.21
Nickel (ppb)		65
Selenium (ppb)		50
Cyanide (ppb)		25
Aldrin (ppb)		0.0014
Alpha-BHC (ppb)		0.13
Benzo(a)Anthracene (ppb)		0.49
Benzo(k)Fluoranthene (ppb)	0.49	
Chlordane (ppb)		0.0008
4,4-DDT (ppb)		0.0059
4,4-DDE (ppb)		0.0059
4,4-DDD (ppb)		0.0059
Dieldrin (ppb)		0.0014
Alpha-Endosulfan (ppb)		0.087
Beta-Endosulfan (ppb)		0.087
Endrin (ppb)		0.023
Gamma-BHC (ppb)		0.63

Heptachlor (ppb)	0.0021
Hexachlorobenzene (ppb)	0.0077
PCBs—total (ppb)	0.0007
Toxaphene (ppb)	0.002
Dioxin compounds (ppt)	0.1

Because of these effluent limitations, the Chevron refinery must also have a self-monitoring program to determine whether it is in compliance with its permit requirements. Under that program, the following chemical parameters must be monitored:

- The pH is monitored daily.
- An effluent sample is analyzed monthly for oil and grease, total organic carbon, arsenic, total chromium, hexavalent chromium, cadmium, copper, lead, mercury, nickel, selenium, silver, zinc, cyanide, and PAHs (e.g., benzo(a)pyrene).
- Quarterly samples are analyzed for ammonia, total phenols, and sulfides.
- Yearly samples are analyzed for dioxins and furans, diazinon, and the remaining organic compounds on the TTO list not previously analyzed.

Because many industrial facilities discharge a complex mixture of both regulated and unregulated toxic chemical pollutants, the USEPA requires a combination of monitoring methods (USEPA, 1992). These methods, which include chemical specific analyses, bioassessments, or whole effluent toxicity tests, are required for most facilities that discharge directly to a water resource. All of these methods have their advantages and disadvantages. For example:

- Chemical Specific Testing is precise but expensive when an effluent contains many toxics. This procedure usually does not analyze for all chemicals and certainly does not consider the interactions between toxic chemicals (as illustrated by the Chevron facility).
- Bioassessments measure ecological effect but do not define the specific chemical source that may be responsible for any damage.
- Whole Effluent Toxicity Testing evaluates the toxicity of the "chemical soup" but offers no information about human health impacts.

Of these methods, only the Whole Effluent Toxicity Testing evaluates the potential synergistic effects of chemical mixtures. This method also evaluates the potential effect of unregulated chemicals independent of whether or not they are identified using chemical specific testing methods. As a result, many industries conduct USEPA-mandated studies to reduce the toxicity of their effluent based on the results of Whole Effluent Toxicity Testing.

The wide application by industry of whole effluent testing is an indirect acknowledgment by the USEPA that a toxic threat does not necessarily occur from one chemical but rather from a mixture of chemicals that may be composed of both regulated and unregulated compounds. This admission is further proof that individual water quality standards are not reliable. Industry use of this method is illustrated in the following examples (USEPA, 1989).

A multipurpose specialty chemical plant operating under a NPDES permit in Virginia was discharging approximately 1.3 million gallons of effluent per day into a surface water resource. This facility manufactured and packaged a number of pesticides. As a result, its effluent contained a mixture of these toxic chemicals. Effluent toxicity testing conducted by the USEPA revealed that the effluent was highly toxic to a number of aquatic organisms. To reduce the toxicity, detailed chemical-specific tests were conducted to identify the compounds responsible for the effluent's toxicity. This testing revealed that the organic chemicals responsible for this toxicity were alkyl diamine, dicyclohexylamine, and piperonyl butoxide.

Tosco's Avon Refinery in Martinez, California produces refined petroleum products, primarily gasoline and diesel fuel. Routine NPDES monitoring requirements were limited to pH, ammonia, oil and grease, chromium, zinc, sulfur, chlorine, and dissolved oxygen (DO). Because the effluent was found to be toxic, it was tested to identify all organic priority pollutants and major nonpriority pollutants. These test data revealed the occurrence of the following compounds in the effluent: ketones, toluene, amines, and dibenzofuran.

The Glen Raven Mills in North Carolina used dyes and surfactants. The effluent from its wastewater treatment plant was evaluated using whole effluent toxicity testing methods and found to be toxic to aquatic organisms. With further analysis for specific chemical constituents, it was determined that surfactants, linear alcohol ethoxylate, and sodium dodecylbenzenesulfonate were responsible for the toxicity of the effluent. These three examples clearly show that industrial facilities do release unregulated toxic compounds into the environment and that there are no set standards for their control. These tests also show that the toxicity of an industrial effluent is a function of more than one chemical constituent. Both of these facts suggest that water quality cannot be protected by using water quality standards as presently promulgated. Furthermore, the use of whole effluent toxicity tests alone are not appropriate as an indicator of water quality, as these tests have no direct relationship to human health effects.

More important, these examples are significant because the compounds responsible for the effluent toxicity are not listed as toxic organic compounds under the NPDES program (see Appendix 2-2) nor regulated as toxic compounds in drinking water (see Table 1-9). This fact illustrates, once again, the failure of existing water quality standards to protect the environment.

These sources of pollution are a potential health threat to our water resources. One very good example is the release of inorganic perchlorate salts that are not regulated under any USEPA program.

Perchlorates are used in a variety of products that include electronic tubes, car air bags, leather tanning, fireworks, and solid fuel propellants for rockets. Large-scale production of perchlorate began in the 1940s and expanded along with the growth of rocket propelled ordnance. Since the 1950s, the primary manufacturing location of perchlorate was at the American Pacific and Kerr-McGee plants (i.e., 1945-1989) outside Las Vegas, Nevada. During their years of production, these facilities discharged their wastewater into Lake Mead. Today, low levels of perchlorate are found in the entire Colorado River system below Las Vegas, which affects the drinking water of 22 million people and water used for irrigation.

Thus, unregulated organic and inorganic chemicals continue to pollute our water resources. This problem has been accurately summarized by the United States' General Accounting Office (1994) review of EPA's toxic substances control programs (Exhibit 2-7).

However, polluted effluent is easily collected and can be treated before its discharge into the environment. Nonpoint sources of pollution, by their very nature, are dispersed into the environment, are difficult to collect, and cannot be easily controlled. As a result, the ability to treat polluted water arising from a nonpoint source is very difficult at best and creates a pollution problem that has few practical solutions.

Exhibit 2-7. *An Assessment of EPA's Control of Toxic Substances*

The General Accounting Office (GAO) reviewed the pollutants discharged from 236 facilities from three industrial sectors (pulp and paper, pharmaceutical, and pesticides manufacturing). The vast majority (77 percent) of the toxic pollutants identified by the GAO were not controlled through the permit process. Priority pollutants (see Appendix 2-2) made up 33 percent of the pollutants, and the remaining chemicals were unregulated, nonpriority pollutants. The GAO concluded that, "Although most of these toxicants are 'nonpriority' pollutants, they do pose risks to both human health and aquatic life. GAO tried to examine the implications of uncontrolled pollution cases identified in the facility sample population, but the majority of cases could not be evaluated because of a lack of criteria for assessing the health risks posed by the discharges."

Nonpoint Sources of Water Pollution

Nonpoint sources of pollution are distributed throughout both urban and rural environments. The major sources of potentially toxic nonpoint pollution are:

- Pesticide pollution from agriculture land. Pesticides applied to the land can be (1) transported by wind to pollute surface water resources, (2) carried in surface water runoff to adjacent surface water resources, or (3) carried in percolating water through soil into groundwater resources.
- Naturally occurring perchlorates and nitrates from geological sources contribute to nonpoint pollution. But unlike other toxic inorganic elements that are used in industrial processes (e.g., chromium, arsenic, mercury), perchlorate and nitrates are also manufactured compounds. Thus, these pollutants are discussed independently of other naturally occurring trace elements.[23]
- Mining of coal and metal sulfides ores (e.g., gold, silver, copper, lead, zinc) from the earth can disturb thousands of acres of land. Because of this disturbance, trace metals and acid will pollute surface and groundwater resources.
- Metallurgical processing of base metals (e.g., copper, lead, cadmium, zinc).
- Chemicals released from commercial, industrial, and waste disposal facilities pollute surface and groundwater.[24]
- Urban environments are subject to the accumulation of chemicals on buildings, streets, impervious surfaces, soils and lawns, in storm channels and sewers, and areas of waste storage or disposal. Common pollutants are pesticides, metals, solvents, and petroleum products. During rainfall events, these chemicals will be carried in the runoff to adjacent bodies of water.
- Rural environments also produce human waste disposal problems from septic systems and surface runoff from feedlots.

Because these sources tend to provide fairly constant and long-term release of chemicals into the environment, they will always be a threat to water resources and our drinking water quality. As a result, USEPA has supported the development of methods to control nonpoint sources of pollution. However, most of these programs are focused only on controlling pollution arising from sediment and fertilizer releases. Chemical and pharmaceutical pollution remains virtually uncontrolled. The impact of

---

[23]Cyanide (CN⁻) is also naturally occurring but is associated with food, plants, and bacteria as opposed to geologic sediments.
[24]Although these sources are usually considered point sources, many of these sites have multiple points of unreported and unregulated chemical releases that are distributed over a wide area.

nonpoint pollution on water quality and our ability to control it is illustrated next.

Agricultural Pollution

Agricultural practices impact water resources as the result of applying pesticides to crops and the management of livestock. These practices pollute both surface water and groundwater resources.

Pesticides

The pollution of water resources by (1) the direct application of pesticides to bodies of water to control aquatic plants, (2) the indirect pollution of surface waters by air transport and/or surface water runoff of pesticides, and (3) the pollution of groundwater by pesticide seepage through soil has been well documented since the 1940s. Since then, more pesticides have been developed and spread across both agricultural and urban landscapes. Thus, it is no wonder that much of our water resources are polluted by pesticides or the residues of degraded pesticides.

For example, a study conducted by the USGS and the U.S. Department of Agriculture (USGS, 2000) found that the pesticide diazinon, which is used in agricultural areas of California's central valley, is also present in amphibians that live in pristine mountain ponds and streams. In fact, more than 50 percent of the frogs and tadpoles tested at Yosemite National Park had measurable levels of diazinon. As a result, diazinon is suspected of being responsible for the drastic decline in amphibian populations in the Sierra Nevada. Since the use of diazinon is restricted to agricultural areas, the diazinon found in the Sierra Nevada ponds, lakes, and streams is probably transported on prevailing summer winds from the intensely agricultural San Joaquin Valley.

This type of pollution is virtually impossible to control, as the application of pesticides in the form of an aerosol or powder makes it susceptible to wind transport. Even after a pesticide has been applied, pesticides adsorbed on fine soil particles can be transported along with the soil eroded by the wind. This type of pollution knows no state or country boundaries. Thus, pesticides banned in the United States but sold to a foreign country may find their way back into the waters of the United States by wind transport.

Pesticides can be applied to a soil or vegetation in the form of a spray, powder, gas, or dissolved in irrigation water. Once a pesticide has been applied to a soil, it can be transported by surface water runoff[25] to adjacent surface bodies of water. As early as 1956 (Middleton and Rosen, 1956), a study on organic pollution reported that concentrations of DDT in the range of 1 to 5 ppb were found in the drinking water of several cities that used rivers as their sources of supply. This study also

---

[25]As the result of either rainfall or irrigation.

concluded that drinking water pollution by organic chemicals was a serious problem because the concentrations at which there might be physiologic effects were not then known.

Another early study conducted by the California Department of Water Resources in 1963 characterized the pollution of surface water drainage in the San Joaquin Valley (Anonymous, 1969). This study identified the following pesticides in agricultural surface waters: aldrin, BHC (hexachlorocyclohexane), chlordane, CIPC (chlorprophan), DDE (dichlorodiphenyldichloroethylene), DDT (dichlorodiphenyl-trichloroethane), dieldrin, endrin, heptachlor, heptachlor epoxide, lindane, methoxychlor, tedion, thiodan, and toxaphene. Approximately 30 years later, the list of pesticides that are found in agricultural and urban water resources has increased despite the banning of a number of pesticides. A water quality study conducted by the USGS (1998) from 1992 to 1996 collected surface water data in 20 of the nation's major hydrologic basins and analyzed for pesticides and pesticide degradation products. The results of this study are summarized in Appendix 2-3. Of the 83 pesticides and pesticide degradation products for which samples were analyzed, 74 were detected. Eleven of the pesticides were detected in more than 10 percent of all surface water samples.

These data demonstrate the widespread distribution of pesticides in our water resources. Given the extensive origins of the pollution, the only effective way to prevent the pollution of drinking water by pesticides is to (1) not allow any runoff into adjacent bodies of water, (2) collect all agricultural runoff for treatment before its release, (3) only allow the use of pesticides that biodegrade rapidly so that there is no runoff threat, or (4) ban the use of pesticides completely. Sadly, none of these options is practical for large areas. Thus, pollution of receiving waters from pesticides in agricultural runoff will continue to be a long-term management problem, particularly since large numbers of pesticides have been detected in surface waters, yet there are very few pesticides included in drinking water standards.

All pesticides will dissolve to some degree in water and be degraded by microorganisms in the soil root zone. Pesticides can also be adsorbed by soil as the pesticide-laden water percolates downward into the groundwater. Thus, the ability to predict which pesticides are a groundwater pollution threat depends on the specific properties of each pesticide and the soil environment in which it is used. Nonetheless, it is obvious that a wide range of pesticides currently pollute our groundwater resources and will continue to do so.

Compared to surface water pollution, pesticide pollution of groundwater is an even greater problem because once it occurs, it is a long-term hazard, tends to impact large areas, and is expensive to remediate.[26] This impact is typified by the pollution that was created by

---

[26]Remediation does not necessarily mean the return to the original water quality.

dibromochloropropane (DBCP) in California. DBCP was added to irrigation water for application to agricultural crops in California's Central Valley from the late 1950s through the mid-1970s (University of California at Davis, 1994). Studies on the application of DBCP in irrigation water concluded that DBCP moved readily with the water and was distributed to wherever the water moved (O'Bannon, 1958), including downward.

As early as 1961, DBCP was reported as being toxic to animals (Torkelson et al., 1961). However, not until 1977, when DBCP was shown to cause sterility in male workers, did California ban its use. Two years later, groundwater pollution by DBCP was reported in California's Central Valley. By 1985, it was determined that almost 2,500 wells throughout the Valley had been polluted. It took the California Department of Health Services another 4 years to establish a maximum contaminant level for DBCP. By 1992, several cities in the Central Valley had to treat the water at their well heads (i.e., at the point of groundwater extraction) to remove DBCP.

This example illustrates that the public was exposed to DBCP in drinking water resources for decades before any actions were taken. Even though a maximum contaminant level was established, this pollutant is still allowed in drinking water. Today, another similar contaminant is being investigated in groundwater resources. The compound is trichloropropane (TCP), which was an intermediate in the synthesis of soil fumigants and also used in the manufacture of epichlorohydrin (i.e., a polymer flocculent used to treat drinking water). Relative to the manufacture of epichlorohydrin, the use of TCP suggests that drinking water treated with epichlorohydrin may also contain trace concentrations of TCP. Yet, TCP is not a regulated compound so it is not monitored. The toxicology and exposure routes for TCP are further discussed by the World Health Organization (2003).

The extent of groundwater pollution by pesticides is also illustrated in the earlier referenced USGS study (1998). The results of this study of groundwater pollution is summarized in Appendix 2-4. Of the 83 pesticides and pesticide degradation products analyzed, 59 were detected in groundwater and 5 were detected in more than 10 percent of all groundwater samples. These studies clearly point out that whether we look at surface waters or groundwaters, once a resource is polluted, there remains a long-term management problem that is not addressed by current drinking water standards.

Livestock Management

The primary pollutants produced from livestock are nitrates, hormones, and pharmaceuticals that occur in animal waste. Unless a farm or ranch has management facilities to collect liquid waste from livestock operations (e.g., collection facilities beneath areas of animal habitation), the urine and feces of livestock are deposited on bare soil. Under these conditions, water-soluble chemicals will percolate through soil into groundwater or run off the soil during rainfall events into adjacent bodies of water. Thus, farming or ranching activities can pollute groundwater and/or surface water resources depending on the number of animals being managed and site-specific climatic, soil, geological, and hydrologic factors.

The livestock operations that pose the greatest pollution threat are defined by the USEPA as Concentrated Animal Feeding Operations (CAFO).[27] For example, the USEPA can designate any animal operation as a CAFO, but usually a CAFO is a livestock operation that has either 1,000 animals (e.g., cows, horses, swine, turkeys) confined at a facility or has more than 300 animals that are confined at a facility that discharges pollutants into surface waters. If a facility is designated as a CAFO, it will need an NPDES discharge permit. In general, the NPDES permit can require the control and/or treatment of polluted runoff as well as requiring impermeable liners between stored animal wastes and groundwater.

Perchlorates and Nitrates[28]

Although perchlorate pollution of surface water and groundwater resources has been primarily associated with rocket fuels (USEPA, 1999), the USEPA found in 1999 that garden fertilizers contained perchlorate at concentrations ranging from 1,500 ppm to 8,400 ppm (Renner, 1999). The source of the perchlorate in fertilizer probably resulted from the use of Chilean nitrates, which have been known since the 1800s to contain perchlorate. Because perchlorate is very soluble in water, fertilizers applied to soil can be a source of perchlorate to both surface water and groundwater. The extent of perchlorate pollution from fertilizers is still unknown, as widespread monitoring for perchlorate has not yet been required.

Based on the occurrence of perchlorates in Chilean nitrates, it is also possible that perchlorates occur naturally in nitrate deposits in the United States given that nitrogen occurs naturally in sedimentary rocks (Stevenson, 1962). For example, a study of natural nitrate pollution in the California's Central Valley (Sullivan et al., 1979) identified high nitrate concentrations in the parent materials of the Valley's soils. This study suggested that irrigation of these soils contributed to natural nitrate groundwater pollution. Given this condition, perchlorates could, if present, also have been leached into the groundwater. As a result, perchlorate may also pollute groundwater in regions that contain high nitrate rocks and soils.

---

[27]Regulations pertaining to Concentrated Animal Feeding Operations are codified under CFR 40, Part 122.
[28]Nitrate pollution is usually associated with nonpoint runoff from agricultural crops, parks, and lawns and groundwater pollution from human/animal waste disposal practices (i.e., land-spreading of sludge, septic systems, cesspools, pits, ponds, and lagoons).

This condition may exist in Texas as well as in other western states. Researchers at Texas Tech University have reported perchlorate contamination in milk that is associated with the water resources in western Texas (Kirk et al., 2003). Given the widespread distribution of perchlorate in groundwater in Texas and 23 other states (Anonymous, 2003), perchlorate is suspected to be anthropogenic.

Mining and Pollution

When iron sulfide minerals are exposed to the atmosphere and water, the water becomes acid and enriched in iron that eventually leaves yellow and reddish deposits in its wake. Other metal sulfides also react with the atmosphere and water to release toxic metals such as arsenic, cadmium, cobalt, copper, nickel, and zinc. Although this process has been known since the early 1800s (Grammar, 1819), the occurrence of toxic metals, such as zinc, copper, arsenic, and antimony in acid mine water, was not widely publicized until the 1870s (Williams, 1877). By the early 1900s, the hazards of mine waters to both surface and groundwater were well known. For example, a report on surface water pollution from mining observed that "the appearance of the stream polluted by mine water is striking and somewhat uncanny, for all vegetable and animal life is destroyed, and the bright, clear waters splash forbiddingly over the bed, which is stained yellow by the iron" (Parker, 1907). The types of mines that are responsible for highly acid water and toxic levels of trace metals include all mines that contain metal sulfides (e.g., iron, lead, zinc, copper). Therefore, mines that extract aluminum, coal, copper, gold, lead, silver, and zinc will usually pollute surface and groundwater resources with acid and toxic metals. As a result, virtually every state in the nation has some mine-related pollution.

Because the process that creates acid and trace metal pollution is the result of excavation (i.e., bringing metal sulfides to the earth's surface), the minerals that cause the formation of acid can be exposed to the atmosphere for hundreds to thousands of years. As a result, the creeks and rivers that drain mining areas will be polluted far into the future. In addition to polluting surface water, the infiltration of polluted mine waters through fractured rock and soil also impacts groundwater.

The mining process by its vary nature disturbs hundreds of acres of land usually intersected by numerous creeks and rivers. Furthermore, the areas from which acid water is discharged to surface water can occur over numerous locations in a watershed. This fact makes the collection and treatment of the water to reduce acidity and metal concentrations more difficult. In most mining environments, some combination of water management and treatment can be implemented to reduce the pollution. Again it should be stressed that mined lands are only reclaimed, not truly remediated. At best they can be "restored" to some acceptable level of pollution. Thus, even after remedial actions are implemented to reduce pollution, low concentrations of toxic metals can and do occur in downstream water resources.

Metallurgical Pollution: Point Source Pollution to a Nonpoint Source

The first decade of the 20th century was pivotal in the recognition of atmospheric pollution from metallurgical smelters in the United States and damage caused to both land and water. These conditions were brought to light in the numerous litigations seeking damage from smelting air pollution (Johnson, 1917). Because smelter smoke was known to contain metals such as lead, zinc, arsenic, copper, and cadmium (Egleston, 1883), the industry was trying to remove these known toxic metals (Seiffert, 1897) from smelter smoke at least as early as 1874 (Egleston, 1883). Yet, at this time period, little attention was given to the contribution of these constituents to atmospheric pollution and the resultant injury to plants, animals, and water resources.

The egregious pollution from four infamous smelting sites in the United States (i.e., Ducktown, Tennessee; Anaconda, Montana; Salt Lake City, Utah; and Selby, California) provided a worldwide focus on airborne emissions from such operations. For example, at Anaconda, Swain (1949) reports that:

> According to investigations made at the time, there were discharged into the atmosphere from this stack daily over three billion cubic feet of gases, measured at stack conditions, 2300 tons of sulfur dioxide, 30 tons of arsenic, 3 tons of zinc and over 2 tons each of copper, lead and antimony trioxide.

An investigation published by the U.S. Department of Agriculture (Haywood, 1908) also reported on the Anaconda situation:

> Finely divided particles of the ore are discharged from the smelter stack and settle on the surrounding country. These fine particles of flue dust may contain sufficient amounts of copper, lead, and zinc to injure vegetation, because of their toxic action through the medium of soil and because of their toxic action on the foliage.

The Ducktown smelting operation left a trail of damage from airborne particulate emissions for some 30 miles, extending across state lines into Georgia (Swain, 1949). Haywood's study in 1908 reported massive deforested areas attributed to smelter emissions. In 1913, the Selby Commission was appointed by the court to research, suggest, and enforce the use of technology(s) for reducing airborne emissions that were damaging vegetation and poisoning livestock.

In October 1948 (Schrenk et al., 1949), "The whole nation was shocked when 20 persons died and several thousand more became ill during the smog that enveloped the town of Donora, Pa., during the last week of October, 1948." The Donora incident demonstrated that metal contaminant toxicity was responsible for both acute and long-term injury. The contribution of metal particulates such as arsenic, antimony, lead, zinc, cadmium, iron, and manganese to this problem was found to be considerable.

An investigation (Jordan, 1975) of a zinc smelter revealed elevated concentrations of zinc and cadmium in the soil horizon up to 25 km east and 16 km west of a zinc smelting operation. The study reported zinc deposition as high as 187 to 561 pounds per acre for the manufacturing year 1969 alone; the operation had been in existence since 1898. This report concludes that:

> Today the once thickly forested slopes are sparsely vegetated or completely barren over an area of about 485 ha. Dry, sun-bleached logs litter the rocky slopes, giving the area a desolate, devastated appearance. . . . Thus it appears that soil Zn content alone is sufficient to account for the stunting of seedlings observed on soils collected within 5 km of the east plant smelter. . . . One fact is certain—without human intervention to ameliorate the metal toxicity, it is likely that the denuded areas will remain barren for decades or centuries to come.

This brief history of the nonferrous smelting industry illustrates how a "point source" of pollution can generate such a widespread plume of toxic metals. Once these massive quantities of metallic elements are deposited on soil and in surface water, they can serve as nonpoint sources of pollution to our water resources. For example, the metals of concern (i.e., lead, zinc, arsenic, cadmium, copper) are all sufficiently soluble in water to pollute both surface water and groundwater with every rainfall. In some respects, the situation parallels acid mine drainage. In other words, a massive accumulation of metals exposed at earth's surface will be a continuing source of metal pollution for decades for the foreseeable future.

Industrial, Commercial, and Waste Disposal Sites

Virtually every industrial facility in the United States that manufactured or used toxic chemicals has historically polluted air, land, and water resources. This pollution is primarily the result of improper waste and chemical handling practices or accidental releases to the environment (Liang, 2001). Such pollution also occurs at smaller commercial facilities that use or sell chemicals as part of their operations. Good examples of the pollution caused by smaller businesses are dry cleaning facilities that use solvents (e.g., tetrachloroethylene or perchloroethylene) and gasoline stations. Thousands of dry cleaning facilities across the nation have created widespread pollution of groundwater resources with prechloroethylene (PCE) and discharged PCE to sewers where it can then reach surface waters. Similarly, the gasoline additive, methyl tertiary butyl ether (MTBE), is a serious problem in many states. In California, it is a particularly widespread problem. The California Department of Health Services reported in August 2001 that the following sources of drinking water are polluted with MTBE: 0.6 percent of groundwater sources,

4.4 percent of surface water sources, and 1.9 percent of public water systems. Although there is no federal water quality standard, the State of California has set a primary drinking water standard for MTBE of 13 parts per billion (ppb).

Another problem with MTBE is that it is difficult to remove from water by standard treatment methods. Fortunately, a new treatment method involving both ozone and hydrogen peroxide has been able to remove MTBE below even California's secondary drinking water standard of 5 ppb (Sullivan, Agardy, and Traub, 2001). Furthermore, the oxidation by-products of MTBE (e.g., t-butyl alcohol, t-butyl formate, isopropyl alcohol, and acetone) are more biodegradable than MTBE itself.

In addition to facilities that either manufacture or use chemicals, waste disposal sites (e.g., sanitary landfills and hazardous waste landfills) that received toxic chemicals have also been significant contributors to the pollution to surface and groundwater. Unlike industrial facilities that manufacture a relatively few number of chemicals, waste disposal sites can contain a vast array of chemicals that can pollute water resources. Thus, the ability to monitor chemical pollution from a landfill is more difficult than monitoring for pollution from industrial facilities because, at an industrial facility, regulators generally know what chemical to monitor. As a result, waste disposal sites can also be a source of unregulated chemical pollutants that go undetected by regulatory agencies because they concentrate on regulated priority pollutants. An example is the study of trace organic chemicals in landfills (Lang et al., 1987). A list of organic chemicals found in landfill leachate and gases (i.e., which dissolve in water) is given in Appendix 2-5.

Regardless of the source, when chemicals pollute the environment, the regulations that exist usually require that the extent of the pollution be characterized and remedial measures be implemented. Pollution from industrial, commercial, and waste disposal sites around the nation is a significant threat to the environment. It has been estimated that more than 300,000 polluted sites within the United States require some degree of soil and/or groundwater cleanup (National Research Council, 1994). The degree to which chemical pollution is remediated is usually based on risk evaluations. Thus, as long as soil pollution is no longer contributing to surface or ground waters, regulatory agencies usually allow the soil to be capped and left in place. Likewise, polluted sediments in streams, rivers, and estuaries may not be remediated if (1) the pollution from these sources is not considered a health risk by the regulatory agency or if (2) dredging a river may well resuspend and distribute more pollution downstream. As a result, remediation of pollution at most industrial and commercial sites usually focuses on groundwater.

Groundwater pollution at industrial and commercial facilities is a significant problem because pollution from these facilities tends to be concentrated and there is little, if any, dilution. A plume of polluted groundwater can spread for miles away from the source of the pollution. In many cases it is impossible or impractical, even when using the best available technology, to reduce pollutants to background levels. The most common groundwater pollutants from industrial and commercial facilities tend to be the USEPA's priority pollutants (Montgomery and Welkom, 1990) because they are regulated compounds.

Examples of chemical contaminants that have polluted and continue to pollute groundwater include the typical dry cleaning solvents, trichloroethylene (TCE) and PCE. Common use of these solvents began in the 1940s when there was a war-driven need for degreasers. Although manufacturers of these chemicals recognized their toxic properties, users typically addressed their disposal by either discharging waste solvents to nearby soil or to sewage and storm water systems. Those more cost conscious recycled their spent solvents. As a result of less-than-adequate disposal practices, many areas of the country are suffering significant clean-up costs so as to remove these solvents from soil and groundwater. It is almost axiomatic that where there were or are dry cleaners, there will be groundwater contamination. Another example is 1,4-dioxane, a solvent and solvent stabilizer that is also found in ordinary household products like shampoos, liquid soaps, baby lotion, and cosmetic products. It is also used in a variety of manufacturing processes including electronics, metal finishing, paper manufacturing, and pharmaceuticals, as well as in the manufacture of foam insulation. Today, it can usually be found as a groundwater pollutant associated with chlorinated solvent releases such as TCE.

When groundwater needs to be remediated, most federal and state laws commonly use cleanup goals based on water quality standards. However, the USEPA has recognized that the ability to attain drinking water standard levels using current treatment methods[29] is not feasible at many sites[30] (National Research Council, 1994). Furthermore, there is no long-term institutional structure in place to ensure that polluted groundwater will not be unknowingly used.

Since groundwater pollution persists for a very long time and "pump and treat" systems have been shown to be inadequate at fully addressing aquifer restoration, alternative methods may need to be used to reduce pollutant levels in the groundwater. Commonly used methods include:

- Mixing water from unpolluted wells with the polluted well water until the level of the pollutant meets water quality standards
- Bringing in a new source of drinking water for those areas affected
- Treating the polluted water at the well head upon its extraction and before its distribution to consumers

---

[29]The most widespread method used is called *pump and treat*. In other words, pump the polluted water out of the aquifer and treat it before it is discharged back into the environment.

[30]This is especially true for groundwater polluted with chlorinated solvents (i.e., trichloroethylene, perchloroethylene).

• Treating the polluted water at the location of its use (i.e., residence or business)

Regardless of the methods used, the polluted groundwater remains a nonpoint source of questionable quality. Even for those few sites where polluted groundwater can be cleaned up to drinking water standards, a low lingering level of pollution will still be present. As a result, industrial and commercial sites that have polluted groundwater can and do serve as long-term sources of low levels of pollution to receiving waters that may be used as a source of drinking water.

Given all of the chemical pollution of our groundwater resources that is the direct result of allowing low levels of organic and inorganic chemicals to be discharged into the environment, there is no federal program that controls toxic pollutants (i.e., GAO study of 1994). This fact can be amply illustrated by comparing those chemical compounds that are required to be monitored in groundwater around hazardous waste sites under the Resources Conservation and Recovery Act to other primary lists of EPA-regulated organic chemicals (see Exhibit 2-8). The "toxic dichotomy" illustrated by this comparison shows that there is no consistent scientific basis for determining which chemicals should be

---

Exhibit 2-8. *The Toxic Chemical Dichotomy*

The chemicals listed as priority pollutants in Appendix 2-2 are clearly hazardous to both the operation of a publicly owned treatment works (POTW), as well as to the aquatic environment that receives polluted wastewater. This fact is obvious because the USEPA actively regulates these compounds. Yet, of the 124 chemicals on the priority pollutant list, only the following chemicals have established primary drinking water criteria:

Benzene, benzo(a)pyrene, carbon tetrachloride, chlordane, chlorobenzene, dioxin, endrin, heptachlor, heptachlor epoxide, hexachlorocyclohexanes (BHC), hexachlorobenzene, 1,2-dichloroethane, dichlorobenzene, 1,1-dichloroethylene, dichloromethane, 1,2-trans dichloroethylene, 1,2-dichloropropane, ethylbenzene, di(2-ethylhexyl)phthalate, pentachlorophenol, polychlorinated biphenyls (PCB), tetrachloroethylene, toluene, 1,2,4-trichlorobenzene, 1,1,1-trichloroethane, trichloroethylene, toxaphene, and vinyl chloride.

The remaining compounds on the list do not have established drinking water criteria (i.e., are unregulated). This omission is significant, as priority pollutant compounds, if present in wastewater, will contaminate receiving waters, which in turn may be a source of drinking water.

(continued)

Exhibit 2-8. (continued)

Furthermore, of the 56 organic chemicals listed in the primary drinking water standards in Table 1-9, the following chemicals are not listed on the priority pollutant list: acrylamide, alachlor, atrazine, carbofuran, 2,4-D, dalpon, DBCP, cis-1,2-dichloroethylene, diaquat, endothall, epichlorohydrin, ethylene dibromide, glyphosate, hexachlorocyclopentadiene, methoxychlor, oxamyl, picloram, simazine, styrene, silvex, and xylenes. These compounds are obviously toxic, yet they would not even be monitored by a POTW.[31]

The USEPA has promulgated regulations that have defined a group of chemicals as being toxic. One group is considered toxic in drinking water, and the other is considered detrimental to POTW operations and the aquatic environment. This dichotomy exists because the cost impact of adding more chemicals to the drinking water standard list is perceived to be unmanageable.

When the Resource Conservation and Recovery Act was implemented to regulate hazardous waste, it was required that groundwater be monitored for industrial chemicals that were listed in CFR 40, Part 264, Appendix IX. This list of potential groundwater pollutants is compared to USEPA-regulated compounds under existing drinking water standards, priority pollutant list, and the ambient water quality criteria (see Appendix 2-9). This comparison shows that of 216 chemicals to be monitored in groundwater around hazardous waste sites, only 29 chemicals are regulated as a drinking water standard, priority pollutant, or ambient water quality criteria. Clearly, there are a significant number of chemicals that are perceived to be both toxic or harmful to human health or the environment; yet they remain unregulated.

The toxic dichotomy also exists between nations. For example, there are major differences between Tables 1-9 (USEPA Drinking Water Standards) and Table 1-10 (European Union and World Health Organization Drinking Water Recommendations) with regard to the allowance of pesticides and industrial chemicals. Once again, there is no consensus regarding which chemicals should be regulated.

---

[31]These compounds occur in urban storm water runoff and from household discharges that are treated by POTWs.

regulated. As a result, allowing unregulated chemicals in our water resources cannot be scientifically justified.

Urban Runoff

When storm water scours an urban environment, accumulated debris, residues, and wastes are commonly flushed into adjacent bodies of water. Today in urban communities, the amount of pollution from industrial facilities and construction sites must be controlled. In larger metropolitan areas (i.e., an urban area with a population greater than 100,000), runoff is required to be treated along with domestic wastes either at the regional treatment plant or during storm periods, collected and stored in large containment structures, and subsequently treated before release. Some communities that do not have the funds necessary to expand wastewater treatment or construct storm water storage capacity have investigated the process of injecting aluminum sulfate into storm water conduits to remove pollutants (Herr and Harper, 2000). These studies have shown an 80 to 90 percent removal rate of heavy metals.

In addition to the more commonly found waste constituents of urban runoff, chemicals that have accumulated in the environment (i.e., roads, roofs of buildings, residential lawns, parks, golf courses) are carried along in runoff. In some sections of the country, urban runoff is viewed as a *water resource* of value and is therefore directed to percolation basins that allow it to percolate through the soil and into groundwater, thereby recharging the groundwater aquifers. For this process to be effective, the soil must retain the pollutants so that groundwater resources are not degraded.

However, recharging aquifers with polluted surface water simply shifts the pollution from surface water to the groundwater. For example, residential drinking water wells have been polluted by pesticides commonly used in urban environments (Eitzer and Chevalier, 1999). This type of pollution typifies the hazard of leachable chemicals commonly found in the urban landscape.

Rural and Developing World Pollution Problems

In many of the less industrialized nations, human wastes, commonly called "night soil" have been and continue to be used as fertilizer. The consequences of this method of fertilization are almost too numerous to mention. For example, it has been responsible for parasites such as schistosomiasis (blood flukes), which have created health problems worldwide. Although in today's environment focus has been placed on esoteric chemical contaminants, the age-old parasite continues to be a major source of illness associated with the disposal of human waste. Other water-borne diseases include amoebic dysentery, bacillary dysentery, cholera, hepatitis A, paratyphoid, typhoid, and polio. The World Health Organization (WHO) has reported that 40 percent of the world's population lack basic sanitation services and at least 1 billion people do not have access to safe (pathogen-free) drinking water. As a consequence, in excess of 2 million die from diarrhea alone (Brown, 2002). On a worldwide basis, millions to billions of disease cases can be related to these pollutants, with many related to polluted water.

Recognizing that in developing countries pathogens pose the greatest risk to drinking water safety, there are steps which can be taken to reduce risk and hopefully eliminate the microbiological impact (Cloete et al., 2004):

- Watershed protection
- Coagulation-flocculation
- Preozonation
- Filtration
- Chloramination
- Pressurized distribution systems (24-hour)

Although microbial pollution is always addressed in the industrialized nations, the current concern and, therefore, the greatest interest and activity focuses on the broad range of inorganic and organic chemicals currently identified both in water resources and drinking waters, the developing nations are still plagued with the more pressing problems of controlling bacteria and viruses, whose health effects are so immediate that they predate and mask the lesser understood and longer term health effects of industrial chemicals. In fact, considering the rather rapid adaptation of so many bacteria and viruses (to which there is often no immediately available medical response) it may well be that pathogen adaptability may eventually pose a greater risk to industrialized nations that the industrial chemicals receiving such attention today.

Water-Borne Incidents in "Affluent" Nations

It is reasonable to assume that affluent nations have and do use the most modern of treatment works to support and sustain high-quality drinking water. Although this assumption is generally valid, it ignores a rather large percentage of populations served by local wells and water systems, many of which receive only minimum if any treatment. All things considered, one would expect that many of the problems that plague developing nations have their parallel in the rural areas of affluent nations (Hrudey and Hrudey, 2004).

Since 1990, incidents of pathogens in drinking water were reported in the United States, Australia, Canada, Japan, Ireland, Denmark, Scotland, Sweden, Finland, Italy, and New Zealand. A more comprehensive description of these recorded events, including events going back through the past three decades, can be found in *Safe Drinking Water* (Hrudey and Hrudey, 2004). The events cited are small in number of persons impacted. The events are also isolated in time, but they did occur and continue to occur. In the final analysis, under our current systems, even affluent nations are not immune to the illnesses suffered in developing nations. They differ only in magnitude.

Basin Water Quality Management

In an attempt to improve national water quality, the Clean Water Act (CWA) established regional performance-based programs to manage the water quality of regional watersheds (e.g., NPDES program, storm water management programs, and the control of nonpoint pollution). As part of these programs, the CWA also established the Total Maximum Daily Load Program (TMDL). Under this program, a given watershed may need to place a TMDL on specific organic pesticides in a watershed that does not meet the established water quality criteria. In general, this program requires states to (1) identify those bodies of water that do not meet applicable water quality standards and (2) provide a road map for efforts to attain and maintain state water quality standards. This usually means that pollution controls need to be improved to meet existing water quality standards.

The ability to identify the pollution parameters that need to be controlled is usually not a problem. The current causes of chemical water quality impairment include nutrients, metals, dissolved oxygen, toxic organics, mercury, and pH. Almost all of these pollutants do come from nonpoint pollution sources (USEPA, 2001). The real problem associated with this program is the ability to quantitatively determine to what extent each pollution source needs to be controlled (i.e., what is the maximum daily pollutant load allowable from each source), the technical ability to reduce the contributions from nonpoint sources, and the ability to impose potentially economically crippling costs on both agricultural and municipal communities in order to comply with the CWA.

Although it is important to control the amount of pollutants discharged into the environment, the TMDL program still relies on performance-based pollution standards. In other words, it is an acceptable practice to pollute the environment up to specified water quality standards and not worry about unregulated chemicals. As a result, we will always have polluted water resources that we may ultimately want to use as sources of drinking water.

Pollution Sources and Water Quality

In today's chemically dependent society, pollutants discharged to the environment cannot be assumed to be assimilated to a degree so that there will be no impact on water quality. Since the turn of the 20th century, pollution has become increasingly complex, with a diversity of chemicals that have been shown to persist in the environment. When this fact is combined with a multiplicity of pollution sources found in both urban and rural landscapes, the potential for the pollution of drinking water resources will continue to increase. The boundaries between wastewater and drinking water have blurred to the point where the drinking water supplies of most Americans originate from sources polluted by wastewater (Maxwell, 2001). Our drinking water resources are threatened by a vast array of chemicals. This threat exists because of the following set of conditions:

- Chemicals released into the environment are not necessarily removed from water by natural processes. Yet, scientists and engineers depend on these natural processes to protect drinking water without specific proof that surface and groundwater systems can safeguard our water resources from the vast array of organic chemicals that are introduced into the environment.
- Point sources that discharge wastewater under the NPDES permit system are allowed to discharge regulated pollutants at specified levels. Thus, low concentrations of pollution are legally discharged to receiving waters.
- Point sources that discharge wastewater under the NPDES permit system are also allowed to discharge unregulated pollutants to receiving waters.
- The removal of unregulated pollutants from POTW discharges can most likely be accomplished only by using microfiltration and reverse osmosis technologies (Sedlak et al., 2000). Use of these technologies is more frequent in water-restricted states such as California and Arizona (Freeman et al., 2002), but these technologies have had limited use in the rest of the nation because of their cost.
- Nonpoint sources are a continuing source of pollution to receiving waters. These sources of chemical pollution will continue to be a significant problem for water resource managers and regulatory agencies.
- The list of pollutants that are not regulated in drinking water and have already been identified in our water resources is extensive.[32] This list, which is presented in Appendix 2-6, includes a large number of compounds; however, these compounds represent only a fraction of the compounds that are probably in our water resources.
- No consistent list of pollutants that should be regulated has ever been compiled. For example, the State of California under Proposition 65 was required to publish a list of chemicals known to cause cancer or reproductive toxicity. This list is given in Appendix 2-7.[33] When the chemicals listed in Appendices 2-4 and 2-5 are compared to toxic compounds regulated in wastewater (Appendix 2-2), the toxic pesticides[34] regulated in food (Appendix 2-8), the Primary Drinking Water

---

[32]This list was compiled from the references in this chapter.
[33]It is not known, however, which compounds are sufficiently soluble in water to be detected.
[34]A significant number of these compounds have limited water solubility, however, once they pass through the root zone, they are less susceptible to being degraded.

Standards (Table 1-9), and the chemicals in Appendix 2-9, it is absolutely clear that the current list of drinking water standards is meaningless given the potential number of compounds that can occur in our water resources.

Because of these conditions, the chemical purity of our drinking water sources cannot be guaranteed. Unfortunately, the extent and magnitude of the problem will never be known, as only a handful of unregulated pollutants in drinking water are actually monitored.

Therefore, depending on where one lives, people should be aware that the source of their drinking water may be significantly polluted. For example, people who obtain drinking water either downstream or downgradient from the following activities should be concerned about consuming chemical pollutants (in a decreasing order of severity):

- Surface waters that transect crop-producing areas that use pesticides.
- Surface waters that receive treated industrial discharges or mine runoff.
- Surface waters that receive discharges from a POTW.
- Surface waters that receive urban runoff.
- Groundwater that receives effluent from sanitary leach fields.
- Groundwater that is recharged with treated municipal or industrial wastewater effluent.
- Groundwater that is under the influence of surface waters that receive agricultural runoff, industrial, or POTW effluents.

Unfortunately, these conditions exist over most of the United States and the world. Communities that receive their drinking water primarily from natural mountain rainfall and snow melt have the potential of obtaining the least polluted water. However, it should be remembered that even mountain lakes and rivers are polluted by pesticides transported hundreds of miles in air.

Summary

When one brings together issues of population growth, pollution generation, water resource availability, and water supply, it is clear that increasing population leads to increased pollution. These forces are pitted against a finite quantity of water resources that are being polluted at rates approaching totality. In other words, the pure water that is everyone's desire and presumably everyone's goal is rapidly turning from reality to wishful thinking. One must face the fact that water, regardless of location, has been and continues to be impacted by human activity. Pollution and contamination, whether biological or chemical, of the world's water resources are here to stay, creating a critical need to protect and purify our drinking water.

References

Anonymous, 1952, *Underground Water Pollution Resulting from Waste Disposal*, State of California, Los Angeles and Santa Ana Regional Water Quality Control Boards.

Anonymous, 1969, California Department of Water Resources, "San Joaquin Valley Drainage Investigation," Bulletin No. 127.

Anonymous, 1976, *Operation of Wastewater Treatment Plants*, Water Pollution Control Federation, Washington, D.C.

Anonymous, 2001, "WE&T Waterline, Tracking Estrogen Pollution," *Water Environment and Technology*, Vol. 13, No. 11, page 22.

Anonymous, 2003, "Perchlorate Emerging as a Likely Arizona Water Quality Issue," Arizona Water Resources, Vol. 12, No. 3 (November-December, 2003).

Barber, L.B. et al., 1997, "Organic Constituents that Persist During Aquifer Storage and Recovery and Reclaimed Water in Los Angeles County, California." In *Conjunctive Use of Water Resources-Aquifer Storage and Recovery*, ed. D.R. Kendall, 261–272, American Water Resources Association, Herndon, Virginia.

Brown, K. 2002, "Water Security: Forecasting the Future with Spotty Data," *Science*, Vol. 297, pages 926–927.

Bruce, Mark et al., 2001, "Mercury," *Water Environment and Technology*, Vol. 13, No. 11, pages 34–38.

Bouwer, H., P. Fox, and P. Westerhoff, 1998, "Irrigating with Treated Effluent," *Water Environment and Technology*, Vol. 1.

Clark, C.E., 1885, "Report on the Main Drainage Works of the City of Boston," *Annual Reports of the Board of Health*.

Cloete, T.E., J. Rose, L.H. Nel and T. Ford, eds., 2004, *Microbial Waterborne Pathogens*, IWA Publishing, London.

Crook, James and William Vernon, 2002, "A Clear Advantage, Membrane Filtration is Gaining Acceptance in the Water Quality Field," *Water Environment and Technology*, Vol. 14, pages 16–21.

Daughton, Christian and Thomas Ternes, 1999, "Pharmaceuticals and Personal Care Products in the Environment: Agents of Subtle Change?" *Environmental Health Perspectives*, Vol. 107, Supplement 6, pages 907–938.

Day, D., ed., 1922, *A Handbook of the Petroleum Industry*, John Wiley & Sons, New York.

Dickey, R. et al., (1949), *Report of the Interim Fact-Finding Committee on Water Pollution*, California State Assembly.

Dow Chemical Company, 1938, *Dow Industrial Chemicals and Dyes*, Midland, Michigan.

Egleston, T., 1883, "The Method of Collecting Flue Dust at EMS on the Lahn," *American Institute of Mining Engineer*, Vol. XI, pages 379–411.

Eitzer, B.D. and A. Chevalier, 1999, "Landscape Care Pesticide Residues in Residential Drinking Water Wells," *Bulletin of Environmental Contaminant Toxicology*, Vol. 62.

Frankland, P. and Silvester, H., 1907, "The Bacterial Purification of Sewage Containing a Large Proportion of Spent Gas Liquor," *Journal of the Society of Chemical Industry*, Vol. 26, No. 6, pages 229–237.

Freeman, Scott, G.F. Leitner, J. Crook and W. Vernon, 2002, "A Clear Advantage, Membrane Filtration Is Gaining Acceptance in the Water Quality Field," *Water Environment and Technology*, Vol. 14, pages 16–21.

General Accounting Office, 1994, "Water Pollution: Poor Quality Assurance and Limited Pollutant Coverage Undermine EPA's Control of Toxic Substances," General Accounting Office, Chapter Report, 02/17/94, GAO/PEMD-94-9.

Goodell, Edwin B., 1904, "A Review of the Laws Forbidding Pollution of Inland Waters in the United States," *U. S. Geological Survey, Water-Supply and Irrigation* Paper No. 103.

Grammar, John, 1819, "Account of the Coal Mines in the Vicinity of Richmond, Virginia," *The American Journal of Science*, Vol. I, pages 125–130.

Harlow, I.F., T.J. Powers and R.B. Ehlers, 1938, "The Phenolic Waste Treatment Plant of the Dow Chemical Company," *Sewage Works Journal*, Vol. 10, pages 1043–1059.

Hairns, P., 1884, "Report on the Obstacles Interposed by Coal Tar Deposits on Dredging of the Potomac River," *The Sanitary Engineer*, Vol.10(1), page 17.

Haywood, J.K., 1908, "Injury to Vegetation and Animal Life by Smelter Wastes," *US Department of Agriculture, Bureau of Chemistry*, Bulletin No. 113.

Herr, J.L. and H.H. Harper, 2000, "Reducing Nonpoint Source Pollutant Loads to Tampa Bay Using Chemical Treatment," *Water Environment Federation Conference Proceedings*, Technical Paper.

Hillebrand, W.F. , "Zinc-Bearing Spring Waters form Missouri," *U. S. Geological Survey*, Bulletin No. 113 (1893).

Hrudey, Steve E. and E.J. Hrudey, 2004, *Safe Drinking Water: Lessons from Recent Outbreaks in Affluent Nations*, IWA Publishing, London, UK.

Hunt, A.E. and G.H. Clapp, 1889, "The Impurities of Water," *Transactions of the American Institute of Mining Engineers*, Vol. XVII, pages 338–355.

Hunt, Tara A., 1998, "Water Quality: Studies Indicate Drugs in Water May Come from Effluent Discharges," *Water Environment & Technology*, Vol. 10, pages 17–22.

Johnson, Douglas W., 1905, "Relation of the Law to Underground Waters," *U.S. Geological Survey Water-Supply and Irrigation*, Paper No. 122.

Johnson, L., 1917, "History and Legal Phases of the Smelting Smoke Problem -I," *Engineering and Mining Journal*, Vol. 103, No. 20, pages 877–880.

Jones, J.W., 1897, "Ferric Sulphate in Mine Waters, and its Action on Metals," *The Proceedings of the Colorado Scientific Society*, Vol. VI, pages 46–55.

Jordan, M.J., 1975, "Effects of Zinc Smelter Emissions and Fire on a Chestnut-Oak Woodland," *Ecology*, Vol. 56, pages 78–91.

Kirk, A.B., E.E. Smith, K. Tian and T.A. Anderson, 2003, "Perchlorate in Milk," *Environmental Science and Technology*, Vol. 37, No. 21, pages 4979–4981.

Lang, R., T.A. Herrera, D.P. Change and G. Tchobanoglos, 1987, "Trace Organic Constituents in Landfill Gas," California Waste Management Board, State of California, University of California Davis.

Levine, B.B. et al., 2000, "Treatment of Trace Organic Compounds by Ozone-Biological Activated Carbon for Wastewater Reuse: The Lake Arrowhead Pilot Plant," *Water Environmental Research*, Vol. 72, No. 4, pages 388–396.

Liang, Sun et al., 2001, "Treatability of MTBE-contaminated groundwater by ozone and peroxone," *Journal of the American Water Works Association*, Vol. 93, page 11.

Magbanua, Benjamin S. et al., 2000, "Quantification of Synthetic Organic Chemicals in Biological Treatment Process Effluent Using Solid-Phase Microextraction and Gas Chromatography," *Water Environmental Research*, Vol.72, No. 1, pages 98–104.

Mason, William P., 1896, *Water-Supply*, John Wiley & Sons, New York.

Maxcy, Kenneth F., 1955, *Preventive Medicine and Public Health*, 8th ed., Appleton-Century-Crofts, Inc., New York.

Maxwell, Steve, 2001, "Ten Key Trends and Developments in the Water Industry," *Journal of the American Water Works Association*, Vol. 93, No. 4, page 5.

Middleton, F. and A. Rosen, 1956, "Organic Contaminants Affecting the Quality of Water," *Public Health Reports*, Vol. 71, No. 11, pages 268–274.

Montgomery, J. and L. Welkom, 1990, *Groundwater Chemicals Desk Reference*, Lewis Publishers, Chelsea, Michigan.

National Research Council, 1994, *Alternatives for Ground Water Cleanup*, National Academy Press, Washington D.C.

Nisbet-Latta, M., 1907, *Handbook of American Gas-Engineering Practice*, Van Nostrand, New York.

O'Bannon, J., 1958, "Application of Emulsifiable Digromochloropropane in Irrigation Water as a Freeplanting Soil Treatment," *Plant Disease Reporter*, Vol. 42, No. 7, pages 857–860.

Ongerth, Jerry E. and Stuart Khan, 2004, "Drug Residuals: How Xenobiotics Can Affect Water Supply Sources," *Journal of the American Water Works Association.*, Vol. 96, page 8.

Parker, Horatio N., 1907, "Stream Pollution, Occurrence of Typhoid Fever, and Character of Surface Waters in Potomac Basin," *U.S. Geological Survey Water-Supply and Irrigation* Paper No. 192.

Peckston, T., 1998, "A Practical Treatise on Gas Lighting, Hebert: London (1841)," *Population Reports*, 1998, Volume XXVI, Number 1.

Pottstown Gas Company v. Murphy, 39 Pa. 257 (May 6, 1861).

Powers, Thomas, 1945, "The Treatment of Some Chemical Industry Wastes," *Sewage Works Journal*, Vol. 17, pages 331–337.

Renner, Rebecca, 1999, "Study Finding Perchlorate in Fertilizer Rattles Industry," *Environmental News, Environmental Science and Technology*, Vol. 33, No. 19, page 3280.

Roefer, Peggy et al., 2000, "Endocrine-Disrupting Chemicals in a Source Water," *Journal of the American Water Works Association*, Vol. 92, No. 8, pages 52–58.

Rudolfs, W., 1953, *Industrial Wastes: Their Treatment and Disposal*, Reinhold Publishing Co, New York.

Salvato, Joseph A., Nelson L. Nemerow, and F.J. Agardy, 2003, *Environmental Engineering*, 5th ed., John Wiley & Sons, Inc., New York.

Saunders, Robin G., 2001, "Letters to the Editor, What You Don't Think You Wanted to Know About Recycled Wastewater," *Scientific American*, Vol. 284, No. 6, page 16.

Schrenk, H.H., H. Heimann, G.D. Clayton, W.M. Gafafer and H. Wexler, 1949, "Air Pollution in Donora, Pa.: Epidemiology of the Unusual Smog Episode of October 1948," *Public Health Bulletin* No. 306, Federal Security Agency, Public Health Service, Bureau of State Services, Division of Industrial Hygiene, Washington, D.C.

Schwarzenbach, R.P. and J. Westfall, 1981, "Transport of Nonpolar Organic Compounds from Surface Water to Groundwater, Laboratory Sorption Studies," *Environmental Science and Technology*, Vol. 15, No. 11, pages 1360–1367.

Sedlak, David, James Gray and Karen Pinkston, 2000, "Understanding Microcontaminants in Recycled Water," *Environmental Science and Technology*, Vol. 34, pages 508–515.

Seiffert, Dr., 1897, "Health of Zinc Works Employees," *The Engineering and Mining Journal and Coal*, Vol. LXIV, No. 20, page 579.

Seiler, Ralph, Steven Zaugg, James Thomas and Darcy Howcroft, 1999, "Caffeine and Pharmaceuticals as Indicators of Wastewater Contamination in Wells," *Groundwater*, Vol. 37, No. 3, pages 405–410.

Stevenson, F.J., 1962, "Chemical State of the Nitrogen in Rocks," *Geochim et Cosmochim Acta*, Vol. 26, pages 797–809.

Sullivan, Patrick J., Garrison Sposito, S. M. Strathouse and C. L. Hansen, 1979, "Geologic Nitrogen and the Occurrence of High Nitrate Soils in the Western San Joaquin Valley, California," *Hilgardia*, Vol. 47, pages 15-47.

Sullivan, P., F. Agardy and R. Traub, 2001, *Practical Environmental Forensics, Process and Case Histories*, John Wiley & Sons, New York.

Swain, R.E, 1949, "Smoke and Fume Investigations, A Historical Review," *Industrial and Engineering Chemistry*, Vol. 41, pages 2384–2388.

Torkelson, T, Sader S. and C. Rowe, 1961, "Toxicologic Investigations of 1,2-Dibromo-3-Chloropropane," *Toxicology and Applied Pharmacology*, Vol. 3.

University of California at Davis, 1994, "Groundwater Quality and Its Contamination from Non-point Sources in California," Centers for Water and Wildland Resources, Water Resources Center Report No. 83.

USEPA, 1989, "Generalized Methodology for Conducting Industrial Toxicity Reduction Evaluations (TREs)," Risk Reduction Engineering Laboratory, Cincinnati, Ohio, EPA 600-2-88-070.

USEPA, 1992, "Introduction To Water Quality-Based Toxics Control for the NPDES Program," Office of Water, Washington , D.C., EPA 831-S-92-002.

USEPA, 1998, "Development Document for Final Effluent Limitations, Guidelines and Standards for the Pharmaceutical Manufacturing Point Source Category, Office of Water, EPA-821-R-98-005.

USEPA, 1999, "Region 9 Perchlorate Update," USEPA Region 9, San Francisco, California.

USEPA, 2001, "The National Costs of the Total Maximum Daily Load Program (Draft Report)," Office of Water, Washington, DC, EPA 841-D-01-003.

USGS, 1998, "Pesticides in Surface and Ground Water of the United States: Summary of Results of the National Water Quality Assessment Program, United State Geological Survey."

USGS, 2000, "USGS Research Finds that Contaminants May Play an Important Role in California Amphibian Declines," News Release of 12-7-00 from USGS, MS119 National Center, Reston, VA 20192.

USPHS, 1946, "Public Health Service Drinking Water Standards," *Public Health Reports*, Vol. 61, No. 11.

Vilbrandt, F., 1934, *Chemical Engineering Plant Design*, McGraw-Hill, New York.

Wauchope, R.D., T.M. Butler, A.G. Hornsby, P.W.M. Augustijn-Beckers and J.P. Burt, 1991, "The SCS/ARS/CES Pesticides Properties Database for Environmental Decision-Making," *Reviews in Environmental Contaminant Toxicology*, Vol. 123.

Williams, Chas. P., 1877, "Analysis of Mine Water from the Lead Region of South-west Missouri, *American Chemist*, Vol. 7, pages 246–247.

World Health Organization, 2003, "1,2,3-Trichloropropane," Concise International Chemical Assessment Document 56, Geneva.

CHAPTER 3

Water Protection

"To be wholesome, water must be ... free from poisonous substances. The possibility of their presence in water supplies is sometimes unsuspected."
Gordon Fair and John Geyer, Water Supply and Wastewater
Disposal, 1954

Water for domestic consumption has been collected, stored, treated, and distributed to thirsty consumers as far back as recorded history. Today nothing has changed except that the consumer, even in developing nations, has a greater selection of water products from which to choose than ever before. Drinking water is now available from domestic water resources taken from surface and underground supplies, bottled water from both domestic and worldwide distributors, and in-home treatment systems for selective use. Given the number of water products available, one would assume that an informed consumer would be able to get water that is sparkling clear, good tasting, and free from "poisonous substances." Unfortunately, consumers have no real assurance that the water they drink does not contain known or "unsuspected" pollutants that may in fact be poisonous.

The 1996 amendments to the Safe Drinking Water Act require that all water utilities conduct source water assessments to determine the potential of an identified drinking water source susceptibility to chemical pollution. This process will involve (1) defining the drinking water source area that could receive pollutants, (2) documenting all activities in the source water area that could discharge pollutants to the source area, and (3) based on this information, hydrogeological basin factors, and the control measures in place, determine the vulnerability of the water source to be polluted. Based on this type of analysis, it is then possible to implement source protection programs to reduce or possibly eliminate the pollution of drinking water resources. Obviously, source protection should be the first line of defense for protecting the public health. In many surface water and groundwater environments, however, source protection may not be possible owing to existing pollution practices and standards. Given the widespread occurrence of polluted water resources, particularly in developing nations, the consumer needs to be aware of the quality of the water products that are available and the technologies that are used to eliminate or reduce pollutants in drinking water supplies. One must never forget that on a global scale there are huge differences regarding source water quality, treatment technology, hygiene, and basic public knowledge regarding water quality.

The Basics of Water Supply

Drinking water supplies are obviously derived from moving bodies of surface water such as creeks and rivers, stored water resources in the form of either surface bodies of water held in lakes and reservoirs, and in groundwater aquifers. As discussed in Chapter 1, it is important to remember that surface water percolates into groundwater aquifers and that groundwater moving in aquifers can emerge naturally at the earth's surface as a spring, an artesian spring (i.e., under pressure), or as discharge unseen into a river, lake, or ocean (Figure 3-1).

The diversion of surface water resources by individuals and communities can be a convenient source of drinking water. This assumes, however, that the surface water occurs throughout the year in sufficient quantities to serve users' needs and that users have a legal right to the amount of water they are diverting. This is usually not a problem in

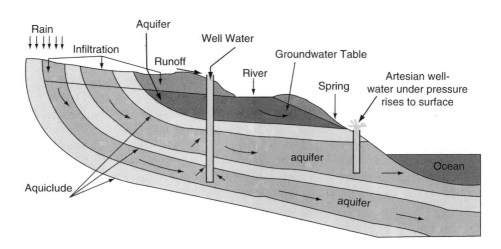

FIGURE 3-1. Geohydrologic terms.

the humid climatic regions of the world. In the United States these regions generally lie east of the Mississippi River and in the Pacific North West. When the demand for water resources exceeds surface water supplies, groundwater supplies can usually be obtained by installing groundwater supply wells. In today's industrialized societies, the users must usually have a permit to install a groundwater extraction well; this is certainly not the case in developing nations. Depending on an aquifer's capability to produce a usable amount of water, a user may need multiple wells at various depths. Therefore, any given geographic region may rely solely on surface water, only groundwater, or a mixture of both.

The one critical issue when using any water resource for drinking water is whether or not that resource is polluted. When considering the likelihood of pollution, there are several general constants (refer to Chapter 2 and Figure 3-1):

- Surface water resources are used as a reservoir that dilutes both permitted and unpermitted discharges of waste, receives polluted runoff from urban and agricultural areas, receives air pollutants that are deposited on the water's surface and if soluble, enters the water resource, and in some cases may also interact with polluted groundwater. In other words, only the most pristine regions of the world have the greatest likelihood of having unpolluted surface water resources. However, even these resources are not totally beyond the potential reach of chemical air pollution and biological hazards endemic to a region's wildlife. Thus, surface water resources that are used for drinking water should be considered as being polluted until demonstrated otherwise.
- Surface waters that receive numerous wastewater discharges from publicly owned treatment works (POTWs) and industrial discharges along the course of inland rivers (e.g., Mississippi, Missouri, Ohio) are particularly egregious because they expose numerous communities to an increasing load of pollutants as water is diverted upstream from a community for drinking water and treated sewage is discharged downstream of each community. This process can continue on for hundreds of miles. If this example holds for industrialized nations, imagine what the situation must be like in countries where no wastewater treatment is provided. Without question, the receiving water is usually undrinkable owing to biological hazards unless it is disinfected. Nevertheless this is often the only source of drinking water, health hazards not withstanding.
- Groundwater resources should contain less pollution than surface water if: (1) there are no sources of pollution into the groundwater (e.g., leaking sewer lines above or in groundwater; chemical spills and leaks from industrial sites, commercial facilities, and fueling facilities; leaching of pesticides through soil in agricultural areas; leakage from land disposal facilities; pollution

from septic systems and cesspools; pollution from mining and metallurgy operations) and there is no direct recharge of treated wastewater into the groundwater, (2) there is no polluted surface water that is recharging groundwater, (3) the origin of the groundwater has at least one or more relatively impermeable and contiguous geologic units[1] between the earth's surface and the aquifer that will be used to extract drinking water, (4) the aquifer in which the groundwater is traveling is composed of sediments containing natural organic matter and clays that may retard the transport of pollutants (e.g., stratified silts, sands, gravels), and (5) where the travel time or residence time between the point of groundwater recharge and extraction is long (i.e., decades to centuries) there is a greater probability of not containing manufactured pollutants. Rarely do groundwater resources possess all of these characteristics.

- Some mineral groundwaters have a lower probability of containing manufactured chemicals owing to their geothermal origin and/or long residence time in an aquifer. They may, however, contain elevated concentrations of trace metals such as arsenic.
- Groundwater that is being recharged with polluted surface water will degrade the quality of the resource. Therefore, a well, located in such a region, will not usually have the same water quality as other groundwater wells that are not under the influence of surface water inputs.

Because either chemical and/or biological pollution of a substantial amount of our water resources is likely, most communities in the industrial nations provide basic water treatment for drinking water. For the most part, the collection, treatment, and distribution of drinking water are provided by local governmental agencies (i.e., city or regional governments). However, there is an increasing trend for cities to purchase their water from both public-owned water agencies and privately owned water companies. In either case, some basic form of water treatment is provided for the majority of a community's population. This does not include individuals who live outside the service range of a water provider.

Individuals who are not serviced by a water provider (i.e., those individuals outside the provider's service range or communities/individuals in countries that do not have the economic capability to treat their water) must locate water resources that are not impacted with biological or chemical pollutants. This normally means avoiding surface water resources that are easily polluted by human activities. Therefore, communities or individuals turn to groundwater as a source of drinking water because they falsely assume that it is inherently pure. The fallacy of this

---

[1] A geologic or stratigraphic formation that retards the flow of water downward.

belief has been most recently demonstrated by the groundwater arsenic poisonings in Bangladesh. A safe water supply continues to be a critical problem for underdeveloped countries. According to United Nations data, more than 1 billion people cannot obtain safe drinking water, and more than 2.2 million people, primarily children, die annually from water-related diseases.[2] Is it any wonder that the global consumption of bottled water has increased by more than 80 percent between 1996 and 2002.[3]

With today's ever increasing population and resulting pollution, it is, for the most part, wiser to assume that our water resources are polluted and should be treated before consumption. This means that individuals must rely on (1) water that has been collected, treated, and distributed to a community, or (2) use a water treatment system on their individual source of drinking water (i.e., either surface water or groundwater). Given that most individuals in the major industrial nations receive their drinking water from a public agency, the basics of municipal water treatment will be discussed first.

Basic Water Treatment

There is clear evidence that ancient societies were concerned with water quality as far back as 2000 BC. These societies improved water quality by allowing the natural action of soil, sand, and course gravel to filter water; by boiling water; and by allowing water to settle in reservoirs and basins in order to remove suspended solids. Today at modern water treatment plants, essentially the same technology is employed, although it is refined and much better understood. Whether we look back 4,000 years or look forward into the future, certain truths appear to be consistent about water quality:

- If water has a bad taste or odor, people will not usually drink it.
- If water shows evidence of turbidity (i.e, it is muddy) and has a color, people will not usually drink it.
- If water makes people sick, they will stop drinking it.

Thus, basic water treatment usually strives to produce water that does not have bad taste, odor, color, or turbidity and is free from bacterial threats.

The basic elements of a typical large water treatment plant[4] today include most, if not all, of the following processes to meet the previously

---

[2]Water, Environment & Technology, WE&T Brief, July 2004, p14.
[3]Beverage Marketing Corporation, 2003.
[4]See definitions of water treatment plants in the section An Issue of Equality.

mentioned objectives: pretreatment, prefiltration, filtration, chemical treatment, and disinfection[5] (Figure 3-2).

- Pretreatment of raw water usually consists of (1) removing floating debris such as weeds, leaves, and other detritus by using screens; (2) aeration, such as often seen in fountains, is used to remove many of the more volatile chemicals that can add taste and odor to water and at the same time add oxygen to the water (i.e., low oxygen in water often leaves a "flat taste"); (3) sedimentation to remove dirt and other materials heavier than water; (4) filtration through sand beds to remove remaining particulates; and (5) occasionally, chlorination to both partially disinfect the water and oxidize and remove some organic chemicals from the water.
- Prefiltration usually consists of adding chemicals to the water to flocculate[6] and filter out suspended particles. However, the added chemicals (acrylamide and epichlorohydrin) can also be carried along with the treated water through the distribution system.
- Filtration is generally accomplished by the use of sand beds. Water passing through sand will remove the vast majority of the suspended and colloidal materials found in water.
- The final water treatment step usually consists of chlorinating the water to eliminate bacterial pollution and to prevent the regrowth of bacteria in the distribution systems carrying the treated water to the consumer. In addition to chlorination, the water may also be treated to reduce hardness (i.e., remove excessive amounts of calcium and magnesium) or the pH may be adjusted to prevent corrosion or scaling of the water distribution mains.

These treatment methods, however, are not specifically designed to remove trace metals and organic compounds. It is only by chance that some trace metal and organic compounds are removed from the raw water by the oxidation, coagulation, and filtration processes. Thus, trace metals and organic compounds can and do pass through the water treatment plant and into the distribution system, ultimately reaching the consumer.

Although treatment is necessary to improve the quality of drinking water, the process also creates its own pollution problems. The standard methods of water treatment used by water utilities can add suspected carcinogens to treated water by (1) the addition of organic flocculents to remove suspended solids and (2) by the use of chlorine, chloramine, or

---

[5]Presently, only surface water systems and systems using groundwater under the direct influence of surface water are required to disinfect their water supplies. However, the USEPA has proposed monitoring requirements to determine if disinfection is necessary.
[6]A process that causes very fine particles to combine into larger particles. The early chemical flocculents were iron and aluminum alums.

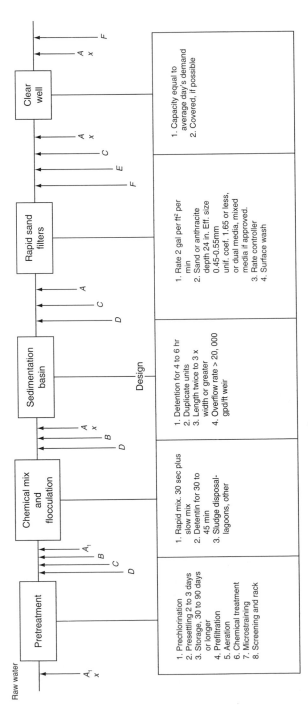

Raw water

Pretreatment → Chemical mix and flocculation → Sedimentation basin → Rapid sand filters → Clear well

Pretreatment

1. Prechlorination
2. Presettling 2 to 3 days
3. Storage, 30 to 90 days or longer
4. Prefiltration
5. Aeration
6. Chemical treatment
7. Microstraining
8. Screening and rack

Chemical mix and flocculation

1. Rapid mix. 30 sec plus slow mix
2. Detentin for 30 to 45 min
3. Sludge disposal-lagoons, other

Sedimentation basin

1. Detention for 4 to 6 hr
2. Duplicate units
3. Length twice to 3 × width or greater
4. Overflow rate > 20,000 gpd/ft weir

Rapid sand filters

1. Rate 2 gal per ft² per min
2. Sand or anthracite depth 24 in. Eff. size 0.45-0.55mm unif. coef. 1.65 or less, or dual media, mixed media if approved.
3. Rate controller
4. Surface wash

Clear well

1. Capacity equal to average day's demand
2. Covered, if possible

Design

Conventional rapid sand filter plant flow diagram. Possible chemical combinations:

A: Chlorine. A1 Eliminate if THMs formed.

B: Coagulant; aluminum sulfate (pH 5.5-8.0). 10 to 50 mg/1; ferric sulfate (pH5.0-11.0), 10 to 50 mg/1; ferrous sulfate (pH 8.5-11.0), 5 to 25 mg/1; ferric chloride (pH 5.0-11.0); sodium aluminate, 5 to 20 mg/1; activated silica, organic chemicals (polyelectrolytes).

C: Alkalinity adjustment; lime, soda ash, or polyphosphate.

D: Activated carbon, potassium permanganate.

E: Dechlorination; sulfur dioxide, sodium sulfite, sodium bisulfite, activated carbon.

F: Fluoridation treatment.

X: Chlorine dioxide, ozone, chlorine-ammonia.

Note that the chlorinator should be selected to postchlorinate at 3 mg/1. Provide for a dose of 3 mg/1 plus chlorine demand for groundwater. Additional treatment processes may include softening (ion exchange, lime-soda, excess lime, and recarbonation), iron and manganese removel (ion exchange, ozone oxidation, sequestering), organics removal (activated carbon, superchlorination, ozone oxidation), and demineralization (distillation, electrodialysis, reverse osmosis, chemical oxidation and filtration, freezing).

FIGURE 3-2. Typical schematic of a water treatment plant.

bromine to destroy biological hazards and the subsequent creation of disinfection by-products. Again, focusing on the developing nations, the threat of disinfection by-products is a secondary concern, the fundamental issue being protection against biological hazards.

Flocculents

To remove small suspended particles in water, a flocculent is used to coalesce small particles into a larger particle so that these particles can be picked up by a filter. The two chemical polymers used for this purpose are *acrylamide* and *epichlorohydrin*. Both of these compounds are listed as "reasonably anticipated to be human carcinogens" in the United States Public Health Service Ninth Report on Carcinogens. These compounds also have established limits in drinking water (see Table 1-9). In other words, drinking water treated with these compounds may routinely contain low levels of these anticipated carcinogens.

Disinfection By-Products

Chlorine, chloramine, and bromine are used in water treatment plants as a disinfectant to destroy biological hazards. Unfortunately, chlorine and bromine can undergo complex chemical reactions with either natural or synthetic organic chemicals and create newly chlorinated or brominated organic compounds in the finished water. The compounds of concern are halogenated methanes, haloacetic acids, haloketones, halonitriles, and nitrosamines.

When natural waters are either chlorinated or brominated, more than 100 potentially toxic halogenated compounds can be created (Gray et al., 2001). Of these potentially toxic compounds, only chloroform, bromoform, bromodichloromethane, and dibromochloromethane have established Primary Drinking Water Standards. These compounds as a group are known as the total trihalomethane (TTHM) compounds. When TTHMs were first regulated, the maximum contaminate level (MCL) was 100 ppb in drinking water. Of the TTHM compounds, only chloroform and bromodichloromethane are listed as "reasonably anticipated to be human carcinogens" in the USPHS Ninth Report on Carcinogens.

However, the Stage 2 Disinfection By-Product Rule issued in May 2002 (1) lowers the TTHMs MCL to 80 ppb; (2) establishes an MCL of 60 ppb for five haloacetic acids (HAA5), including monochloroacetic acid, dichloroacetic acid, trichloroacetic acid, monobromacetic acid, and dibromoacetic acid; (3) establishes an MCL of 10 ppb for bromate; and (4) establishes an MCL of 1,000 ppb for chlorite. Even with these modifications, a recent article in the *Journal of the American Water Works Association* (Roberts et al., 2002) reports that there should be standards for the nine haloacetic acids (HAA9) instead of the HAA5 compounds. This article states that "given the HAAs are thought to pose greater health risks than TTHMs do, the possible widespread occurrence of HAA9 should be of concern to water suppliers and regulators alike." This finding should also be a concern to the average consumer.

According to the report entitled, "Consider the Source, Farm Runoff, Chlorination Byproducts, and Human Health" by the Environmental Working Group, halogenated methanes and haloacetic acids represent a serious health threat to the American Public and specifically pregnant women (Gray et al., 2001). Some of the report's main conclusions are that more than 137,000 pregnancies each year are at increased risk of miscarriage and birth defects, and, of the 50 communities studied, the top five with the most risk were Montgomery County, Maryland; Washington, DC; Philadelphia, the suburbs of Pittsburgh; and San Francisco.

Furthermore, the Environmental Working Group report contends that the U.S. Environmental Protection Agency's (EPA) ability to link one health effect (i.e., bladder cancer) to halogenated by-products illustrates how the nation's health tracking programs force decisions based on just a fraction of the public health data on environmental pollutants. To better quantify the health impacts of halogenated by-products, the group recommends the creation of a national health tracking system. In other words, not only does the USEPA lack the ability to establish human health effects for specific chemicals, it also lacks the ability to track environmental impacts on individuals and communities.

Even though the Stage 2 Rule established new drinking water standards for the halogenated disinfection by-products, low levels of a handful of suspected halogenated compounds will be allowed in drinking water, while virtually hundreds of potentially toxic halogenated compounds go unregulated. Moreover, to meet just the new HAA5 standards, new treatment technologies will most likely have to be installed by many community water systems make a switch to chloramine. In fact, this has been the case.

The obvious solution to this problem would be to either remove the dissolved organic matter that could be halogenated or not to disinfect drinking water with halogens. European countries have been using ozone and ultraviolet (UV) light to disinfect drinking water for decades. These proven technologies are both reliable and cost effective. However, most existing water treatment systems in the United States have not chosen to make the transition to ozone and UV light treatment for two reasons. First, ozone and UV light treatment provides little residual disinfection once the water is pumped into the distribution system. Second, it would be more costly to make the conversion. Therefore, some water utilities that have had problems with halogenated disinfection by-products in their treated drinking water have switched or plan to switch to chloramine. The switch to chloramine is logical, as it is not as aggressive as chlorine and bromine at creating a TTHM or HAA compound in drinking water. Thus, water utilities will be able to meet the lower TTHMs and HAA5 drinking water standards, but these compounds will remain in the treated water.

One set of disinfection by-products not addressed by the Stage 2 Rule is nitrosamines. Research suggests that nitrosamines are formed as the result of the disinfection process (Najm and Trussell, 2000). Nitrosamines have also been identified in drinking water distribution systems as a result of chlorination. Nitrosamine formation within the drinking water system is a significant threat because these compounds are potential carcinogens. One nitrosamine, N-nitrosodimethylamine, is of particular concern because of its carcinogenicity and widespread distribution (USPHS, 2001).

In addition to the occurrence of water treatment chemicals and disinfection by-products in community water systems, fluoride is often added to the finished water product. The debate over the benefits of the fluoridation of drinking water to reduce dental cavities versus its carcinogenic and toxic characteristics has persisted for several decades. It is not our intent to continue this debate other than to point out that fluoride is a toxic substance. According to the 1960 Merck Index (Stecher et al., 1960), sodium fluoride[7] is an insecticide used to kill roaches and ants. Thus, in the realm of polluted drinking water, fluoride is just another toxic compound among many that can be found in drinking water. Moreover, fluoridation at a concentration not to exceed 4.0 ppm of fluoride is consistent with government policies that allow drinking water to contain (be polluted with) toxic chemicals relying on "standards" by which to maintain safety and protect human health.

The use of water treatment technologies to purify water has proven to be a significant factor in protecting the public health. For waters that come from protected watersheds,[8] chemical pollution is far less of a concern than water from urban or agricultural watersheds. In addition to the chemical pollutants already present in water resources, basic water treatment technologies can add a number of potentially toxic manufactured chemicals to a community water supply. These concerns were raised in a *Consumer Reports* (Harris and Brecher, 1974) article entitled, "Is the Water Safe to Drink?" This report concluded that (1) no substance capable of causing cancer in any animal species at any dosage should remain in the water we drink if a feasible method is available for removing it, and (2) the cost of promptly improving our drinking water is reasonable enough to justify the added protection that would be gained. Sadly, almost 28 years later these recommendations have yet to be implemented, even though various treatment technologies are capable of removing pollutants to as close to zero as possible, while studies continue to show that our drinking water is polluted (Heavner, 1999; Olsen et al., 1999; Coupe and Blomquist, 2004). As a result, other methods of

---

[7]The specific chemical compound used to fluoridate water.
[8]Such watersheds have little or no human activity so that chemical and human pollution is minimized, although not absent.

water treatment may be required to reduce or eliminate pollutants from these resources.

Beyond Basic Water Treatment

The ability of a water treatment facility to deliver a wholesome and high-quality product to its consumers usually occurs when advanced treatment methods are used. Some of the more advanced technologies used today focus on the ability to use instrumentation to control the optimal quantities of chemicals needed in process operations (i.e., coagulation, flocculation, sedimentation, and filtration). Such instrumentation is also used to continuously monitor water quality throughout the water treatment process. In addition to instrumental controls, numerous advanced treatment methods can be used to reduce both biological and chemical pollutants.

The use of advanced instrumentation in water treatment systems is an obvious benefit to a water utility because these technologies save money on chemicals and labor. As a result, these advance instrumentation methods have received acceptance at several water treatment plants. The use of advanced treatment methods, however, does increase the operating cost of water treatment systems. Because cost is often the limiting factor for the use of advanced treatment methods, their use tends to be restricted to those few facilities that can afford the added expense or have no other alternative to meet discharge requirements. For the facilities that can pay the added costs, the following advanced treatment methods are available:

- In the United States, as well as in other countries, activated carbon or granulated activated carbon (GAC) has been used to remove organic chemicals from water since the early 1900s.
- Ion exchange resins have been employed to remove inorganic metals for decades.
- Ozonation has been successfully used in Europe, instead of chlorine and bromine, to both disinfect water and oxidize organic constituents. Such systems are still being tested in the United States and as yet have not received wide acceptance. Ozonation in combination with hydrogen peroxide (i.e., advanced oxidation) has been used in the United States to remove persistent compounds such as methyl tertiary butyl ether (MTBE) (Liang, et al., 2001).
- Ultraviolet light, which has also been used in Europe, is being evaluated at a number of municipal treatment facilities in the United States for bacterial and viral control.
- As far back as 1985, the French have made use of membrane separation technology (i.e., reverse osmosis) for the removal of organic and inorganic compounds. Today, a number of reverse osmosis installations are operating in the United States and other industrialized

nations. Reverse osmosis membranes can hold back a wide range of micropollutants including pesticides and pathogenic organisms and can replace the need for both ozonation and activated carbon filtration. A variation of reverse osmosis, called nanofiltration, which operates at lower pressures than reverse osmosis, can remove compounds in the 300 to 1,000 molecular weight range.

- Advances in the development of organic polymers for improved flocculation are referred to as enhanced coagulation.
- Source protection, although not strictly in the category of treatment, can play a critical role in supporting the quality of water supplies. As of 1998, the city of New York spent approximately $1 billion to conserve and protect catchment areas in upper New York State. The cost tradeoff was to have spent an estimated $5 billion on a state-of-the-art filtration plant with an associated additional $300 million in annual operating costs.[9]

Because of the age of most community water supply systems (50 to 100 years old), local state and federal funds tend to go toward existing maintenance problems[10] instead of advanced treatment technologies. For example, to meet seismic codes, the City and County of San Francisco was recently faced with a major upgrade of its transmission pipelines bringing water from the Hetch Hetchy reservoir to local storage reservoirs. The estimated upgrade and retrofit costs will run into billions and will necessitate rate increases for many years. Yet, this upgrade deals only with the main transmission system.

Although advanced technology does exist, it is infrequently used on most of the public water systems, and rarely on the smaller rural systems. Nevertheless, a state-of-the-art treatment plant that used all available technologies to remove disease causing constituents such as bacteria and viruses, removed all naturally occurring and synthetic organic chemicals, limited the amounts of naturally occurring metals and radioactive elements, and could effectively protect human health as well as maintain the integrity of the distribution system. Contrast this approach with programs in developing nations where the crudest of technologies are used to meet limited but critical objectives such as the reduction of water-related diseases. In Bangladesh hand pumps and health education often represent the available "technology" by which to control diarrheal diseases, while in northern Brazil and sections of Iran, latrines, communal taps, hand pumps, and public standpipes represent the best methods by

---

[9]"Consequences of Overuse and Pollution," The Population Information Program, Center for Communication Programs, The Johns Hopkins School of Public Health, Volume XXVI, No. 1, September 1998.
[10]Older systems suffer from clogged reservoirs, which reduce their storage capacity, and leaking reservoirs. This leads to extensive water losses and deteriorating treatment facilities and transmission pipes.

which to control schistosomiasis and ascariasis.[11] Many nations of the global community simply do not have the financial resources or the technology to even approach the most rudimentary technology of the industrialized nations.

In the United States, the state-of-the-practice is to "patch up" what is broken or in need of repair, update existing treatment units, and, when forced by either consumer complaint or regulatory requirement, add additional technology in hopes of maintaining or keeping up with regulatory compliance. The magnitude of the facility maintenance problem was recently described in an USEPA study[12] that found an estimated expenditure of more than $1 trillion will be needed between 2010 and 2020 to meet current objectives of upgrading and extending the operational capabilities of our nation's water and wastewater systems. Quoting the USEPA, "currently capital spending is not adequate, new investment is flat, the long term financial solvency of many systems is doubtful and more households are having problems affording services."

Government funding and regulatory requirements aside, a community water system could increase the cost of its product to its consumers. All that is needed is political approval from the appropriate regulatory agency (e.g., mayor, city supervisors, public utility board). When there is a clear threat, community water systems do use these advanced water treatment technologies. The problem is that there is no established process for determining that a threat of chemical pollution justifies the cost of implementing advanced treatment technologies. Furthermore, even if a threat were identified, not all community water systems would be politically and financially able to respond. For example, privately owned water utilities might be more likely to implement advanced treatment technologies simply because they operate only one business. This situation is in sharp contrast to a municipally owned water utility where the owner (a city or county) has competing budgets and funding priorities.

In the western United States, the cost of water treatment for both biological and chemical components is approximately 15 percent of a community water system budget. To install the best available technologies to remove chemical pollutants to a level as close to zero as possible would only increase these budgets by 15 to 25 percent (personnel communication, Jerome Gilbert, Past President of the American Water Works Association, 2002).

Given the actual cost to the consumer and the quality of water received, many communities would probably be agreeable to such an increase in the cost of water. This conclusion assumes that either (1) some level of governmental support will be required to subsidize low income water users and/or (2) the increased costs are weighted toward

---

[11]Tables of Solutions for a Water-Short World, Population Reports, Series M, No. 14, 2000.
[12]Taken from the Federal Water Review process of September 2000.

those consumers who have the greatest consumption. If community water systems are not willing to implement all the technologies required to provide a water product with an absolute minimum concentration of chemical pollutants, then the necessary technology will have to be phased in over a number of years. For example, compliance with the Stage 2 Disinfection Byproduct Rule will take approximately 6 to 12 years. Under this program, the community water systems that cannot meet the TTHMs and HAA5 Maximum Contaminant Levels will have to use the best available technology, such as enhanced coagulation, GAC, or nanofiltration, if the water treatment system uses chlorine as the primary and residual disinfectant.

Under all of these conditions, it is clear that many community water systems will have to add advanced treatment technologies to comply with the Stage 2 rule. However, it is not clear how many community water systems would be willing to upgrade their treatment system to the point that their product has the minimum concentration of chemical pollutants possible without governmental funding. Regardless of funding issues, the economies of scale would suggest that the larger municipalities would have a greater probability of installing advanced treatment methods for their consumers than small systems (e.g., those serving less than 100,000 persons).

An Issue of Equality

The size and type of a water treatment system determine two important characteristics that can influence the quality of the water a customer consumes. These characteristics are the extent to which advanced treatment technologies are used and the type and frequency of water quality monitoring conducted.

Because the implementation of advanced treatment technologies is generally a function of local, state, and federal funding, many water systems, lacking such financial support, provide only basic water treatment. In general, the larger the water provider, the greater probability that advanced treatment technologies may be used. Thus, consumers who receive their water from a provider that serves a population greater than 100,000 (i.e., large system) are more likely to have their water treated to the highest quality. As of 1993 (USEPA, 2000), approximately 45 percent of the population of the United States was served by large water systems, while roughly 2.5 percent were served by small water systems (i.e., those that serve fewer than 500 people). These figures, however, do not include the approximately 22 million individuals who obtain their water from private water resources. Thus, a significant proportion of the population receives a water product that could and should be treated to a greater degree.

The other significant factor that can influence whether a customer will knowingly or unknowingly consume polluted drinking water is the extent and frequency of "required" water quality monitoring. Federally mandated monitoring requirements are tied to a water provider's classification and size. Therefore, it is necessary to define the USEPA's public water system classification scheme before monitoring requirements can be defined. The USEPA has defined a public water system as a facility that provides piped water for human consumption to either a minimum of 15 service connections or one that serves at least 25 persons for a minimum of 60 days a year. Based on this definition, the USEPA has further defined the following three types of public water systems:

- Community Water System (CWS): A public system that supplies water to the same population year-round (e.g., individual residences or businesses)
- Non-Community Water System (NCWS): A public water system that regularly supplies water to at least 25 of the same people at least 6 months per year, but not year-round (e.g., schools or hospitals)
- Transient Non-Community Water System (TNCWS): A public water system that provides water in a place where people do not remain for long periods of time (i.e., a campground or highway rest area)

Because CWS systems generally provide a water product to locations where individuals permanently reside, the quality of the water is measured by drinking water standards that are supposed to protect against potential heath effects that may occur from long-term exposure.[13] Thus, for water quality standards to serve as the means for protecting human health, the chemicals included within drinking water standards must be regularly monitored.

Chemical Monitoring and Warnings for Regulated Pollutants

The Federal Safe Drinking Water Act that was passed into law in 1974 attempts to ensure the quality of drinking water by setting both the standards (see Table 1-9) and the frequency of sampling drinking water for analysis. In other words, by setting water quality standards, monitoring will be conducted to ensure that a water resource does not exceed any standard and, if the standard is exceeded, these data can be used to (1) address any lack in treatment capability and (2) advise the consumer, thereby protecting the public health. Conceptually this goal sounds reasonable and the monitoring requirements developed by the USEPA can

---

[13]The NTNCWS and TNCWS systems mostly serve individuals who have another primary source of drinking water. As a result, the USEPA generally allows these systems to only monitor for pollutants that may have an acute or immediate health effect on the consumer (e.g., microbiological hazards).

be reviewed in Appendix 3-1. Unfortunately, these monitoring regulations have the following defects:

- Only chemicals that are listed in Table 1-9 are required by law to be monitored on a frequent schedule (i.e., more than once a year).
- If a specific chemical occurs in drinking water above a set water quality criteria, the public is only warned that the specific chemical is in the drinking water.
- The offending water provider must then increase its monitoring frequency and continue to notify consumers of the violation for as long as the violation continues.
- No treatment to remove the offending pollutant is required. However, most state water agencies will set a specified period to reach compliance with drinking water standards. The time to comply, depending on monitoring data and the reason for the pollution, can take years.
- When violations occur, and until they are rectified, it is recommended that consumers either purchase bottled water or treat the water using in-home treatment systems.

An example of these regulatory defects is typified by arsenic pollution in Fallon, Nevada. Fallon became the focus of the U.S. Geological Survey and the national media (USGS, 2001) in April 2001 because of a leukemia cluster that involved 12 children and the occurrence of arsenic in the community drinking water. The concentration of arsenic in the regional groundwater aquifer is approximately 90 ppb, which is almost double the drinking water standard that was in effect at the time. Drinking water is supplied in Fallon by individual groundwater wells and a municipal system that taps this regional groundwater aquifer. Because arsenic was not removed from its drinking water, the community has only been warned of the hazard for the last decade. By the middle of 2003, the City of Fallon had to begin to treat drinking water to remove arsenic. Unfortunately, Fallon is not the only community that has been affected. According to Congress (Congressional Report, 2000), approximately 15 million Americans are currently exposed to unsafe levels of arsenic. As a result, treatment will be required to reduce their arsenic levels to accepted standards. In anticipation of a compliance date of early 2006, a number of communities in Arizona, Nevada, and Maine have installed pilot programs employing household point-of-use (POU)/point-of-entry (POE) systems to determine both their efficiency in reducing arsenic to below regulatory levels and the relative cost of such systems.[14] To quote the authors of the article, "the POU treatment for arsenic removal for

---

[14]Kommineni, Sunil et al., "Point-of-Use/Point-of-Entry Treatment for Arsenic Removal: Operational Issues and Costs," California-Nevada Section, American Water Works Association, 2004 Spring Conference, Las Vegas, NV.

small systems appears to be promising." As with most economic models, the greater the volume treated, the lower the unit cost.

Existing monitoring and warning requirements for regulated chemicals in drinking water clearly demonstrate the foolishness of regulatory programs that rely on standards to protect human health. Because pollution events can occur at anytime but monitoring occurs only during an extremely short period, how can the public health be protected? Individuals who can afford to buy an alternate source of drinking water that is not polluted can avoid part of the problem. However, even if a consumer obtains an alternate drinking water source, polluted water for bathing still provides a route of exposure.

Because consumers usually demand a safe source of drinking water, most state regulatory agencies eventually force water utilities to come into compliance with established drinking water standards. The time period required to come into compliance, however, can take years to accomplish if (1) the standard for a chemical of concern is not consistently exceeded or (2) there is political resistance for immediate change because of economic considerations. Until change comes, consumers are left to find their own source of safe drinking water. Consumers who want to know the results of chemical monitoring for their water utility can contact the utility directly and request a copy of their annual drinking water quality reports. Examples of the type of information provided in these annual reports is given in Exhibit 3-1. These reports were selected to illustrate variability between water providers and water sources.

Exhibit 3-1. Water Quality Report Examples

Water utilities sample for the same set of regulated compounds but only report those compounds that are detected.

2003 Report for the Anaheim Public Utilities, California
Seventy-five percent of the source water comes from local groundwater wells (400-500 feet deep) and 25 percent from the Southern California Coastal Plain surface water and Colorado River water. This utility has two water treatment plants. Therefore, data are given here only for the highest constituent concentrations detected in either water treatment plant.

Detected Constituents	Highest Detected Concentration	Comments
Barium	112 ppb	
Copper	220 ppb	
Fluoride	0.41 ppm	

(continued)

Exhibit 3-1. (continued)

Nitrate	1.4 ppm	
Alpha activity	2.5 pCi/L	
Beta activity	5.9 pCi/L	
Uranium	2.6 pCi/L	
Total trihalomethane (TTHM)	63.3 ppb	
Haloacetic acid (HAA)	30.0 ppb	
Total coliform	1.0%	
Boron	160 ppb	Unregulated
Perchlorate	5.0 ppb	Unregulated
Vanadium	3.4 ppb	Unregulated

Only one unregulated pesticide, Diuron, was analyzed for but was not detected.

2003 Report for the City of Lodi, California
All of the water for the City of Lodi comes from 25 groundwater wells. No information is available on the depth of the wells or other geologic/hydrologic information. There is no routine disinfection of the water source.

Detected Constituents	Highest Detected Concentration	Comments
Arsenic	9 ppb	
Barium	230 ppb	
Copper	550 ppb	
Lead	5.2 ppb	
Fluoride	0.23 ppm	
Nitrate	36 ppm	
Alpha activity	15.15 pCi/L	Over MCL
Radium (combined)	568 pCi/L	
Uranium	11.7 pCi/L	
Tetrachloroethylene	1.3 ppb	
1,1-Dichloroethylene	0.81 ppb	
Trichloroethylene	3.0 ppb	
Dibromochloropropane	0.35 ppb	Over MCL
Total coliform	1.3%	
DCPA	1.6 ppb	Unregulated
Trichloropropane	37 ppb	
Unregulated		
Vanadium	0.05 ppb	Unregulated

2003 Report for the city of Waxahachie, Texas

The City of Waxahachie uses surface water for their drinking water.

Detected Constitutes	Highest Detected Concentration	Comments
Barium	45.9 ppb	
Copper	715 ppb	
Lead	3.2 ppb	
Fluoride	0.6 ppm	
Nitrate	0.78 ppm	
Atrazine	0.72 ppb	Over MCL
TTHMs	91.5 ppb	Over MCL
HAA5	73.5 ppb	Over MCL
Total coliform	1.0%	
Chloroform	62.9 ppb	TTHM compound
Bromodichloromethane	7.9 ppb	TTHM compound
Chlorodibromomethane	21.8 ppb	Unregulated

2003 Report for the City of Muncie, Indiana

The source water for Muncie and surrounding communities relies on surface water (White River and Prairie Creek Reservoir) and one well field with three wells. There is no geologic information of the groundwater resource. The water treatment plant uses chloramine as the disinfectant.

Detected Constitutes	Highest Detected Concentration	Comments
Barium	96 ppb	Average value
Copper (90th percentile)	164 ppb	Average value
Chromium	13 ppb	Average value
Cyanide	62 ppb	Average value
Lead (90th percentile)	3.0 ppb	Average value
Selenium	3.0 ppb	Average value
Fluoride	0.9 ppm	Average value
Nitrate	2.78 ppm	Average value
Atrazine	0.2 ppb	
TTHMs	78.7 ppb	
HAA5	24.0 ppb	
Chloramine	2.28 ppm	
Chloroform	12.5 ppb	TTHM compound
Bromodichloromethane	5.3 ppb	TTHM compound
Dibromochloromethane	1.0 ppb	TTHM compound

(continued)

Exhibit 3-1. *(continued)*

2003 Report for the City of Allentown, Pennsylvania

The City of Allentown uses four source waters, which include two surface waters (Little Lehigh Creek and the Lehigh River) and two groundwaters. The groundwater sources include Crystal Spring and the Schantz Spring.

Detected Constituents	Highest Detected Concentration	Comments
Barium	45.9 ppb	
Copper	715 ppb	
Lead	3.2 ppb	
Fluoride	1.2 ppm	
Nitrate	4.9 ppm	
Chlorine	0.24 ppm	Data from 2001
TTHMs	27.8 ppb	
HAA5	14.0 ppb	
Zinc	15.0 ppb	Unregulated

2003 Report for the Town of Burlington, Massachusetts

The Town of Burlington uses surface water from the Mill Pond Reservoir and groundwater wells located in the Vine Brook Aquifer that range in depth from 30 to 100 feet deep.

Detected Constituents	Highest Detected Concentration	Comments
Antimony	3 ppb	
Arsenic	5 ppb	
Barium	300 ppb	
Beryllium	1 ppb	
Cadmium	0.5 ppb	
Chromium	30 ppb	
Copper	210 ppb	
Cyanide	10 ppb	
Lead	3 ppb	
Mercury	0.5 ppb	
Selenium	5 ppb	
Thallium	0.1 ppb	
Fluoride	1.0 ppm	
Nitrate	0.44 ppm	
Nitrite	0.04 ppm	
Radium (combined)	0.2 pCi/L	
TTHMs	18.8 ppb	
HAA5	8.0 ppb	
Nickel	20.0 ppb	Unregulated

Although flawed, the water quality standard system has provided monitoring data that can be used to assess the general extent of polluted drinking water in the United States (i.e., the same data collected and reported in water utility annual drinking water quality reports). After collecting years of chemical monitoring data on public water supplies, the USEPA began in August 1999 to assemble this information into a national database. This national database contains the monitoring results for regulated and unregulated chemicals found in drinking water sources that have been distributed by community water systems. Most, if not all, industrialized nations have developed similar programs and make available similar information.

The National Drinking Water Contaminant Occurrence Database

As previously noted, not all public water systems are required to monitor for chemical pollutants. The extraordinary cost of chemical analyses also limits the collection of chemical data from smaller community water systems. Therefore, it should be no surprise that of the 168,690 community water systems in the United States only 7.6 percent have actually monitored for those chemicals with established primary Drinking Water Standards (i.e., the regulated pollutants). For the community water systems that participated in the chemical monitoring, regulated chemical pollutants were detected 19,111 times, and in many cases standards were exceeded. These data are summarized in Appendix 3-2. Although the data as summarized by the USEPA database cannot be used to determine the number of community water systems that provided polluted drinking water to their customers, the data suggest that the occurrence of regulated chemicals in drinking water is a common problem. Of even more concern is the occurrence of unregulated chemicals in drinking water.

The USEPA database also contains monitoring data on 46 unregulated chemical pollutants detected in community water systems. These monitoring data appear in Appendix 3-3. For the community water systems that participated in the chemical monitoring program, unregulated chemical pollutants were detected 5,601 times. Once again, the data as summarized by the USEPA database cannot be used to determine the number of community water systems that provided unregulated pollutants in the drinking water distributed to their customers. More important, this database only provides monitoring information on a small number of unregulated chemicals that could potentially pollute drinking water. This information, however, does provoke a more meaningful question. What other unregulated chemicals are also in drinking water? This question can be answered only by even more extensive monitoring programs.

Unregulated Pollutants and Monitoring Regulations

It should be obvious that the occurrence of any known toxic compound in water distributed to consumers should be addressed as a (potentially) serious problem. However, the federal government clearly does not consider unregulated chemical pollution of drinking water a concern based on existing unregulated monitoring requirements. For example, in January 2001 the USEPA implemented the Unregulated Contaminant Monitoring Regulations (USEPA, 2001) that required 2,800 large public water systems[15] and 800 of 66,000 small public water systems to conduct assessment monitoring during any continuous 12-month period[16] for the following chemicals: DCPA mono-acid, DCPA di-acid, 4,4-DDE, 2,4-dinitrotoluene, 2,6-dinitrotoluene, EPTC, molinate, MTBE, nitrobenzene, terbacil, acetochlor, and perchlorate. In addition to these chemicals, the USEPA also required a random selection of 300 large and small public water systems to monitor for the following chemicals: alachlor, diazinon, 2,4-dichlorophenol, 2,4-dinitrophenol, 1-2-diphenylhydrazine, disulfoton, diuron, ESA, fonofos, linuron, 2-methyl-1-phenol, polonium-210, prometon, RDX, and 2,4,6-trichlorophenol. If any of these chemicals are detected, each customer will receive a notice each year by July 1 that will identify the chemical pollutants in their drinking water. A unique aspect of these regulations is that USEPA is limited to monitoring no more than 30 pollutants in any 5-year monitoring cycle (USEPA, 1999).

Given the immense number of potential chemicals that can and do pollute our water resources, this monitoring program lacks credibility. In general, however, it is not unexpected that the number of chemicals being monitored makes up a short list, as monitoring costs increase for every chemical analyzed. For example, it is estimated that the annual cost (nationally) just to monitor for the first set of chemicals is $8.4 million. Because of this cost, we may never have a comprehensive monitoring program for unregulated pollutants.

The effectiveness of a monitoring program to protect the public health is obviously tied to the likelihood of a chemical occurring in a water resource, the chemical's perceived risk, and its inclusion into routine monitoring programs. Under the current system of water protection, the selection of which chemicals to monitor is a critical process. Unfortunately, the probability that a substantial number of new chemicals will be added to the Primary and Secondary Drinking Water Standards list or to the unregulated monitoring list, in real time, is highly unlikely.

---

[15]CWS and NTNCWS systems that serve more than 10,000 persons. No transient water systems were included.

[16]During this period, quarterly samples must be taken for surface water sources, whereas only biannual samples are required for groundwater sources.

Setting New Drinking Water Standards

Because of the vast number of chemicals that can occur in drinking water, the 1996 Amendments to the Safe Drinking Water Act required the USEPA to develop a list of unregulated pollutants that may pose risks in drinking water, and to determine which should be added to the Primary Drinking Water Standards. As previously noted, USEPA's consideration of economics has, and probably always will, drive policy decisions that pertain to setting any environmental standard. This reality, combined with the enormous number of potential drinking water pollutants, has forced the USEPA to identify an economically manageable list of chemicals to determine which of these compounds should be added to the Primary Drinking Water Standards List. It is worth emphasizing, "an economically manageable list of chemicals." This group of chemicals is called the Drinking Water Contaminant Candidate List (CCL).

The first CCL was developed by the USEPA and published in 1998 (USEPA, 1998). The first CCL included all of the identified toxic or hazardous compounds already regulated by various environmental programs (e.g., Clean Water Act; Resource Conservation and Recovery Act; and the Comprehensive Environmental Response, Compensation and Liability Act, etc.). This list contained 391 chemicals. This group of chemicals was then evaluated to estimate which of the compounds had the greatest probability of actually occurring in drinking water. For example, USEPA tried to determine which compounds were produced in the greatest quantities and whether these chemicals would persist in the environment long enough to pollute drinking water resources. Based on these criteria, a final list of 50 chemicals was selected (Table 3-1).

This list of 50 chemicals is an incredibly small number when one considers that approximately 72,000 chemical substances are listed in the Toxic Substance Control Act inventory of commercial chemicals. Because the USEPA limited its selection process to only those chemicals that were already regulated, the National Research Counsel (NRC, 1999) made the following observation:

> This approach, while useful for developing a CCL in a short time period, is like "looking under the lamp post" because it overlooks potential chemical contaminants not previously identified through inclusion on one of the selected lists. For example, the first CCL development process did not collect and evaluate data on radionuclides, most degradation products of known contaminants, or even all classes of commercial chemicals (such as pharmaceuticals).

Obviously, limiting the selection to only 50 chemicals is not realistic and does the average consumer a great disservice. Using USEPA's own monitoring data for unregulated chemicals as reported in Appendix 3-3, it can be shown that the following chemicals actually occur in drinking water, are known to be toxic, yet are not on the CCL list: carbaryl, cyanazine, aldicarb, aldicarb sulfone, aldicarb sulfoxide, bromochloromethane, bromomethane, butachlor, butylbenzene,

TABLE 3-1
USEPA Drinking Water Contaminant Candidate Chemical List

Initial 1998 List	CCL 2 (2004)
Acetochlor	x
Alachlor ESA (+degradation products)	x
Aldrin	
Aluminum	x
Boron	x
Bromobenzene	x
DCPA non-acid degradate	x
DCPA di-acid degradate	x
DDE	x
Diazinon	x
1,1-dichloroethane	x
1,1-dichloropropene	x
1,2-diphenylhydrazine	x
1,3-dichloropropane	x
1,3-dichloropropene	x
2,2-dichloropropane	x
2,4-dichlorophenol	x
Dieldrin	
2,4-dinitrophenol	x
2,4-dinitrotoluene	x
2,6-dinitrotoluene	x
Disulfoton	x
Diuron	x
EPTC (s-ethyl-dipropylthiocarbanate)	x
Fonofos	x
Hexachlorobtadiene	
p-isopropyltoluene (p-cymene)	x
Linuron	x
Manganese	
Methyl bromide	x
Methyl-t-butyl ether (MTBE)	x
2-Methyl-phenol (o-cresol)	x
Metolachlor	x
Metrobuzin	
Molinate	x
Naphthalene	
Nitrobenzene	x
Organotins	x
Perchlorate	x
Prometon	x
RDX	x
Sodium	
Sulfate	
1,1,2,2-tetrachloroethane	x
Terbacil	x
Terbufos	x
Triazines (and degradation products)	x

TABLE 3-1 *(continued)*

Initial 1998 List	CCL 2 (2004)
2,4,6-trichlorophenol	x
1,2,4-trimethylbenzene	x
Vanadium	

sec-butylbenzene, tert-butylbenzene, o-chlorotoluene, p-chlorotoluene, dia-camba, dibromomethane, m-dichlorobenzene, dichlorodifluoromethane, trans-1,3-dichloropropene, 3-hydroxycarbofuran, methomyl, propachlor, n-propylbenzene, toxaphene, 1,2,3-trichlorobenzene, trichlorofluoromehtane, 1,2,3-trichloropropane, trifluralin, 1,2,4-trimethylbenzene, and 1,3,5-trimethylbenzene. This example illustrates that the selection of the CCL chemicals based on predefined criteria and economic constraints fails to address the real problem.

That drinking water is polluted by known toxic but unregulated chemicals is not in question. However, many of the identified chemicals are not considered because they do not meet bureaucratic and economic objectives. The National Drinking Water Advisory Council Report on the CCL Classification Process to the USEPA (May 19, 2004) is suggested reading for those who wish to review the details on how chemicals will be selected for the CCL. This tome clearly demonstrates that selecting specific chemicals that are a potential human health risk can be justified administratively, but the process is not practical, nor does it result in the selection of validated standards.

This result is preordained since the Safe Drinking Water Act directs the USEPA to identify only those contaminants that pose the greatest public health concerns. Thus, even when a drinking water resource is polluted, only those compounds that are considered to pose the greatest risk need be evaluated. Such an approach is particularly egregious considering that the true human health effects of many chemicals remains largely unknown.[17] Furthermore, the CCL list was updated with the draft CCL 2 list (Table 3-1) published in the *Federal Register* on April 2, 2004 (Volume 69, Number 64), but no new chemicals were added and eight compounds were removed from consideration. It would appear from this modification that the objective of the National Drinking Water Advisory Council and the USEPA is to reduce the number of chemicals in consideration rather than its expansion.

The selection process for the CCL is significant because it is from this list that the USEPA must select any chemical that it wants to regulate in the future. More specifically, the 1996 Amendments to the Safe Drinking Water Act require the USEPA, within 5 years after the final

---

[17]For example, the USEPA's existing chemical toxicity testing and assessment program is only focusing on the approximately 3,000 high-production volume chemicals (i.e., those produced at levels over 1 million ponds per year). The USEPA has a long way to go considering that only 550 of these chemicals have been evaluated since 1979.

CCL is published, to make a decision on whether to regulate at least five of these chemicals. Even with this deadline, it will be years before any new standards are promulgated. This standard setting procedure is discussed in more detail in Exhibit 3-2. Given the level of pollution that exists in our nation's surface and groundwater, this rate of regulation is clearly inadequate.

It is a reasonable assumption that regulatory programs have been designed to ensure that the human health is protected but, in fact, it cannot be demonstrated that these programs actually minimize the risk associated with consuming polluted drinking water, in a quantifiable sense. Therefore, it is up to the consumer to decide whether they are comfortable drinking water (presumably validated as safe by present government regulations) that might possibly contain a number of unspecified pollutants. The consumer should be concerned because studies continue to show that drinking water resources are polluted with a vast array of manufactured chemicals.

Why Consumers Should Be Concerned

Since the 1940s, the State of California has been an acknowledged leader in environmental protection. Therefore, it was not unexpected that in 1999, the California Public Interest Research Group (Heavner, 1999) published a major report on pesticides in drinking water.[18] This report showed that the majority of California counties had detectable concentration of pesticides in drinking water sources that originated from both surface and ground waters. Table 3-2 shows the number of drinking water sources where pesticides were detected in more than 10 sites since 1990. Contrary to general theories of pesticide pollution, these data show that pesticide detections in groundwater were significantly greater than in surface water sources.

Furthermore, many of these pesticides continue to be unregulated. Of the 600 water supplies where pesticides were detected, the utilization of advanced treatment systems would have the capability of removing the pesticides. The City of Fresno, the City of Modesto, the City of San Bernardino, the Contra Costa Water District, and the Monterey District all installed granulated activated carbon treatment systems. The City of Riverside estimated that the cost to install and maintain an activated carbon system would raise a Riverside's residence water bill approximately $80 a month. This is a small price to pay, considering that cost of bottled water can run as high as several thousand times the cost of municipal supplies—not the best of economic tradeoffs.

---

[18]It is unfortunate that more states have not conducted similar studies.

Exhibit 3-2. *You Do Know That Science Has Nothing to Do with Policy: How Drinking Water Standards Are Set in the United States*

According to the USEPA,[19] the United States enjoys one of the best water supplies in the world. The United States has also spent hundreds of billions of dollars to build drinking water treatment and distribution systems and currently spends $22 billion per year to operate and maintain that system. Why is it necessary to spend these large amounts on drinking water? USEPA believes that "all sources of drinking water contain some naturally occurring contaminants."

The USEPA controls the levels of contaminants that are allowed in public drinking water systems by setting National Primary Drinking Water Regulations (NPDWR). Drinking water standards, or maximum contaminant levels (MCLs), have been set for more than 80 contaminants. MCLs are established based on known or potential health effects, the availability of technologies to remove the contaminant, their effectiveness, and the cost of treatment. MCLs are supposed to be set at levels that protect public health. State public health and environmental agencies have the primary responsibility for ensuring that these federal drinking water quality standards, or more stringent ones required by an individual state, are met by each public water supplier.

Before 1974 each state set the standards that had to be met at the local level. Drinking water quality and protection standards differed from state to state. In 1974, the Safe Drinking Water Act (SDWA) was passed to ensure that a consistent level of protection and quality existed throughout the United States. As part of the 1996 SDWA amendments, USEPA is now required to publish a list of contaminants that are not currently subject to an NPDWR and are "known or anticipated" to occur in public water systems. This list, the Contaminant Candidate List (CCL), is supposed to set research priorities and help in the development of guidance from USEPA and the selection of contaminants for making regulatory determinations and/or monitoring by the States.

(continued)

[19]USEPA (2001). "Water on Tap: A Consumer's Guide to the Nation's Drinking Water." URL: http://www.epa.gov/safewater/wot/introtap.html.

Exhibit 3-2. (continued)

The CCL currently consists of 50 chemical and 10 microbiological contaminants. Contaminants on the list include industrial solvents, metals, pesticides, explosives, rocket fuels, biocides, and common elements. Exposure to the listed chemicals are known to lead to a host of significant health effects including cardiovascular, pulmonary, immunologic, neurologic, and endocrine (e.g., diabetes) effects, cancer, and even death. The SDWA required USEPA to determine whether to regulate not less than five of the contaminants (but not all of the contaminants) from the CCL by 2001. The USEPA revised the CCL by February 2003 and can defer the selection of a final list until August 2006, eight years after the first CCL was created. If a chemical is deferred until 2006 for regulation, then a proposed NPDWR for that chemical must be issued no later than August 2008. The final rule for any chemical on the CCL is required 18 months later (February 2010) and can be delayed an additional 9 months (November 2010). Under the SDWA, water systems have 3 years to comply with an NPDWR and may take an additional 2 years if capital improvements to the water system are necessary. Therefore, it is possible that a chemical recognized by USEPA in 1998 as a potential health threat will not be removed from a drinking water system until 2013! After spending up to 15 years evaluating the potential threat from a contaminant and making improvements to aging water distribution systems, the public may finally get relief from exposure to that contaminant!

The delay in the decision-making process is understandable when that time is spent gathering critical data on the potential health effects of exposure to contaminants. What is not understandable is the amount of time, money, and energy spent by various advocacy groups to delay the decision-making process. Consider a set of recent decisions by USEPA on the proposed MCL for arsenic, a known human carcinogen. On January 22, 2001, USEPA published a "final" ruling on a proposed MCL for arsenic, reducing it from 50 micrograms per liter (μg/L) to 10 μg/L (Federal Register, 66 FR 6976). This drinking water standard was based on the standard set by the United States Public Health Service in 1943. The current World Health Organization (WHO) standard is 10 μg/L. USEPA staff was interested in setting the standard at 5 μg/L, but changed it to 10 μg/L after intense criticism that USEPA's cost-benefit ratio was unrealistic and their estimated cost for mitigation was excessively low. The "final" rule was to become effective March 23, 2001 and compliance with the rule required by January 2006. On March 20, 2001, USEPA withdrew the January 22, 2001 standard for arsenic and sought an independent review of the science behind the standard and the cost estimates to communities that would have to implement the rule. On April 23, 2001, USEPA announced its intention to put in place an arsenic MCL that would require compliance with that standard by 2006. USEPA also asked the National Academy of Science to review a range of 3 to 20 µg/L for the new drinking water standard.

While the USEPA spent considerable time and effort developing a proposed MCL for arsenic, a known human carcinogen, the MCL was tabled not by concerns over the science but rather over the cost of implementing that standard! As a result, the prospect for developing MCLs for the 50 contaminants on the CCL in a timely manner are limited at best. As an attorney for an industrial advocacy group against the new standard pointed out "science has nothing to do with policy."

TABLE 3-2
Pesticides Detected in More Than Ten Sites Used for Drinking Water in California Since 1990

Pesticide	Pounds Used in 1997	Groundwater Wells with Detections	Surface Water Sites with Detections
DBCP	Banned	918	19
Simazine	764,586	580	34
Diuron*	1,228,114	343	8
Atrazine	46,568	178	11
Bromacil*	82,424	179	
EDB	Banned	129	10
1,2-Dichloropropane	Banned	63	
Prometon*	20	28	4
Bentazon*	1,907	16	7
Methyl Bromide*	15,663,832	16	4
Diazinon*	955,108	4	14
2,4-D	609,039	5	13
Chlorthal*	342,000	8	7
Dalapon	2	7	6
Metolachlor*	212,714	5	7
Carbaryl*	753,801	5	7
Heptachlor	Banned	9	2
Norflurazon*	212,621	10	1
Carbon disulfide*		10	1
EPTC*	579,245	2	9
Thiobencarb*	894,287	1	10
Chlorpyrifos*	3,152,564	1	9
Trifluralin*	1,433,999	2	8
Cyanazine*	470,838	4	6
Molinare*	1,170,699	2	8

* = Pesticides not regulated by primary drinking water standards.

TABLE 3-3
United State Geological Survey of Drinking Water Resources, Pesticide Mixture
Frequency of Detection (%)

Pesticide mixtures	Agricultural Land Use	
	Surface water	Ground water
Atrazine, DEA	35%	26%
Atrazine, DEA, Metholachlor	33%	4%
Atrazine, DEA, Prometon, Metholachlor	17%	3%
Atrazine, DEA, Prometon, Metholachlor, Simazine	14%	2%
Atrazine, DEA, Prometon, Metholachlor, Simazine, Alachlor	10%	

	Urban Land Use	
	Surface water	Ground water
Simazine, Peometon	66%	7%
Prometon, Atrazine, Simazine	50%	5%
Atrazine, Simazine, Prometon, Diazinon	38%	
Atrazine, Simazine, Prometon, Metholachlor, Diazinon	23%	
Atrazine, Simazine, Prometon, Metholachlor, Diazinon, Chloropyrifos	15%	

A study by the United States Geological Survey (Patterson and Focazio, 2001) on contaminants in drinking water sources reported that pesticides were reported in both surface water and groundwater in areas dominated by agricultural and urban land use. For example, Table 3-3 lists the frequency (%) of detection at or above 0.01 ppm that various mixture of pesticides were found. Contrary to what might be expected, surface water in urban areas had a greater frequency of pesticide pollution than surface water in agricultural areas. This report also showed that in areas where MTBE was used in gasoline, approximately 20 percent of the groundwater wells tested contained detectable concentrations of MTBE.

Canada has registered 550 pesticide active ingredients under the Pest Control Products Act, and registers 10 to 15 new actives each year.[20] Canada currently lacks a systematic, coordinated, interjurisdictional system for monitoring pesticides in aquatic systems (water and sediment).

In 1999, the Natural Resources Defense Council (Olson et al., 2003) reviewed the water quality compliance of various cities around the United States and ranked their performance. The rating of each city for

---

[20]Maguire, R.J. et al., "Pesticides," National Water Research Institute—Environment, Canada.

data collected in 2001 is given in Table 3-4. Their summary is consistent with the overall data provided in Appendix 3-2 and Appendix 3-3. The database just identifies specific cities. All of these reports and data clearly demonstrate that the occurrence of manufactured chemicals in our drinking water resources is ubiquitous throughout the nation. The Natural Resources Defense Council study also reports that most cities' water supply systems need immediate repair and upgrading. When old pipes leak and break, bacteria and other contaminants can get into the water. Furthermore, outdated drinking water treatment plants (i.e., those without advanced treatment systems) allow many contaminants "to slip through."

TABLE 3-4
Natural Resources Defense Council Water Quality Compliance Assessment Summary

City	Rating	Assessment
Albuquerque	Poor	Violated radon and arsenic standards
		Exceeded national health goals for gross alpha radiation, thallium (TTHMs) and haloacetic acid (HAA) compounds, total coliforms
Atlanta	Fair	Violated turbidity standard (can influence microbial levels)
		Reported high levels of lead and HAA
		Poor pipe maintenance and distribution system
Baltimore	Good	Violated turbidity standard
		Reported high levels and spikes of lead and HAA
Boston	Poor	Violations for lead
		Reported elevated levels of TTHMs and the presence of *Cryptosporidium*
Chicago	Excellent	Reported low levels of lead, TTHMs, and HAA
Denver	Good	Reported moderate levels of TTHMs and HAA but high levels of lead
Detroit	Good	Reported high levels of TTHMs, HAA, lead, and total coliform
Fresno	Poor	Reported high levels of pesticides, chlorinated solvents, lead, arsenic, radon, and gross alpha radiation
Houston	Fair	Violated radon and HAA standards
		Reported high levels of TTHMs, arsenic, and total coliform
Los Angeles	Fair	Violated proposed standard for perchlorate
		Reported high levels of TTHMs, HAA, arsenic, radon, and nitrates
Manchester	Good	Reported high levels of lead and TTHMs
		Reported low levels of MTBE and trichloroethylene

(continued)

TABLE 3-4 *(continued)*

City	Rating	Assessment
New Orleans	Good	Reported high levels of TTHMs and HAA Reported moderate levels of atrazine
Newark	Fair	Violated lead standards Reported high levels of TTHMs and HAA
Philadelphia	Fair	Reported high levels of TTHMs, HAA, lead, and the occurrence of *Cryptosporidium* and *Giardia* Reported low levels of industrial chemicals, metals, and pesticides
Phoenix	Poor	Reported high levels of arsenic, TTHMs, HAA, nitrates, chromium, and di(2-ethylhexyl)phthalate Exceeded draft perchlorate standard
San Diego	Fair	Exceeded draft perchlorate standard Reported high levels of TTHMs, HAA, ethylene dibromide, lead, MTBE, gross alpha and beta radiation, uranium, and total coliform
San Francisco	Fair	Violated TTHMs and HAA Reported high levels of lead and the presence of *Cryptosporidium* and *Giardia* Reported cross-connection risk between potable and nonpotable sources
Seattle	Fair	Violated lead standard Reported high levels TTHMs and HAA and presence of *Cryptosporidium*
Washington, D.C.	Fair	Violated TTHMs Reported high levels of HAA, cyanide, and total coliform Reported moderate levels of radioactive contaminants

Unlike the previous studies, a report on water-soluble pesticides in the finished water of 12 community water systems by Coupe and Blomquist (2004) looked at both large and small water supplies.[21] This approach is critical, as small community water systems have less stringent monitoring requirements than large community water systems that were reported by Olsen et al. (2003). Coupe and Blomquist (2004) reported that the most frequently detected pesticides were atrazine, cyanazine, prometon, simazine, acetochlor, alachlor, and metolachlor. Most important, atrazine, metolachlor, prometon, and simazine were detected in the source and finished water of every system sampled, whereas deethylatrazine[22] was detected in the source and finished water of every system but one.

---

[21]Systems with less than 10,000 consumers.
[22]A degradate of triazine herbicides.

One of the community water systems that was sampled by Coupe and Blomquist (2004) was in Tensas Parish, Louisiana. The water system operated by the Tensas Water Distribution Association, Inc., takes a majority of their source water from Lake Bruin. The water treatment system begins with a clarification step using chlorine, aluminum, and soda ash and is followed by filtration. Water samples were taken from April 1999 through December 2000. The chemical results for Tensas Parish are given in Table 3-5. These data show that for the two regulated pesticides, atrazine and simazine, their concentrations in the finished water were below primary drinking water standards. The remaining compounds found in the finished water are currently unregulated.

Contact with the mayor's office in the City of St. Joseph, Louisiana, the largest community serviced by the Tensas Water Distribution Association, indicated that no one in the community was even aware that their drinking water had been sampled and analyzed for pesticides (personal communication, 2004) . In fact, their first notification of any water sampling was from the USEPA, when the Tensas Water Distribution Association was found to be in violation of exceeding the HAA5 standard (63 ppb, 120 ppb, and 143 ppb, respectively, for the first, second, and third calendar quarters of 2004). The violation notice in the November 27, 2004 issue of the Tensas Gazette, St. Joseph, LA contained the following information for the residents of St. Joseph's: "EPA does not consider this violation to have any serious adverse health effects on human health as a result of short-term exposure; however, continued long term exposure to HAA5 levels above the standard (e.g., 20 years of exposure) has the potential to have serious adverse effects on human health. . . . Some people who drink water containing HAA5 in excess of the MCL [60 ppb] have an increased risk of getting cancer."

This is an interesting response considering that in the same legal notice it is stated that "Compliance with the HAA5 standard for public water systems serving less than 10,000 individuals initially became

TABLE 3-5
Median Pesticide Data for Source Water and Finished Water, Tensas Parish, Louisiana (1999-2000)

Pesticide	Source Water Concentration (ppb)	Finished Water Concentration (ppb)	Removal Efficiency	MCL (ppb)
Atrazine	0.885	0.733	17%	3
Cyanazine	0.179	0.117	35%	
Deethylatrazine	0.156	0.134	14%	
Metolachlor	0.015	0.013	13%	
Prometon	0.007	0.004	43%	
Simazine	0.124	0.102	18%	4
Tebuthiuron	0.007	0.003	57%	

effective and enforceable on January 1, 2004." In other words, the system had never been monitored for HAA5 compounds until that year. Since Tensas Parish has used chlorination for disinfection for decades, and assuming the level of dissolved organic matter in Lake Bruin has not changed, the residents of the community have most likely been exposed to HAA5 compounds in excess of the standard for a very long time. This example illustrates the problem small community water systems have in regard to knowing what chemicals are in their water. This obviously does not include those individuals with private source waters (where the water quality could actually be worse rather than better—without testing no one really knows).

Drinking water has been found to contain a wide array of chemical contaminants. To date, the most common compounds found in drinking water are the disinfection by-products and pesticides; metals, perchlorate,[23] fuel compounds, and chlorinated solvents have been detected less often. Yet, there is virtually no information on pharmaceuticals and chemicals associated with personal care products in our drinking water sources. Because these are unregulated chemicals, we may never know their true impact on drinking water sources.

The previous data demonstrate that chemical monitoring for specific chemicals is not a credible approach for managing chemical risks in drinking water. Furthermore, it is impossible to monitor for all potential chemical pollutants in drinking water on a national level because (1) the laboratory performing the analysis needs to know what specific chemicals to look for in the water; (2) most laboratories require standard methods of analysis for each specific chemical, but these methods are not routinely available for unregulated pollutants; and (3) it would be economically impracticable even for large community water systems. Thus, the extent of existing chemical pollution in drinking water cannot be monitored, and consumers will continue to unknowingly consume manufactured chemical pollutants. Given this condition, consumers need to be aware of the pollutants they may be consuming and the approaches to mitigate that consumption.

Approaches to Mitigate Chemical Exposure

The current system of regulating chemicals in drinking water is simply not credible. As a result, it is reasonable that individuals exercise some degree of caution with regard to the water they drink. Many consumers

---

[23]Interestingly, perchlorate was noticed as a contaminant as far back as the early 1950s when the California State Water Pollution Control Board restricted the discharge of perchlorate from the AeroJet facility in the Sacramento, California area. Subsequently, in an annual summary addressing groundwater quality published by the American Water Works Association; also in the 1950s, the presence of perchlorate in groundwater in the Sacramento area was noted, presumably from the same source addressed by the state.

have been doing so by drinking bottled water. A survey (Anonymous, 1993) conducted by the American Water Works Association in 1993 reported that 35 percent of people were worried about tap water safety and that over half of all Americans drink bottled water. Another study (Anadu and Harding, 2000) of the risk perception that drives bottled water use found that among other reasons "public perceptions of chemical risk are also influenced by the level of distrust the public holds for government and industry." This perception could be one of the reasons that sales of bottled water rose to $5.2 billion in 1999 and are expected to increase by a compound rate of 15 percent. In terms of consumption, this equates to approximately 3.4 billion gallons annually or more than 12 gallons per person. These data suggest that consumers believe that drinking bottled water is one solution to reducing the consumption of chemical pollutants. On a global scale, consumption of bottled water is growing faster than 10 percent per year, with sales estimated to have reached $100 billion per year.[24]

Bottled Water

In general, the quality of bottled water is expected to be better than the quality of tap water. After all, the product is in a bottle, it looks good, and it is expensive relative to tap water. Because of this perception, providers of bottled water should be held to a higher standard of purity.[25] Unfortunately, they are not held to a higher standard, nor do they usually provide analysis of their products for manufactured pollutants. In fact, bottled water is not necessarily better than tap water and can in some cases be just as polluted as tap water. Furthermore, bottled water is not regulated by the USEPA but rather by the U.S. Food and Drug Administration (FDA)[26] under Title 21, Part 165.110 of the Federal Code of Regulations.[27] Under these regulations, bottled water must comply with the National Primary and Secondary Water Quality Standards[28] (see Table 1-9). The major difference between USEPA and FDA oversight lies in biological purity standards.

For example, the standards for biological purity required by the USEPA and the FDA are compared in Table 3-6. USEPA standards are more comprehensive than FDA standards.[29] Even with this dichotomy,

---

[24]Gleick, Peter H. et al., "The World's Water 2004-2005," Island Press.

[25]There are no regulations that require producers of bottled water to assess or disclose the environmental quality of their product.

[26]Bottled water in the United States is also regulated by individual states.

[27]Some states have standards that are stricter than the federal standards.

[28]Chemical monitoring frequency is essentially the same between the USEPA and FDA, neither of which are adequate.

[29]Since bottled water is classified as a food by the FDA, why do FDA regulations on pesticides (see Appendix 2-8) in food not apply to water? For some reason (probably economic), the FDA believes that chemicals, which are banned from foods, are not found in water.

TABLE 3-6
Comparison of Biological Purity Standards between the USEPA and FDA for Bottled Water

Parameter	USEPA Tap Water	FDA Bottled Water
Disinfection Required	Yes	No
E. coli and Fecal Coliform Banned	Yes	No
Testing Frequency for Bacteria	Hundreds/month	Once/week
Filtration to Remove Pathogens or Have a Protected Source	Yes	No
Must Test for Viruses	Yes	No

the good news concerning bottled water is that for the past 37 years, there have been no confirmed reports in the United States of illnesses or disease linked to bottled water. However, the potential for microbial pollution of bottled water still exists.

The bad news is that a survey conducted over 4 years by the Natural Resources Defense Council (Olsen et al., 1999) found that one-third of the 103 brands of bottled water tested contained elevated levels of bacteria, inorganic chemicals, and/or organic chemicals. In this study, the specific chemicals that were tested for varied widely between laboratories. In general, however, the chemicals looked for were those with established primary drinking water standards. The chemicals found in bottled water include acetone, n-butylbenzene, 2-chlorotoluene, dichloroethane, ethylbenzene, p-isopropyltoluene, methylene chloride, styrene, trichloroethylene, toluene, and xylene. This list of chemicals is consistent with those compounds found in drinking water sources throughout the United States (see Appendix 3-2 and 3-3). The occurrence of these compounds in bottled water suggests that (1) source water contained manufactured chemical pollutants or (2) they were potentially added during the bottling process (e.g., toluene, xylene, ethylbenzene). As a result of these data, the Natural Resources Defense Council report concluded that bottled water was not necessarily safer than tap water.[30]

Domestic bottled water can also contain disinfection by-products if it comes from community water systems.[31] This water could contain any number of the TTHMs and HAA5 compounds. Because the Stage 2 Rule amended the Primary Drinking Water Standards for the TTHMs and HAA5 compounds, bottled water must ultimately meet the same standards (Federal Register, March 28, 2001).

---

[30]Both government and industry estimate that between 25 and 40 percent of all bottled water is actually tap water.
[31]Bottled water providers use ozone or UV light for disinfection; thus, there is no threat of halogenated disinfected by-products.

In the United States, there are different types of bottled water products. These bottled water definitions are listed in Table 3-7 with related groundwater characteristics shown in Figure 3-1. In the European Union, Natural Mineral Water corresponds to a "microbiologically wholesome water, originating in an underground water table or deposit and emerging from a spring tapped at one or more natural or bore exits"[32] that must originate from a protected source. This product is not sterile and can contain natural microorganisms. In addition to these standards, all bottled water imported into the Unites States must meet FDA and state regulations.

Unfortunately, none of these descriptive terms or regulations really give the consumer any quantitative indication that the actual bottled water does not contain manufactured chemical compounds, microorganisms, or trace elements. In other words, the true environmental characteristics or quality of bottled water are unknown. At most, the only

TABLE 3-7
Different Types of Bottled Water (U.S. Definitions)

Water Product	Source Water
Artesian Water	Well water from a confined aquifer that pushes the water above the level of the land surface
Mineral Water	Water containing not less than 250 parts per million as total dissolved solids
Spring Water	Water derived from an underground formation from which water flows naturally to the surface of the earth
Well Water	Water from a hole bored, drilled, or otherwise constructed in the ground and tapping into an aquifer
Sparkling Water	Water that contains natural or added carbon dioxide in the same concentration that it had at the point of emergence from its source
Purified Water	Water that has been produced by distillation, deionization, reverse osmosis, or other processes that meets the most recent definition of purified water in the United States Pharmacopeia
Distilled Water	Water that has been produced by vaporizing and then condensing the water during the process of distillation
Drinking Water	Bottled water that is obtained from an approved source, meets all applicable federal and state standards, and has undergone a minimal treatment process consisting of filtration and some type of disinfection

Notes:
1. Label Statement: Any water that comes from a community water system must be labeled as such with a label that states: "from a community water system" or "from a municipal source."
2. "Mineral" and "Sparkling" waters originate from artesian wells, groundwater wells, or springs.

[32]Council Directive 80/777/EEC of July 15, 1980.

information provided on a bottle of water are the concentration of the major cations and anions (i.e., Na, K, Ca, Mg, Cl, HCO_3, and SO_4),[33] and potential marketing statements concerning the purity of the product. Some bottled water producers have web sites that will provide chemical analysis of the major cations and anions and more details on the geologic origin of the source water and the potential for manufactured chemicals to pollute the source. For the vast majority of these web sites, there is a fundamental lack of environmental information. The type of environmental information that is recommended to be provided by bottled water producers should include:

- Trace element analyses that include radioactive elements
- Chemical monitoring data on microbial analysis, pesticides, industrial chemicals, and fuel and lubricant compounds
- Details on the geologic characteristics of the aquifer (e.g., type of aquifer material, depth of aquifer, type and number of impermeable geologic units between the aquifer and earth's surface, the land use of the recharge areas, the land use of the source area, the travel time from the recharge area and the source area)

Until this type of information is made available, the consumer is at the mercy of the bottled water producer's marketing hype.

The basic inorganic chemistry of specific bottled waters can be obtained for most bottled waters around the world from web sites that sell or promote these products. Examples[34] of the inorganic water chemistry of mineral waters (total dissolved solids greater than 250 ppm) are given in Appendix 3-4; examples for nonmineral waters are given in Appendix 3-5. In addition to this basic information, the amount of potential natural radioactivity associated with these bottled waters was calculated using the equation developed by Sparovek et al. (2001) and converted to picocuries (pCi/L). This equation (3-1)[35] was developed from the analysis of 12 mineral waters with a range of total ions from 112 to 6,157 ppm. The data from their study are summarized in Table 3-8.

$$pCi/L = (6.85Cl^- - 24.6K^+ \ 4.68)(0.027) \qquad (3-1)$$

Trace element data that were available for some of the waters in Appendix 3-4 and 3-5 were collected by the Institute of Plant Nutrition and Soil Science, Federal Agricultural Research Center (FAL) Braunschweig, Germany, in 2003. These data (with concentrations in ppb) are presented in Appendix 3-6.

---

[33]Sometimes the level of arsenic is indicated on the label.
[34]These examples do not include all commercially available bottled waters. Waters were selected to illustrate the range of available information.
[35]The r^2 for this regression equation is 98 percent.

TABLE 3-8

Data on Radioactivity Originating from Ra 226 and Uranium (U) in Mineral Waters

Mineral Water	Radioactivity from Ra 226 + U (pCi/L)	Radioactivity from U (%)	Total Ions (mg/L)	Cl^- (mg/L)	K^+ (mg/L)
Bismark	0.17	0	305	14.9	4.2
Wittenseer	0.18	0	310	17.8	1.3
Hella	0.20	0	142	29.1	3.2
Volic	0.36	72	112	8.4	5.7
Contrex	1.10	86	2,181	8.6	3.2
Vittel	1.12	44	517	29.4	4.9
Evian	1.53	86	480	4.5	1.0
Perrier	2.10	85	609	21.5	5.1
Apollinaris	3.25	2	2,650	100.0	20.0
Heppinger	8.45	2	4,513	244.7	52.7
Pellegrino	9.98	59	1,132	68.0	2.8
Mlynsky	48.59	1	6,157	591.1	93.0

Adapted from Sparovek et al., 2001.

When looking at the available inorganic chemistry of bottled waters, there is considerable variation in the distribution of the major cations and anions, trace metals, and radioactivity. When considering the environmental consequences of bottled water and the inorganic chemistry, it must be remembered that the human species evolved consuming water with dissolved minerals. In other words, it would be difficult to establish whether a specific mineral water had an identifiable human health risk other than if it exceeded existing primary drinking water standards. Since all bottled waters sold in the United States must meet the primary drinking water standards, the consumer should have little concern.

All of the bottled waters have been sorted by the concentration of their total dissolved solids (TDS) or total ions (i.e., in the absence of TDS data).[36] TDS was selected as a major environmental characteristic in bottled water for several reasons. First, many of the commercially available bottled waters are mineral waters (i.e., water with a TDS greater than 250 ppm). In the United States, TDS has a maximum contaminate level of 500 ppm. Therefore, consumers should be aware of this inorganic water quality standard when selecting a bottled water for every day consumption. Second, sometimes the higher the TDS level the greater the travel time of the groundwater or depth of percolation (i.e., approaching geothermal conditions); this can be an indirect indication of a water that potentially could have a lower probability of containing manufactured

---

[36]For those waters that do not have a measured total dissolved solids (TDS), the sum of the total ions can be compared to other waters with known TDS and total ion values (if all the major cations and anions are measured; Na, K, Ca, Mg, Cl, HCO_3, and SO_4).

pollutants (i.e., in the absence of analytical data).[37] The downside is that this water may also contain elevated concentrations of trace elements.

It is recommended that consumers select bottled waters that minimize their exposure to regulated trace elements. Unfortunately, having access to published trace element data, including radioactive elements, for specific bottled waters is rare, as most bottled water producers do not publish this data. Given this major omission, the existing data in this chapter can generally be used to help evaluate the inorganic chemistry of the bottled waters in Appendixes 3-4 through 3-6. Of all the potentially harmful elements (i.e., for those who require low-sodium diets), at least sodium can be evaluated. Consumers concerned with the level of sodium in bottled water can almost always find this information on the bottle. Consumers concerned about the amount of radioactivity in bottled water can refer to the calculated radioactivity (expressed in pCi/L) in Appendix 3-4 and Appendix 3-5. Keep in mind, however, that these numbers are modeled estimates.

In addition to the bottled water examples given previously, there are also specialty bottled waters that are purified or have added chemical elements or characteristics. Examples of these specialty bottled waters that are sold in the United States are listed in Table 3-9. Given that many bottled waters lack specific inorganic and organic chemical analysis, purified waters offer a good alternative to a water that has not been completely characterized. When considering the selection of a purified bottled water, a bottled water that has been treated with multiple treatment technologies is preferable to a water product that has only been treated with one chemical technology. Water treatment, however, can sometimes add cost to the product. Table 3-10 provides a cost comparison of bottled waters that were available in a store located in Northern California. Clearly, there is a significant price range in the various water products for which the consumer can purchase.

In addition to the chemical pollutants that can be present in natural bottled water, the consumer should also be aware of several issues associated with the plastic bottles in which it is often sold. According to Stover et al. (1996), the migration potential exists for traces of monomers, oligomers, additives, stabilizers, plasticizers, lubricants, and reaction products of polymers and additives from plastics into a packaged product. Examples from this report include (1) the release of measurable amounts of the antioxidant 2,6-di-t-butyl-p-cresol[38] from polyethylene, polystyrene, and polypropylene and (2) the leaching of acetaldehyde[39] from

---

[37]For example, given that the introduction of manufactured synthetic organic compounds were not mass produced until the 1920s and 1930s, any aquifer with a travel time of 100 years should not contain these compounds at the source today. This assumes that contamination of the aquifer did not occur from other environmental conditions.

[38]This compound is not listed as a human carcinogen, but there is evidence of carcinogenicity in animals.

[39]Acetaldehyde (CAS Number: 75-07-0) is a listed hazardous substance and may be a carcinogen in humans.

TABLE 3-9
Examples of Specialty Bottled Waters Sold in the United States

Product	Advertised Characteristics	Unknowns
Angle Fire Water	Oxygenated, 8.6-8.9 pH, mineral enhanced	Source of water, analytical data, minerals added
Aquafina	Purified by reverse osmosis and carbon filtration, non-detect for sodium	Source of water, analytical data
AVO2	Spring, carbon filtered, and reverse osmosis	Analytical data
Cool Wave	Alkaline water, treated to remove impurities; electrolysis to reduce molecular clusters	Source of water, analytical data treatment methods
Dasani	Purified by reverse osmosis, minerals added non-detect for sodium	Source of water, analytical data
Essentia	Alkaline water (pH 9.5) formulated to simulate intracellular fluid electrolytes, ionization to reduce molecular clusters	Source of water, analytical data
eVamor	Alkaline water, containing chromium, vanadium, and selenium	Source of water, total analytical data
Glaceau Smartwater	Vapor distilled and nutrient enhanced	Source of water, analytical data, nutrients added
HiOsliver	Contains 60 ppm oxygen, pH 8.4, 364 ppm TDS, 6.2 ppm Na, 4.4 ppm Ca, 110 ppm Mg, 3.1 ppm NO_3 (Mount Hamilton, California)	Source of water
Life O2	Oxygenated	Source of water Analytical data
Penta	Contain 60-70 ppm oxygen, treated to remove all chemical pollutants (i.e., pesticides, arsenic, MTBE, chromium (VI), etc.), treated to reduce molecular clusters	Analytical data

TABLE 3-10
Bottled Water Price Comparison

Name	Price ($)/Liter	TDS(ppm)	Remarks
Arrowhead, USA	0.66	125	
Calistoga, USA	0.89	130	
Aquafina, USA	0.92	No Data	Treated water
Dasani, USA	1.06	No Data	Treated water
Crystal Geyser, USA	1.09	120	
Fiji, Fiji	1.39	160	
Esker, Canada	1.50	No Data	
Ice Age, Canada	1.52	4	
Smartwater, USA	1.59	No Data	Treated water, minerals added
Essentia, USA	1.66	No Data	Treated water, minerals added
Evian, France	2.19	309	Mineral water
Evamor, USA	2.49	No Data	High pH water
Acqua Panna, Italy	2.49	188	
San Pellegrino, Italy	2.65	1079	Mineral water
Perrier, France	2.65	478	Mineral water
Lurisa, Italy	2.69	No data	
Trinity, USA	2.81	195	Geothermal mineral water, alkaline
Penta, USA	3.98	No data	Treated water, oxygen added
Voss, Norway	4.73	22	

polyethylene terephthalate (PET). PET is one of the most common plastics used for bottled water.[40]

Acetaldehyde imparts both taste and odor to bottled liquids. As a consequence, manufactures of PET monitor the amount of acetaldehyde produced during the production of PET. This is such a common problem that PerkinElmer Instruments made an autoanalyzer just for this product. Studies by Sheftel (2000) showed that acetaldehyde was found in carbonated mineral water and lemonade in concentration ranges between 11 and 7.5 mg/L.

Polycarbonate plastics have also been suggested to contribute Bisphenol A into liquid products under various temperature ranges and bottle characteristics (Takao et al., 1999). The report by the World Wildlife Fund (Lyons, 2000) identifies Bisphenol A as a known endocrine

---

[40]It is also important to note that polyethylene terephthalate (PET) is recycled and can contain organic pollutants from previous uses (Komolprasert, 2001). The FDA has also approved the use of recycled PET in food containers. Recycling of PET is purified and rebuilt at the molecular level, yet the impact on bottled water is unknown.

disruptor. This report also establishes that Bisphenol A has been detected in water and mineral water in the low ppb range.

Under the Test Section of the August 1, 2000, *Consumer Reports* (Anonymous, 2000), bottled water was evaluated for taste. According to this article, "waters bottled in PET plastic generally tasted better than those bottled in HDPE [High Density Polyethylene]." One water, in a PET bottle ". . . imparted a hit of sweet, fruity plastic flavor (imagine the scent when you blow up a beach ball)," while a water bottled in HDPE has a taste ". . . a bit like melted plastic (imagine the smell when you get a plastic container too close to a flame)." All of the previous data show that plastic bottles can and do impart very low concentrations (ppb or parts per trillion range) of chemical compounds into their product. Unfortunately, bottled water producers that use plastic bottles either do not test their stored product for potential compounds known to be leachable, or they do not publish the results. In the absence of specific proof that PET or other plastic bottled water products do not contribute trace chemical pollutants to their products, the prudent consumer should purchase water either in glass bottles or, if plastic containers cannot be avoided, use a disposal carbon filter that is specifically designed for plastic bottles.[41] On a global scale, packaging of bottled water ranges from Brazil packaging 99.9 percent of bottled water in plastic to Germany where glass is used in 97 percent of bottled water packaging.[42]

Drinking bottled water that contains no manufactured chemicals and acceptable ranges on trace and radioactive elements is one more important step toward limiting a person's exposure to pollutants.[43] However, those individuals who wish to minimize their exposure and their family's exposure to potential chemical pollutants in municipal water or their own private water source, should install an in-home water treatment system.

In-Home Water Treatment Systems

The necessary water treatment technologies exist today for consumers to treat their source water. There are no technical barriers to prevent the removal of manufactured chemicals from drinking water. Therefore, consumers who are dissatisfied with a water utility that (1) does not use advanced treatment technologies to remove manufactured chemicals, (2) causes excess residual chlorine or bromine in their water product, or (3) reports detectable levels of disinfection by-products, industrial chemicals, fuel products, or pesticides in their water product, should consider

---

[41]If a carbon filter is used on the bottle, the consumer should make sure that the manufacturer's recommendations on the volume of water that can be treated are followed.
[42]Gleick, Peter H. et al., "The World's Water 2004-2005," Island Press.
[43]Having this level of purity in bottled water then raises the questions as to what is the quality of water used in soft drinks and beer? Obviously, if there are chemical pollutants in bottled water, why not in soft drinks, reconstituted juices, and beer?

installing an in-home water treatment system. Consumers who have their own groundwater or surface water supply should also consider an in-home water treatment system depending on the surrounding land use and source water characteristics.

Although there are various methods of removing pollutants from water, the most commonly used techniques used by in-home water treatment systems are reverse osmosis (RO) and granulated activated carbon (GAC). RO technologies use fine porous membranes to separate inorganic and organic chemicals from water. These units are effective at removing dissolved salts, suspended matter, and a wide variety of dissolved organic chemicals, as well as bacteria and viruses. A typical "claim list" of chemicals removed by an RO system is given in Table 3-11, and Table 3-12 provides an example of RO removal efficiencies. Carbon filters (GAC) can reduce chlorine, many manufactured synthetic chemicals (chlorinated and nonchlorinated) including pesticides, some radiological constituents, fluoride, radon, and some metals (Seelig et al., 1992). An example of removal efficiencies using GAC is given for a set of selected chemicals in Table 3-13.

As can be seen from Tables 3-11 through 3-13, a significant reduction in chemical pollutant levels can be achieved using these technologies; however, chemical removal efficiencies are only known for a small

TABLE 3-11
Reverse Osmosis Chemical Reduction Claim

alachlor	1,2-dichloropropane	styrene
atrazine	cis-1,3-dichloropropylene	1,1,2,2-tetrachloroethane
benzene	dinoseb	tetrachloroethane
carbofuran	endrin	toluene
carbon tetrachloride	ethylbenzene	1,2,4-trichlorobenzene
chlorobenzene	ethylene dibromide	EDB
dibromochloropropane	1,1,1-trichloroethane	DBCP
heptachlor	1,1,2-trichloroethane	o-dichlorobenzene
p-dichlorobenzene	heptachlor epoxide	trichloroethylene
hexachlorobutadiene	trihalomethanes	xylenes (total)
hexachlorocy-clopentadiene	1,2-dichloroethane	lindane
trans-1,2-dichloroethylene	methoxychlor	1,1-dichloroethylene
dichloroethylene	2-4-D	cis-1,2-dichloroethylene
pentachlorophenol	2,4,5-TP (silvex)	simazine
total dissolved solids	barium	cadmium
copper	hexavalent chromium	trivalent chromium
lead	radium	selenium

Source: Eco Home Products Hydro Line 5000 Reverse Osmosis Unit, 7745 Alabama Ave. #11, Canoga Park, CA 91304.

TABLE 3-12
Reverse Osmosis Chemical Removal Efficiencies

Compound	Percent	Compound	Percent
Aluminum	97–98	Polyphosphate	98–99
Bromide	93–96	Pyrogen	99+
Cadmium	96–98	Radioactivity	95–98
Chloride	94–95	Silica	85–95
Chromate	90–98	Silicate	95–97
Chromium	96–98	Silver	95–97
Copper	98–99	Sodium	94–98
Cyanide	90–95	Strontium	96–99
Ferrocyanide	99+	Sulfate	99+
Hardness	95–98	Thiosulfate	99+
Iron	98–99	Virus	99+
Lead	96–98	Magnesium	96–98
Ammonium	85–95	Manganese	98–99
Arsenic	94–96	Mercury	96–98
Bacteria	99+	Nickel	98–99
Barium	96–98	Nitrate	93–96
Bicarbonate	95–96	Orthophosphate	98–99
Borate	40–70	Phosphate	99+
Boron	60–70		

Source: The Good Water Company, 151 N. Main Street, Suite 700, Wichita, KS 67202.

TABLE 3-13
Activated Carbon Performance Test

Chemical	Removal Efficiency (Percent)
Chlorine	98.0
Dichloromethane	98.9
Chloroform	99.5
Trichloroethylene	99.1
Perchloroethylene	99.6
Benzene	99.3
Toluene	99.3
p-Xylene	99.4
Aldrin	98.5
HCH	99.4
p-DDT	98.0
PCB	95.0
Atrazine	99.0
Phenols	99.3
Naphthalene	99.2
Fluoranthene	98.4
Benzo-a-pyrene	94.7

Source: Katadyn Water Filters (Katadyn can be contacted at webmaster@Katadyn).

number of compounds (i.e., the regulated chemicals). Therefore, it must be assumed that other unregulated compounds will also be removed with the same level of efficiency. This assumption is generally valid but not without some degree of risk. For example, compounds such as carbon disulfide, methyl bromide, chloromethane, and dichlorodifluoromethane (Freon 12) would not be substantially removed using GAC. Without actual removal efficiencies for a vast number of unregulated compounds, it will be necessary to combine both RO and GAC into one treatment system to provide a level of treatment that ensures the greatest level of pollutant removal.

The amount of water treatment required largely depends on the source of a home's water supply. Water from a community water supply will require less treatment than water from a private well. Furthermore, the greater the amount of inorganic and organic compounds that are in the source water, the more frequent the maintenance (i.e., change-out of RO membranes and/or GAC filter media). As a result, some waters that contain elevated concentrations of calcium and magnesium (i.e., hard water) should be softened[44] to reduce potential scaling of the treatment system. Iron can also reduce the efficiency of RO membranes. As a consequence, it is recommended that if iron levels in the source water are above 5 ppm, they should be removed before an RO system to reduce RO maintenance.

The maintenance frequency of any water treatment system ultimately depends on when pollutant pass-through or breakthrough occurs. At this point, a new RO membrane or fresh GAC filter media should be installed. How does a home owner determine when pollutants are no longer being removed? This calculation is fairly straightforward for RO technologies. When an RO membrane begins to clog, the pressure difference across the membrane increases. This pressure difference can be monitored electronically to warn the home owner that it is time to change the membrane. In addition, when the efficiency of an RO membrane decreases, the salt content of the treated water will increase. Salt levels can also be monitored using a simple salinity probe. The problem with GAC filter media is that there is no way to determine when breakthrough of organic pollutants occurs. The only way to determine whether low concentrations of a compound are no longer being adsorbed by GAC is by analyzing the water for the occurrence of that compound. Given the wide range of compounds that can occur in drinking water, this option is not a realistic monitoring method. Thus, GAC filter media should be changed out on a predefined maintenance schedule designed to

---

[44]This is usually accomplished using ion exchange technologies where sodium is exchanged for the calcium and magnesium in the water (sodium carbonates do not form solids like calcium and magnesium carbonates).

ensure that sufficient capacity to remove organic pollutants remains.[45] This change-out cycle should be determined on the basis of the amount of organics that will pass through the GAC filter for a given volume of treated water. This filter cycle can be determined in consultation with the filter manufacturer or the filter supplier.

When home owners use a home treatment system, they must assume responsibility for maintaining it. Failure to do so will result in either a decreased pollutant removal efficiency or no pollutant removal at all. If home owners do not want to assume these very important responsibilities, they should engage a water purification service company to provide routine monitoring surveillance and maintenance activities.

Home purification systems fall into two broad categories, point-of-use (POU) systems and point-of-entry (POE) systems. A POU system is installed at the location of the water's use; a POE system is installed to treat all water entering the house. Since POU systems only provide a means of supplying treated water to one location in a household, they are not universally recommended, as they do not treat bath/shower water, dishwasher water, or water used for the washing machine. However, if space limitations or economic considerations do not allow the use of a POE system, a POU system is more than acceptable.

In most homes, POU systems are usually installed at the kitchen sink. The two most common POU systems are GAC filters attached to water tap and under-the-sink units that use RO, GAC, or a combination of both methods. Of these two methods, GAC filters on the tap, although commonly used, are not recommended because under normal household use, water may not come in contact with the GAC filter long enough to effectively remove pollutants. Furthermore, it is almost impossible to keep track of how much water has passed through the filter.[46] As a result, it is highly likely that the adsorbing capacity of the GAC filter will be exhausted (i.e., no longer providing any treatment) without the home owner's knowledge.

Furthermore, potential for bacterial growth in both RO and GAC units can become a serious issue. These treatment units can serve as a potential location for microbial growth. As a result, all under-the-sink treatment systems should include a UV light unit to destroy microorganisms. As is the case with other in-home systems, UV water treatment systems for home use are commercially available. In addition to this problem, if the source of the drinking water has hard water (i.e., high

---

[45]Granulated activated carbon (GAC) can also be an excellent place for bacteria to grow. Thus, GAC should be changed out on a regular basis to minimize bacteria pollution. This may not be necessary if there are KDF pre- and postfilters. KDF is an electrochemical/oxidation reduction process that will oxidize chlorine, remove iron, aluminum, lead, silver, and cadmium and inhibits bacterial growth.

[46]A filter can only treat a specific volume of water, for example, 60 gallons.

levels of calcium and magnesium) or contains more than 3 ppm iron, it should be treated at the POE to remove these materials. Pretreatment for hardness and iron removal will increase treatment efficiency and reduce maintenance.

A good POU system at the kitchen sink may include (1) a prefilter to remove turbidity or suspended solids and chlorine, (2) a canister system employing a high flow RO unit followed by an GAC unit, and (3) a UV light treatment unit. These systems require a 4- to 5-gallon storage tank between the GAC unit and the UV light treatment unit. If the size of the system does not fit under the sink, it can be placed in either a basement or garage with the appropriate plumbing to the POU. In many cases, only a GAC system will be necessary, which will eliminate RO units and bulky storage tanks.

A POE system takes a portion of the total amount of water that enters the home, treats it and stores it for future distribution to all sinks, showers, bath tubs, and household appliances. The remaining untreated water[47] can then be distributed to outside faucets and irrigation systems. Generally, a POE system has the same equipment as a POU system except the individual components are larger and more expensive. For example, a minimum storage tank is usually 100 to 300 gallons and should be glass lined. These systems may require as much as a 50- to 70- square foot space in a garage, basement, or utility room; the exception to this case is when only a high flow (i.e., 10 to 14 gallons per minute) GAC column and UV or KDF system is installed on a municipal water source. These systems require very little space (3 square feet) and are reasonably priced ($2,000 to $2,500).

What any POU or POE system can accomplish is easily summarized by reviewing the respective performance characteristics of both RO and GAC components (see Tables 3.3, 3.4, and 3.5). The prefilter removes turbidity and also helps to delay early fowling of the RO unit. Such filtration is especially important if the source water is not from a community water supply. The RO system removes a wide variety of chemical and biological pollutants. Removal efficiencies approaching 95 percent are routinely obtained, with some chemicals having a 99 percent removal rate. The GAC unit acts as a polishing step, removing constituents that might have passed the RO unit. The GAC unit also operates at efficiencies typically exceeding 90 percent. Thus, using both RO and GAC units in combination should result in excellent water quality. Because the GAC is used as a polishing step, probability of any organic chemical passing through the treatment system is lower when the GAC filter is replaced on a routine basis. GAC units used in this manner are also less likely to have a bacterial problem, as levels of organic compounds, suspended sediments, and nutrients in the filter media are generally low. The UV unit assures disinfection of any bacterial pollutants that may

---

[47]Which is most of the water used by a household.

have passed through the previous units. It should also be noted that UV lamps must be replaced once or twice a year and that water storage tanks should be routinely cleaned on an annual basis.

Because both RO and GAC treatment units have unknown removal efficiencies for unregulated chemicals, it could be necessary to combine them in the same treatment system. This issue aside, it is widely accepted that these treatment units, when properly maintained, will achieve overall removal efficiencies of 90 to 95 percent for dissolved inorganic and organic compounds. At present, in-home treatment systems are not regulated by federal, state, or local laws. However, the industry is self-policed by several organizations including the National Sanitation Foundation (NSF). The NSF, a not-for-profit group, has been accredited by the American National Standards Institute, the Occupational Safety and Health Administration, and the Standard Council of Canada. NSF's program certifies the performance of in-home water treatment units and system components. The NSF does not certify complete systems. It is highly recommended that only NSF certified treatment units and components be used for in-home treatment systems.

Although water treatment systems require routine maintenance to guarantee water quality, this fact should not negate the value of their use. Furthermore, for individuals who are unsure how to maintain a water treatment system or just do not want to be bothered with having to remember maintenance cycles, a service company can be hired to monitor, clean, and replace filters as necessary. Ultimately, as more consumers install in-home treatment systems, reductions in cost of these treatment technologies and their maintenance will probably be realized.

Some municipal water utilities have even considered providing such a service for costumers outside their normal service range. Another benefit to in-home treatment systems is that in the event that a chemical or biological organism is intentionally added to a municipal water system, in-home treatment can serve as a second line of defense.

Potential Pollution by Terrorists

A USEPA-sponsored study was conducted in the early 1970s (Agardy, 1972) to address the threat of the intentional poisoning of drinking water resources. This study concluded that the release of a chemical or biological agent into a drinking water resource could occur in two fundamental ways. Hazardous materials could be either (1) introduced into a water resource reservoir before its treatment and release to the water distribution system, or (2) injected into drinking water that is already within the water distribution system pipelines (i.e., posttreatment). These same conclusions were reached by the American Water Works Association in February 2002 (Regush, 2002).

The pretreatment threat was not considered large because large community reservoirs generally contain tens of millions to billions of gallons of water. Such an enormous volume of water would dilute any toxic chemical or biological agent to the point of being ineffective. To create a hazardous condition, the perpetrator would need to dump truckloads of chemicals or hundreds of pounds of biological agents. However unlikely, it is not impossible to obtain truckloads of toxic chemicals (i.e., as a licensed transporter or by highjacking) in various locations of the United States. Still, it would be difficult for terrorists to produce, steal, or smuggle hundreds of pounds of biological agents into the United States for the purpose of polluting water supplies. Furthermore, once the water was chlorinated at the water treatment plant, most biological agents would be destroyed.

These conclusions were basically supported by a 2001 article in *Water Environment and Technology* (Anonymous, 2001a). However, this same article pointed out that with more than 6,800 public drinking water intakes[48] on rivers in the United States, these intakes "can be considered vulnerable to disruption by accidental or intentional release of hazardous chemicals or biological substances." Given the potential for the intentional pollution of water resources, Congressional hearings in November 2001 on Antiterrorism (Luthy, 2001) concluded that existing water treatment technologies could be put together in a series to provide "multiple barriers" to block the chemical or biological pollution. In other words, the technical solutions exist; they need only be implemented. Still, it was believed that an intentional act of polluting water resources before treatment was a serious threat.

The greatest threat to posttreatment water supplies would be from toxic chemicals introduced into small water storage tanks. Doing so would be a difficult task considering that the terrorist would have to be familiar with the storage system, be able to gain access to the area, and remain undetected during the time necessary to completely pump the chemicals into the tank. The earlier USEPA study also suggests that the posttreatment release of toxic chemicals into the water distribution system could constitute a serious problem. On one hand, the introduction of a biological agent into a posttreatment water supply line is much less of a hazard, as residual chlorine or bromine is usually present (unless UV light is being used).[49] The real threat exists from introduction of a toxic chemical into the distribution system outside the confines of the treatment plant. In many cities, this aim can be accomplished by simply pumping a truckload of a toxic chemical into any fire hydrant. The potential number of individual households or businesses that could be affected would obviously depend on their proximity to the point of injection. These same concerns were mentioned in the November 2001

---

[48]This number of intakes on rivers also indicates the magnitude of public water supplies that can be impacted by upstream pollution without any accident or intentional act.

[49]Some community water systems are switching to ultraviolet light for disinfection to avoid halogenated disinfection by-products. Thus, these distribution systems would be much more vulnerable to biological attack.

article by the American Water Works Association (Anonymous, 2001b).

On October 10, 2001, the Federal Bureau of Investigation (FBI) provided a Statement for the Congressional Record on "Terrorism: Are American's Water Resources and Environment at Risk." The conclusions of the FBI report were generally the same as the 1972 study. Among their conclusions were that (1) pollution of a water supply with a biological agent that causes illness or death is possible but not probable, (2) pollution of a water resource with a biological agent would unlikely produce a large risk to public health, (3) a successful attack would require knowledge of the water supply system, and (4) a successful attack would likely involve a posttreatment injection. As a result, the FBI recommended that water utilities "maintain a secure perimeter around the source, if possible, and the treatment facility. In addition, security should be maintained around critical nodes such as tunnels, pumping facilities, storage facilities, and the network of water mains and subsidiary pipes should be enhanced."

Although the FBI recommended increased security on the posttreatment distribution systems, there were no specific recommendations for locked or secured watermains and fire hydrants or electronic monitoring. Given the cost associated with such a program, along with the FBI's estimate of the low probability of a successful attack or potential for large-scale damage to a significant number of consumers, it is highly unlikely that federal funding would be available for such security measures. In addition to the FBI's assessment, the American Water Works Association issued a press release on October 18, 2001, that proclaimed "Terrorist Threats to Nation's Drinking Water Supply Remote."

These warnings remain important because they confirm that a threat to our drinking water from a terrorist act is possible. No matter how remote this threat may be, it should be remembered that such an attack does not require sophisticated technology or a lot of money. Furthermore, the safety of drinking water cannot be guaranteed because (1) there are multiple points of attack in a posttreatment distribution system, and (2) these systems require extensive and costly monitoring to detect any breach in security.

Water Pollution and Risk

With the amount of chemical pollution that has been identified in our water resources, it is not surprising that our sources of drinking water are also polluted. Given this condition, is there really a health risk that needs to be addressed or managed beyond our reliance on existing drinking water standards? This issue needs to be critically evaluated in light of growing populations and the increasing occurrence of unregulated chemicals in our water. Again it should be recognized that the term *risk* has many meanings depending on one's global location. In the industrialized nations it is a fair observation that a great deal of attention is placed on risks associated with synthetic organic chemicals in the very low ppb range. This is certainly not true in the developing nations where the greatest concern continues to be directed on bacterial contamination. Thus, there is a very wide spectrum as regards pollution and risk.

References

Agardy, Franklin J., 1972, *The Threat . . . from Additions of Chemicals and Biologicals to a Municipal Water Supply*, URS Research Company, San Mateo, California.

Anadu, E.C. and A.K. Harding, 2000, "Risk Perception and Bottled Water Use," *Journal of the American Water Works Association*, Vol. 92, pages 82–90.

Anonymous, 1993, "Consumer Attitude Survey on Water Quality Issues," American Water Works Association Research Foundation Report.

Anonymous, 2000, "It's Only Water, Right," *Consumer Reports*, Vol. 65, Issue 8 (Aug. 1), pages 17–22.

Anonymous, 2001a, "WE&T News Watch, Water-related Bioterrorism Unlikely, Experts Say," *Water Environment and Technology*, Vol. 13, pages 10–14.

Anonymous, 2001b, "Manager to Manager, Are We Prepared?" *Journal of the American Water Works Association*, Vol. 93, pages 28–30.

Congressional Report, 2000, Report to Rep. Henry A. Waxman on "Public Exposure to Arsenic in Drinking Water," Special Investigations Division, Committee on Government Reform for the U.S. House of Representatives (October 4).

Coupe, Richard H. and Joel D. Blomquist, 2004, "Water-soluble Pesticides in Finished Water of Community Water Supplies," *Journal of the American Water Works Association*, Vol. 96, pages 56–68.

Gray, Sean et al., 2001, *Consider the Source, Farm Runoff, Chlorination Byproducts, and Human Health*, Environmental Working Group, Washington, D.C.

Harris, Robert H. and Edward M. Brecher, 1974, "Is the Water Safe to Drink, Part 1: The Problem," *Consumer Reports*, June issue, pages 436–442.

Heavner, Brad, 1999, "Toxics on Tap, Pesticides in California Drinking Water Sources," California Public Interest Research Group Charitable Trust, San Francisco, California.

Komolprasert, V. et al., 2001, " Volatile and Nonvolatile Compounds in Irradiated Semi-rigid Crystalline Poly(ethylene terephthalate) Polymers," *Food Additives and Contaminants*, Vol. 18.

Liang, Sun et al., 2001, "Treatability of MTBE-contaminated Groundwater by Ozone and Peroxone," *Journal of the American Water Works Association*, Vol. 93, pages 110–120.

Luthy, Richard, G., 2001, "Safety of Our Nation's Water," Statement before the committee on Science, U.S. House of Representatives hearing on: H.R. 3178 and the Development of Anti-Terrorism tools for Water Infrastructure (November 14).

Lyons, Gwynne, 2000, "Bisphenol A, A known Endocrine Disruptor," World Wildlife Fund (United Kingdom) European Toxics Program Report.

Najm, Issam and R. Rhodes Trussell, 2000, "NDMA Formation in Water and Wastewater," American Water Works Association, Water Quality Technology Conference, Salt Lake City (November).

NRC, 1999, *"Identifying Future Drinking Water Contaminants,"* National Research Council and National Academy Press, Washington D.C.

Olson, Erik D. et al., 1999, "Bottled Water, Pure Drink or Pure Hype?," Natural Resources Defense Council, New York.

Patterson, Glenn G. and Michael J. Focazio, 2001, "Contaminants and Drinking-Water Sources in 2001: Recent Findings of the U.S. Geological Survey," United States Geological Survey, Open-file Report 00-510.

Regush, Nicholas, 2002, "Questions on Protecting US Water Supplies," *Journal of the American Water Works Association*, Vol. 93, pages 52–53.

Roberts, Megan G., Philip C. Singer and Alexa Obolensky, 2002, "Comparing Total HHA and Total THM Concentrations Using ICR Data," *Journal of the American Water Works Association*, Vol. 94, pages 103–114.

Seelig, B., F. Bergsrud and R. Derickson, 1992, "Treatment Systems for Household Water Supplies: Activated Carbon Filtration," North Dakota State University, NDSU Extension Service, Report AE-1029.

Sheftel, Victor O., 2000, *Indirect Food Additivies and Polymers: Migration and Toxicology*, Lewis Publishers, Boca Raton, Florida.

Sparovek, R., J. Fleckenstein and E. Schnug, 2001, "Issues of Uranium and Radioactivity in Natural Mineral Waters," *Landbauforschung Völkenrode*, Vol. 4, No. 51, pages 149–157.

Stecher, P. G., M. J. Finkel, O. H. Siegmund and B. M. Szafranski (editors), 1960, *The Merck Index of Chemicals and Drugs*, 7th Edition, Merck & Company, Rahway, New Jersey.

Stover, Richard L. et al., 1996, "Report of the Berkeley Plastics Task Force," City of Berkeley, California, April 8, 1996.

Takao, Y. et. al., 1999, "Fast Screening for Bisphenol A in Environmental Water and in Food by Solid-phase Miroextraction," *Journal of Health Science*, Vol. 45, No. 39.

USEPA, 1998, "Announcement of the Drinking Water Contaminant Candidate List; Notice," *Federal Register*, Volume 63, Number 40:10274-10287.

USEPA, 1999, "Final Revisions to the Unregulated Contaminant Monitoring Regulation," Office of Water, Washington, D.C., EPA 815-F-99-005.

USEPA, 2000, "Providing Safe Drinking Water in America," Office of Enforcement and Compliance Assurance, Washington, D.C., EPA 305-R-00-002.

USEPA, 2001, "Unregulated Contaminant Monitoring Regulation: Monitoring for List 1 Contaminants by Large Public Water Systems," Office of Water, Washington, D.C., EPA 815-F-01-003.

USGS, 2001, "The Environment and Human Health," U.S. Geological Survey, Fact Sheet FS-054-01, May 2001.

USPHS, 2001, "9th Report on Carcinogens," U.S. Department of Health and Human Services, National Toxicology Program, Washington D.C. (January).

CHAPTER 4

Living with the Risk of Polluted Water

"Our problems would be much simpler if we needed only to consider the balance between food and population. But in the long view the progressive deterioration of our environment may cause more death and misery than the food-production gap."
Paul R. Ehrlich, The Population Bomb, 1971

As demonstrated in Chapters 2 and 3, our water resources and specifically our drinking water in every developed country of the world is polluted with a vast array of manufactured chemicals. This condition is the direct result of either nonexistent environmental programs or programs that permit low levels of chemical pollutants to be distributed throughout the human habitat. As a consequence of this pollution, drinking water has been and will continue to be polluted by chemicals that are known to be toxic or hazardous at some levels but are allowed in our drinking water as long as their concentrations meet existing standards.

Under this scenario, does having a single chemical or a mixture of chemicals at low concentrations in our drinking water really pose a threat to our health? After all, for the most part, chemical pollution is not an acute threat (i.e., such as pollution from a chemical spill or terrorist act) but rather a chronic condition in the form of a multitude of chemical pollutants that occur at low concentrations ingested over time. The risks to our health of such long-term exposure is difficult to quantify. Such a dilemma can foster inaction and complacency. Furthermore, the burden of proof in establishing the real health risks of drinking polluted water should not be limited only to the occurrence of a "cancer cluster" in a community[1] consuming polluted drinking water. In today's chemically dependent societies, the risk we all live with needs to be much better understood and addressed.

---

[1] For example, Toms River, New Jersey, and Woburn, Massachusetts.

The Burden of Proof

Most people believe, at a gut level, that environmental factors, such as the exposure to chemicals, exposure to airborne smoke and particulates, stress, drug and alcohol use, diet, electromagnetic fields, and radiation, can contribute to or cause health problems. This belief is strongly supported by an article in the July 2000 *New England Journal of Medicine* (Lichtenstein et al., 2000) that studied the medical history of twins. The article established that it was much more likely that the occurrence of cancer was linked to the exposure of environmental factors than to genetics. These fears are further bolstered by the fact-based films such as "A Civil Action" and "Erin Brockovich," which have heightened the public's awareness that chemical pollution of water supplies can have life-threatening consequences. In each film, individuals in two communities suffered from illnesses that appeared to be caused by chemical pollution of their drinking water[2] (i.e., trichloroethylene in one case and hexavalent chromium in the other). As a result of these illnesses, individuals filed lawsuits against the companies that caused the pollution to recover medical costs as well as punitive damages for the chemical insult that was unknowingly introduced into their bodies.

In both cases, the sources of chemical pollution and the companies responsible for the pollution were easily determined. The problem facing the attorneys for the plaintiffs was establishing "beyond a reasonable scientific certainty" that there was a direct linkage between the health problems exhibited by their clients and the chemicals they unknowingly consumed. Unfortunately, the attorneys for the plaintiffs in both cases were unable to prove such a linkage. As a result, both cases were settled by mutual agreement between the plaintiffs and defendants with a cash payment to the plaintiffs. Because these cases settled, the courts never affirmed or denied the plaintiffs' allegations that their health problems were caused by the chemicals in their drinking water. From the standpoint of a citizen concerned about the effects of chemical pollution on their health, this is a totally unsatisfying outcome.

These results, however, were virtually preordained because in a court of law the plaintiffs have the burden of proof. In other words, the plaintiffs must first prove to the court that the company in question actually caused the chemical pollution and then that this pollution caused their illness. This second step is not a simple proof. In a toxic tort case, experts for the plaintiffs would have to establish the following sequence of facts:

- What is the pollutant that allegedly caused their illness? This is the simplest fact to establish.

---

[2]Because pollutants are absorbed through the skin while bathing or inhaled while showering, this water should be as pure as the water we drink.

- How did the pollutant reach the location at which the individual was exposed? In most cases this can be established with a high degree of certainty.
- What was the concentration of the pollutant at that location where the individual was exposed? This is more difficult, but not impossible to determine.
- How long was the individual exposed to the pollutant? While this is not as easily established, reasonable estimates can be made.
- What is the effect of the pollutant on the human body (i.e., its effect at the gene, cell, organ, and system level)? This information is known for only a handful of chemicals. Thus, the likelihood of establishing *effect* is nearly impossible, given the tens of thousands of chemicals manufactured in the United States alone.
- For a specific individual, are the symptoms of their specific illness consistent with exposure to this pollutant? Again, this information is known for only a handful of chemicals for which industrial exposure studies have been completed. Thus, the probability of establishing this relationship in a nonindustrial environment is nearly impossible.

Recognizing these weakness, experts for the defense can confuse the issue by emphasizing the potential causative relationship between an illness of a specific individual and any or all of the following environmental factors: exposure to other chemicals either in the home or workplace, smoking, alcohol, diet, stress, drug use, and medical history. It is ironic that other "environmental factors" are used to invalidate the act of actually consuming a chemical pollutant. Defense lawyers will also argue that there is little, if any, scientific evidence linking a specific chemical to a specific health problem.[3] However, one reason for this is that health data (i.e., occupational epidemiologic studies) are inherently inaccurate as a result of the complexities associated with these studies (NRC, 1999b). Because there usually is no clear relationship between a specific chemical and a human illness, courts often decide to exclude expert testimony on this issue. For example, in Federal Court pursuant to Daubert, expert testimony offered by the plaintiff to demonstrate a link between the illness and exposure to a toxic chemical was found to be inadmissible because it was not scientifically valid (Colvig, 2001). Given this difficult proof, it is no wonder that most of these cases never go to trial.

The failure of plaintiffs to establish the necessary facts in a toxic tort case is the direct result of not having chemical-specific information on the relationship between exposure to known concentrations of a chemical

---

[3]In another Erin Brockovich/PG&E chromium lawsuit in Kettleman, California, reported in the *San Francisco Chronicle* on January 12, 2002, PG&E spokesman Jon Tremayne stated that the defense "will be to question the links between the chromium contamination and the illnesses of plant workers and local residents."

and human health effects.[4] The only chemical for which these relationships are fairly well known is lead and this is only because the health effects of lead on humans have been extensively studied for nearly a century. In today's litigious environment, the ability to develop human-specific chemical toxicity data on the tens of thousands of manufactured chemicals used in the United States would appear to be impossible unless the federal government allows human chemical exposure studies. Because this is highly unlikely, researchers will never be able to evaluate the true impact of specific chemical pollutants on human health, let alone the effect of mixtures of these chemicals that occur in drinking water.

Because our understanding of chemical toxicity is usually based on animal studies, the concentration at which a specific chemical causes a toxic response or cancer in a human is not known. This failing calls into question the ability of the government to establish the concentration, or standard, at which a chemical in drinking water will not impact human health. After all, what is the purpose of establishing a safe drinking water standard if it is not based on actual human exposure data.

If it is not possible to demonstrate a causative link between a chemical and harm to the human body in a court of law, then what can one conclude as regards the safety of drinking water based on arbitrary standards? In spite of this critical deficiency, existing federal laws imply that there is "no harm" if a chemical pollutant is present at a concentration that is less than an existing water quality standard. This dichotomy has perpetuated environmental programs that have legalized pollution[5] while it is advertised as being protective of the public health. Under these programs, it is acceptable to (1) ingest small quantities of chemicals known to have harmful effects to animals at elevated concentrations; (2) ingest any concentration of a known or potentially toxic chemical that does not have a federal or state standard, which is approximately 99.9 percent of all the chemicals currently used in the United States (as well as most developed countries); and (3) drink water containing a mixture of chemical pollutants just as long as each chemical is either unregulated or below its published standard. Clearly, the existence of these environmental programs begs the question, how did we saddle ourselves with this type of pollution control?

Permissible Pollution

The primary objective of the 1925 U.S. Public Health Service (USPHS) drinking water standards was to "safeguard the health of the public"

---

[4]This can be generally referred to as a dose-response relationship. This relationship would quantify the specific chemical concentrations required to induce a specific health response (e.g., hormone imbalance, cancer, death).
[5]Federal and state environmental laws allow industry and/or individuals to release chemicals into water as long as the pollution is below "acceptable" levels.

against the most seriously recognized dangers such as "contamination by disease (e.g., such as typhoid fever and other illnesses of similar origin and transmission)." These standards did, however, warn that "to state that a water supply is 'safe' does not necessarily signify that *absolutely no risk is ever* incurred in drinking it." This statement, which was made more than three quarters of a century ago, is probably one of the most accurate predictions of today's pollution problems.

The USPHS water quality standards further stated that an acceptable water supply should be suitable for drinking, be clear, colorless, odorless, pleasant to the taste, and free from toxic compounds. These standards also established limits for the following toxic metals in drinking water: lead at 0.01 mg/L, copper at 0.2 mg/L, and zinc at 5.0 mg/L. If a water resource exceeded any of these standards, it was to be rejected and an alternate source used in its place. Today, however, unlike 1925, it is more difficult to find an alternate water source that is not polluted. The types of chemicals released into the environment have evolved and increased since the first standards were implemented in 1925. These chemicals are not only more complex, but they are more numerous. It has been estimated that in excess of 72,000 chemicals are produced in the United States (NRC, 1999a), yet the number of these chemicals that are currently regulated is extremely small (see Table 1-9).

Furthermore, of the 168,690 public water systems (USEPA, 2000a) in the United States, only 7.6 percent of these have actually monitored their water for the chemicals in the primary Drinking Water Standards list, and reported the occurrence of these chemicals to the USEPA. These data are compiled in the USEPA's National Drinking Water Contaminant Occurrence Database and can be accessed at http://www.epa.gov. Of the public water systems that did report to the USEPA, approximately 90 percent reported the detection of at least one chemical from the list as being present in their drinking water. In many cases, the standards for these chemicals were exceeded. Unfortunately, the use of standards to protect the environment and public health has been adopted throughout federal programs (USEPA, 2000b).

Every significant pollution control regulation in the United States (e.g., Clean Water Act, Safe Drinking Water Act, Clean Air Act, Resource Conservation and Recovery Act, Toxic Substances Control Act, and the Comprehensive Environmental Response Compensation and Liability Act) allows for chemical pollution up to set limits that are (presumably) established to protect human health and the environment. But do these standards really protect human health? Since virtually all standards are established using toxicity data collected specifically on animals, standards that may be applicable to the environment and the general ecology may not serve the same purpose with regard to humans.

This is a difficult question because there is practically no scientific data showing a direct relationship between the adsorption and ingestion of specific chemical pollutants or mixtures of chemicals by humans and directly observed human health problems (i.e., except for lead). Without having chemical-specific and/or chemical mixtures information collected on humans and correlated with human health-specific information, the scientific basis of drinking water standards cannot be demonstrated.

Furthermore, even when potential human health risks have been identified for specific chemicals, it usually takes the federal government years to recognize, let alone address, the hazard. For some of the most common water pollutants, the gap between the time a compound has been released into the environment, and its recognition as a health hazard by the government can be decades. All of these conditions cast doubt on the ability of drinking water standards to protect human health. Because of this doubt, it is important to explore the process of defining the ways in which adverse health risks are determined by the scientific, regulatory, and legal communities.

This exploration requires an introduction into the toxicology of chemicals (see Appendix 4-1 for a primer on toxicological terminology), quantitative and nonquantitative risk assessments, special populations that must be identified before standard setting, and standard setting.

The Dose Makes the Poison

Toxicology is the study of poisons, or more correctly the study of how chemicals interfere with the normal function of a biological system. According to the 15th century physician and father of toxicology, Paracelsus (1493-1541), "All substances are poisons; there is none which is not a poison. The right dose differentiates a poison and a remedy," or the dose makes the poison.

In today's terms, toxicology is an applied science that is built on other medical sciences including physiology (the study of living organisms), biochemistry (the chemistry of living organisms), pathology (the study of diseases and their progression in the body), pharmacology (the study of drugs on living organisms), medicine (the science of diagnosing, treating, or preventing disease and other damage to the body or mind), and epidemiology (the study of disease patterns). The specialties of toxicology fall into three distinct categories: descriptive toxicology, research/mechanistic toxicology, and applied toxicology (Williams, 2000).

- *Descriptive toxicology* focuses on the testing of chemicals. Descriptive toxicology focuses on identifying the organ toxicities of a test agent under a wide range of exposure conditions. In addition to identifying all possible acute and chronic toxicities, including genotoxic, reproductive, developmental, and carcinogenic potential of the test agent, a complete descriptive toxicologic evaluation seeks to identify the metabolites of the test agent in the body, absorption patterns of the test agent, distribution and accumulation in tissues and organs, as well as the excretion patterns of the compound.

- *Research/mechanistic toxicology* attempts to determine how substances exert deleterious effects on living organisms. Studies for mechanistic toxicology are designed to identify the cellular or biochemical mechanism involved in organ or systemic toxicity. The focus is on the actual interaction of the toxicant or its metabolites on cellular structures.
- *Applied toxicology* is concerned with the use of chemicals outside the laboratory (the real world). Applied toxicologists attempt to determine the physiologic effect or safety of an administered product. A regulatory toxicologist judges whether a substance has low enough risk to justify making it available to the public. The process whereby safe dose or levels of exposure are derived is termed *risk assessment*. This can be a quantitative process or a qualitative process.
- *Forensic toxicology* is essentially a specialty area of analytical chemistry. Forensic toxicology involves the use of analytical chemistry, pharmacology, and toxicology concerned with the medical and legal aspects of drugs and poisons.

These four specialties are all used in *clinical toxicology*, which involves the treatment of individuals poisoned by medicines and over-the-counter drugs, and *occupational toxicology*, which focuses on the exposure of workers and the diseases that result from those exposures. In the last several decades, there has been the development of *environmental toxicology*, which is concerned with chemical exposures in the environment. Obviously, the chemical pollution of our water resources is most closely related to environmental toxicology and our potential exposure to drinking chemical pollutants. These exposures can be from intentional releases (e.g., regulated point source discharges) or unintentional releases (e.g., pesticide/herbicide residue in runoff). In all these areas of toxicology, the central concept is the dose of a specific chemical to which an individual is exposed.

Basic Concepts of Dose

The key principles in toxicology are exposure, dose, and dose-response relationships. The principal concept of toxicology is that after an organism has been exposed to a compound, an effect is elicited in the organism (Figure 4-1). As the dose increases, the effect will become more pronounced, until the organism can no longer tolerate the compound. For example, the probable oral lethal dose for humans is shown in Table 4-1. Dose is the total amount of a substance administered to, taken, or absorbed by an organism.

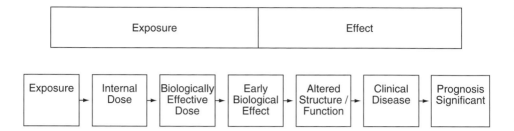

FIGURE 4-1. Exposure of organisms leads to effects.

The dose-effect relationship is an association between dose and the magnitude of a continuously graded effect, either in an individual or in a population or in experimental animals. For individuals there is a graded dose-response relationship, and for populations responses are measured in quantals (IUPAC, 1993). For example, Figure 4-2 illustrates the dose-response for individuals and Figure 4-3 illustrates the dose-response curve for a population.

A *quantal* can be defined as meaning all or none, and comes closest to a classification of whether something is safe or toxic. The quantal dose-response is used to determine the median lethal dose (LDm) and judge what percentage of the population is affected by a dose increase.

Dose-response relationships are defined using the following (IUPAC, 1993):

- Dose-related effect: Situation in which the magnitude of a biological change is related to the dose.
- Dose-response curve: Graph of the relation between dose and the proportion of individuals in a population responding with an all-or-none effect.
- Dose-response relationship: Association between dose and the incidence of a defined biological effect in an exposed population.

TABLE 4-1
Probable Oral Lethal Dose for Humans

Toxicity Rating or Class	Dose (mg/kg)	For Average Adult
1. Practically nontoxic	> 15,000	> 1 quart
2. Slightly toxic	5,000-15,000	1 pint to 1 quart
3. Moderately toxic	50-5,000	1 ounce to 1 pint
4. Very toxic	50-500	1 teaspoon to 1 ounce
5. Extremely toxic	5-50	7 drops to 1 teaspoonful
6. Supertoxic	< 5	< 7 drops

From *American Industrial Hygiene Association Journal*.

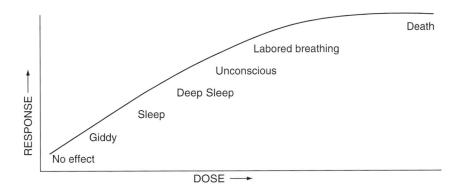

FIGURE 4-2. Dose-response for individual.

The dose-response relationship is normally represented graphically and includes a number of mathematical distributions/values to express the relationship clearly (Figure 4-4). Dosages are often described as *lethal doses* (LD), where mortality is the measured endpoint; *toxic doses*, where adverse effects are measured endpoints; and *sentinel doses*, where minimally-adverse or nonadverse effects are the measured endpoints. Sentinel effects include things such as minor irritation, headaches, and drowsiness, and serve as a warning that further exposure or greater doses may result in more serious effects. Dose-response data are used to make calculations and comparisons. For example, Figure 4-5 shows a comparison of multiple dose responses for studies X, Y, and Z. When discussing the dose concept, dose can be further refined to include (IUPAC, 1993):

- Absolute lethal dose (LD_{100}): The lowest amount of a substance that kills 100 percent of test animals under the exposure conditions.

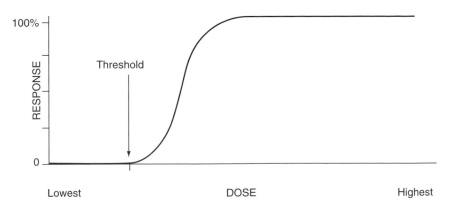

FIGURE 4-3. Dose-response curve for population.

The LD_{100} is dependent on the number of organisms used in the assessment.

- Absorbed dose (of a substance): The amount of a substance absorbed into an organism or into organs and tissues of interest.
- Absorbed dose (of radiation): Energy imparted to matter in a suitably small element of volume by ionizing radiation divided by the mass of that element of volume. The SI unit for absorbed dose is joules per kilogram (J/kg-1) and its special name is gray (Gy).
- Estimated exposure dose: Measured or calculated dose of a substance to which an organism is likely to be exposed, considering exposure by all sources and routes.
- Lethal dose: Amount of a substance or physical agent (radiation) that causes death when taken into the body by a single absorption (denoted by LD).
- Maximum tolerated dose: High dose used in chronic toxicity testing that is expected on the basis of an adequate subchronic study to produce limited toxicity when administered for the duration of the test period. It should not induce (1) overt toxicity, for example, appreciable death of cells or organ dysfunction, (2) toxic manifestations that are predicted materially to reduce the life span of the animals except as the result of neoplastic development, or (3) 10 percent or greater retardation of body weight gain as compared with control animals. In some studies, toxicity that could interfere with a carcinogenic effect is specifically excluded from consideration.
- Median effective dose (ED_{50}): Statistically derived dose of a chemical or physical agent (radiation) expected to produce a certain effect in 50 percent of test organisms in a given population or to produce a half-maximal effect in a biological system under a defined set of conditions.
- Median lethal concentration (LC_{50}): Statistically derived concentration of a substance in an environmental medium expected to kill 50 percent of organisms in a given population under a defined set of conditions.
- Median lethal dose (LD_{50}): Statistically derived dose of a chemical or physical agent (radiation) expected to kill 50 percent of organisms in a given population under a defined set of conditions.
- Minimum lethal dose (LD_{min}): Lowest amount of a substance that, when introduced into the body, may cause death to individual species of test animals under a defined set of conditions.
- Noneffective dose: Amount of a substance that has no effect on the organism. It is lower than the threshold of harmful effect and is estimated while establishing the threshold of harmful effect.

In addition to the qualitative and statistical definitions of dose, there are numerous interactions that can occur within the human body. In an effort to predict chemical interactions in populations, mathematical

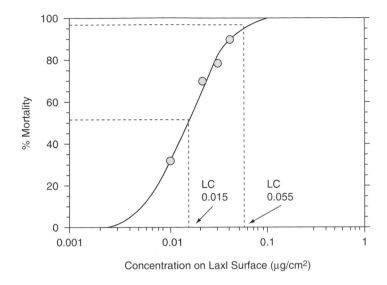

FIGURE 4-4. Comparative dose-response curves.

constructs of existing dose response data or metabolic models have been devised to estimate how a body deals with a specific substance, showing the proportion of the intake that is absorbed, the proportion that is stored and in what tissues, the rate of breakdown in the body and the subsequent fate of the metabolic products, and the rate at which it is

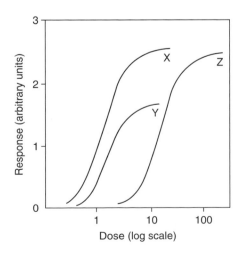

FIGURE 4-5. Comparison of multiple dose responses.

eliminated by different organs as unchanged substance or metabolites. The most frequently used models are briefly described here (IUPAC, 1993):

Multistage Model: Dose-response model for cancer death estimation of the form

$$P(d) = 1 - \exp[-(q_0 + q_1 d^1 + q_2 d^2 + \ldots\ldots + q(k)d^k)] \qquad (4\text{-}1)$$

where $P(d)$ is the probability of cancer death from a continuous dose rate, d, the q's are constants, and k is the number of dose groups (or, if less than the number of dose groups, k is the number of biological stages believed to be required in the carcinogenesis process). With the multistage model, it is assumed that cancer is initiated by cell mutations in a finite series of steps. A one-stage model is equivalent to a one-hit model.

One-Hit Model: Dose-response model of the form

$$P(d) = 1 - \exp(-bd) \qquad (4\text{-}2)$$

where $P(d)$ is the probability of cancer death from a continuous dose rate (d) and b is a constant. The one-hit model is based on the concept that a tumor can be induced after a single susceptible target or receptor has been exposed to a single effective dose unit of an agent. This model is frequently cited as the preferred regulatory construct for determining safe doses of potential carcinogens.

Weibull Model: Dose-response model of the form:

$$P(d) = 1 - \exp(-bd^m) \qquad (4\text{-}3)$$

where $P(d)$ is the probability of cancer death due to a continuous dose rate, d; b and m are constants.

Clearly, the ability to establish a relationship between dose and any effect on a biological organism plays a central role in establishing regulatory standards. Yet the complexity of dose and exposure relationships is also functionally tied to toxicity mechanisms.

Mechanism of Toxicity

Toxicity is variously defined as (1) the capacity to cause injury to a living organism by the administration of a substance, the way in which the substance is administered (inhalation, ingestion, topical application, injection) and distributed in time (single or repeated doses), the type and severity of injury, the time needed to produce the injury, and the nature of the organism(s) affected and other relevant conditions; (2) the adverse effects of a substance on a living organism defined with reference to the quantity of substance administered or absorbed, the way in which the substance is administered (inhalation, ingestion, topical application, injection) and distributed in time (single or repeated doses), the type and severity of injury, the time needed to produce the injury, the nature of the organism(s) affected, and other relevant conditions; or (3) as the measure of incompatibility of a substance with life: this quantity may be expressed as the reciprocal of the absolute value of median lethal dose ($1/LD_{50}$) or concentration ($1/LC_{50}$) (IUPAC, 1993).

The toxicity of a substance is measured using a number of variables including the nature and duration of exposure, the biological systems affected, the stage of development for the biological system, and genetic variables that make individuals more susceptible to adverse reactions from substances. The toxic effect(s) of a compound is the direct result of the interaction of the compound with the substances and not as a side effect. Side effects are defined as nondeleterious, for example, dry mouth. Toxic effects are the undesirable results of a direct effect (IUPAC, 1993). Measures of the nature and duration of exposure include acute, chronic, and subchronic. These common definitions are:

- Acute toxicity
    - Adverse effects occurring within a short time (usually up to 14 days) after administration of a single dose (or exposure to a given concentration) of a test substance or after multiple doses (exposures), usually within 24 hours.
    - Ability of a substance to cause adverse effects within a short time of dosing or exposure.
- Chronic toxicity
    - Adverse effects following chronic exposure.
    - Effects that persist over a long time whether or not they occur immediately on exposure or are delayed.
- Subchronic toxicity
    - Adverse effects resulting from repeated dosage or exposure to a substance over a short period, usually about 10 percent of the life span.
    - The capacity to produce adverse effects after subchronic exposure.

Measures of biological systems affected include systemic toxicity, local or organ toxicity, cellular toxicity, and genotoxicity (IUPAC, 1993). These measures are:

- Local toxicity: Circumscribed change occurring at the site of contact between an organism and a toxicant.
- Systemic toxicity: Consequence that is of either a generalized nature or that occurs at a site distant from the point of entry of a substance: a systemic effect requires absorption and distribution of the substance in the body.
- Genotoxicity: Ability to cause damage to genetic material. Such damage may be mutagenic and/or carcinogenic.

Measures of the stage of development affected include developmental, embryotoxicity, fetal toxicity, and cellular toxicity (IUPAC, 1993). These measures are:

- Developmental toxicity: Adverse effects on the developing organism (including structural abnormality, altered growth, or functional deficiency or death) resulting from exposure before conception (in either parent), during prenatal development, or postnatally up to the time of sexual maturation.
- Embryotoxicity includes (1) production by a substance of toxic effects in progeny in the first period of pregnancy between conception and the fetal stage and (2) any toxic effect on the conceptus as a result of prenatal exposure during the embryonic stages of development. These effects may include malformations and variations, malfunctions, altered growth, prenatal death, and altered postnatal function.
- Fetotoxicity: This is toxicity to the fetus.

Measures of the stage of genetic variables that make individuals more susceptible to adverse reactions from substances include latent effects, allergic reactions, idiosyncratic reactions, and hypersensitivity (IUPAC, 1993). These measures are:

- Delayed/latent effect: Consequence occurring after a latent period after the end of exposure to a toxic substance or other harmful environmental factor.
- Allergic reactions are symptoms or signs occurring in sensitized individuals after exposure to a previously encountered substance (allergen: an antigenic substance capable of producing immediate hypersensitivity) that would otherwise not cause such symptoms or signs in nonsensitized individuals. The most common forms of allergy are rhinitis, urticaria, asthma, and contact dermatitis.
- Antigen: Substance or a structural part of a substance that causes the immune system to produce a specific antibody or specific cells and that combines with specific binding sites (epitopes) on the antibody or cells.
- Idiosyncratic reaction: Genetically based, unusually high sensitivity of an organism to the effect of certain substances.
- Hypersensitivity: State in which an individual reacts with allergic effects after exposure to a certain substance (allergen) after having been exposed previously to the same substance.
- Cell-mediated hypersensitivity: State in which an individual reacts with allergic effects caused by the reaction of antigen-specific T-lymphocytes after exposure to a certain substance (allergen) after having been exposed previously to the same substance or chemical group.

- Cell-mediated immunity: Immune response mediated by antigen-specific T-lymphocytes.

The relative strength of a toxin is normally measured qualitatively. A relative ranking system for characterization of the acute toxicity of a chemical (later) provides a scale by which substances can be measured. *This scale assumes exposure to only one compound at a time.*

For dosing of multiple substances even more qualitative values are used. For mixtures of chemicals, the toxicity may be evaluated in one of two ways. The first is to evaluate the mixture as a whole. This method is preferable, when adequate human and animal studies have been conducted, as it may account for various effects of chemical interaction, including antagonistic, synergistic, and potentiative effects, resulting from one or more of the chemicals. These interactions are given in Table 4-2. Antagonists are chemicals that when combined diminish each other's effects (e.g., 2 + 4 = 3). Synergists are chemicals that have the same toxicity and when combined produce a greater effect than their additive effect (e.g., 1 + 1 = 20). Potentiation occurs when a chemical that does not produce a specific toxicity increases the toxicity of another chemical (e.g., 2 + 0 = 10). It is important to note that this type of information is usually developed from actual toxicity studies.

The second method uses models to evaluate the toxicities of individual chemicals in a mixture when there is not adequate information to evaluate the mixture as a whole. In these cases, indicator chemicals (i.e., usually the most toxic or highest concentration in a mixture) are used to estimate the potential risk. Having measured the gross mechanism of toxicity for a chemical, it is also important to determine the processes that may have contributed unintentionally to the mechanism of toxicity, the body's detoxification process.

Biotransformation and Detoxification

Toxicity occurs in a number of ways, most via a dangerous metabolite of a substance that has been activated by an enzyme, light, or oxygen

TABLE 4-2
Mathematical Representations of Chemical Interactions

Effect	Relative Toxicity (hypothetical)	Example
Additive	2 + 3 = 5	Organophosphate pesticides
Synergistic	2 + 3 = 20	Alcohol and acetaminophen
Potentiation	2 + 0 = 10	Alcohol and carbon tetrachloride
Antagonism	6 + 6 = 8 or 5 + (−5) = 0 or 10 + 0 = 2	Toluene and benzene or caffeine and alcohol or British Antilewiste (BAL) and mercury

reaction in a process known as *biotransformation*. Toxic reactions often depend on how metabolites are processed by an individual's body, how proteins build up and bind at effector sites in the body. Some metabolites destroy liver cells, others brain tissue, and still others operate at the DNA level. Toxic reactions can be classified as one of three (3) reactions (Williams et al., 2000):

- Pharmacologic: For example, injury to the central nervous system.
- Pathologic: For example, injury to the liver.
- Genotoxic: For example, creation of benign or malignant neoplasm or tumors.

If the concentration of toxin doesn't reach a *critical level*, the effects will usually be reversible. Pharmacologic reactions, for example, are of this type. To sustain permanent brain damage, dosages must be above a standard critical level. Pathological reactions can be repaired if discovered early enough, but most liver damage occurs over a period of a few months to a decade. Genotoxic or carcinogenic effects may take 20 to 40 years before tumors develop. Most of the time, toxic metabolites are activated by enzymatic transformation, but a few are activated by light. This means that exposure of the skin to sunlight produces a photoallergic reaction or *phototoxic reaction* within 24 hours. It is important to understand that *the target organ of toxicity is not the site where toxin accumulates*. Lead poisoning, for example, results in an accumulation of lead in bone marrow, but the toxic effect is the creation of lesions on skin and soft tissue. Carcinogenesis is even more complicated, involving the creation of promoter electrophiles, which activate or potentiate the growth of latent tumors given some biological trigger or subsequent environmental attack. Different people, of course, have *chemical allergies* (as well as food allergies), depending on the serology of their allergen-antigen history. In such people, toxic reactions take different forms. Other people have *idiosyncratic reactions*, which means they have certain unique genetic triggers. Furthermore, people exposed to multiple toxins can have *synergistic reactions;* that is, two or more toxins interact at the metabolic level to be greater or less than the effects of the individual toxins (Williams et al., 2000).

Xenobiotics (foreign chemicals) absorbed by the body must normally be transformed to enhance their excretion. Most of the transformations are performed by cellular enzymes in the so called Phase I and Phase II reactions. Phase I reactions use mono-oxygenases (e.g., cytochrome p450), reductases, and hydrolases (for esters and epoxides) to add reactive functional groups to the molecule. Phase II reactions use covalent conjugation (glucuronidation) to make the molecule hydrophilic and more excretable. The reactions are catalyzed by glycosyltransferases and sulfotransferases (for hydroxyaromatics and carboxy groups), glutathione S-transferase (for electrophilic functional groups such as halogens, nitro groups, or unsaturated/conjugated sites), acetyltransferases (for primary amines or hydrazines), and aminoacyltransferases (for forming peptides from carboxy groups using free amino acids). The metabolism of xenobiotics in this manner results in more polar compounds than the parent compound, enhancing excretion.

The information on the interaction between xenobiotics and humans is critical in the establishment of health-based standards for drinking water. The standard setting process involves the synthesis of this complex toxicologic information with the political process of regulation.

Toxicity and Defining Standards

Drinking water standards, according to USEPA, are regulations to control the level of contaminants in the nation's drinking water. The Safe Drinking Water Act of 1974 (SDWA) gives the USEPA the authority to set drinking water standards. This process involves a significant effort to determine what is the "safest" (within the limits of the current level of understanding of toxicology) and most economically feasible level possible, requiring at least a decade from the time the chemical is listed as a contaminant to the time that a standard is promulgated.

The SDWA was originally passed by Congress in 1974 to protect public health by regulating the public drinking water supply. The SDWA was amended in 1986 and 1996 and requires specific actions to protect drinking water and its sources: rivers, lakes, reservoirs, springs, and groundwater wells. The SDWA does not regulate private wells, which serve fewer than 25 individuals. SDWA applies to every public water system in the United States, estimated to be approximately 160,000 at this time.

The SDWA created a multiple barrier approach to drinking water protection, which includes assessing and protecting drinking water sources, protecting wells and collection systems, making sure water is treated by qualified operators, ensuring the integrity of distribution systems, and making information available to the public on the quality of their drinking water. In most cases, USEPA delegates responsibility for implementing drinking water standards to states and tribes.

There are two categories of drinking water standards, National Primary Drinking Water Regulations and National Secondary Drinking Water Regulations. Primary standards are legally enforceable standards that apply to public water systems. Primary standards protect drinking water quality by limiting the levels of specific contaminants that can adversely affect public health and are known or anticipated to occur in water. They take the form of maximum contaminant levels (MCLs) or treatment techniques (TTs).

Secondary standards are nonenforceable guidelines regarding contaminants that may cause cosmetic effects (such as skin or tooth discoloration) or aesthetic effects (such as taste, odor, or color) in drinking water. USEPA recommends secondary standards for water systems but does not require systems to comply. It is left to the state's discretion as to whether it wants to create enforceable standards from the Secondary Standards. The 1996 Amendments to SDWA require USEPA to (1) identify drinking water problems, (2) establish priorities, and (3) set standards. The standards are normally based on peer-reviewed science and data support an intensive technologic evaluation, which includes the occurrence of the contaminant in the environment; potential human exposure and risks of adverse health effects to the general population and sensitive subpopulations; appropriate analytical methods of detection; technical feasibility for the removal of the contaminant; and the impacts of regulation on water systems, the economy, and public health.

In the first step of this process, the USEPA compiles a list of potential contaminants of interest, in the form of the National Drinking Water Contaminant Candidate List (CCL). Identifying drinking water problems is accomplished by determining the health risks and the likelihood that a contaminant occurs in public water systems at levels of concern. It is in this first step that toxicological research plays a key role. The health risks are identified by completing a comprehensive risk assessment of potential CCL compounds, in which scientists evaluate whether fetuses, infants, children, or other groups are more vulnerable to a contaminant than the general population. In this evaluation the physicochemical characteristics, toxicokinetics/toxicodynamics and mode-of-action testing strategy, human health effects data, toxicologic effects in laboratory animal studies, pharmacokinetic models for the mode-of-action, dose-response assessments for human health, and major risk characterizations may be detailed in a comprehensive report. This information is used later to determine the health basis for any drinking water standards. Examples of the types of information gathered in these evaluations are available through the USEPA (http://www.epa.gov) or through advocacy groups such as the Natural Resource Defense Council (NRDC) (http://www.nrdc.org), which in June 2003 published a report of 19 major metropolitan water systems across the United States (NRDC, 2003). Examples of the information are provided in Appendix 4-2.

The CCL, published March 2, 1998, listed contaminants that (1) were not already regulated under SDWA, (2) may have adverse health effects, (3) were known or anticipated to occur in public water systems, and (4) may require regulations under SDWA. The first CCL included 50 chemicals or chemical groups and 10 microbial contaminants. To assist in the identification of contaminants that were known or were anticipated to occur in public water systems, USEPA established a National Drinking Water Contaminant Occurrence Database (NCOD) and an Unregulated Contaminant Monitoring Regulation. The NCOD contains data on the occurrence of both regulated and unregulated contaminants.

On April 2, 2004, USEPA listed 50 contaminants (9 microbiological and 41 chemical contaminants or contaminant groups), for a draft CCL2. With the announcement of the draft CCL2, USPEA is allowed to continue with research and data collection activities related to the list, prepare to make regulatory determinations in the 2006 time-frame using the data collected from these activities, and focus resources on completing ongoing work with the National Drinking Water Advisory Council on an expanded process for classifying drinking water contaminants in the future. The original CCL and CCL2 for chemicals are compared in Table 3-1.

The SDWA requires USEPA to review each primary standard at least once every 6 years and revise them, if appropriate. SDWA specifies that any revision must maintain or increase public health protection. To date, USEPA has conducted detailed contaminant occurrence analyses for 61 regulated contaminants, using data provided by a national cross section of 16 states. Most of the sample data were collected between 1993 and 1997.

The second step in the process is to establish priorities. Contaminants on the CCL are divided into priorities for regulation, health research, and occurrence data collection. USEPA is required by SDWA to select five or more contaminants from the regulatory priorities on the CCL and determine whether to regulate them. To support these decisions, USEPA must determine that regulating the contaminants would present a meaningful opportunity to reduce health risk.

The USEPA will also select up to 30 unregulated contaminants from the CCL for monitoring by public water systems serving at least 100,000 people. Currently, most of the unregulated contaminants with potential of occurring in drinking water are pesticides and microbes. Every 5 years, USEPA will repeat the cycle of revising the CCL, making regulatory determinations for 5 contaminants and identifying up to 30 contaminants for unregulated monitoring. In addition, every 6 years, USEPA will reevaluate existing regulations to determine whether modifications are necessary.

The third step in the process is the establishment of drinking water goals. After reviewing health effects studies, USEPA sets a Maximum Contaminant Level Goal (MCLG), the maximum level of a contaminant in drinking water at which no known or anticipated adverse effect on the health of persons would occur, and which allows an adequate margin of safety. MCLGs are nonenforceable public health goals. Since MCLGs consider only public health and not the limits of detection and treatment technology, sometimes they are set at a level that water systems cannot meet. When determining an MCLG, USEPA considers the risk to sensitive subpopulations (infants, children, the elderly, and those with compromised immune systems) of experiencing a variety of adverse health effects.

For chemicals that do not cause cancer, the MCLG is based on the reference dose (RfD). In the RfD calculations, sensitive subgroups are included, and uncertainty may span an order of magnitude.

- The RfD is multiplied by typical adult body weight (70 kg) and divided by daily water consumption (2 liters) to provide a drinking water equivalent level (DWEL).
- The DWEL is multiplied by a percentage of the total daily exposure contributed by drinking water (often 20 percent) to determine the MCLG.

For chemicals suspected or known to cause cancer in humans, the MCLG is set to zero, as carcinogens are generally considered to have no dose below which the chemical is considered safe. If a chemical is carcinogenic and a safe dose can be determined, the MCLG is set at a level above zero that is safe.

For microbial contaminants that may present a public health risk, the MCLG is set at zero because ingesting one protozoa, virus, or bacterium may cause adverse health effects. USEPA is conducting studies to determine whether there is a safe level above zero for some microbial contaminants. So far, however, this has not been established.

Once the MCLG is determined, USEPA sets an enforceable standard. In most cases, the standard is the MCL, the maximum permissible level of a contaminant in water that is delivered to any user of a public water system. The MCL is set as close to the MCLG as feasible, which the Safe Drinking Water Act defines as the level that may be achieved with the use of the best available technology, treatment techniques, and other means that USEPA finds are available (after examination for efficiency under field conditions and not solely under laboratory conditions), taking cost into consideration.

When there is no reliable method that is economically and technically feasible to measure a contaminant at particularly low concentrations, a TT is set rather than an MCL. A TT is an enforceable procedure or level of technological performance that public water systems must follow to ensure control of a contaminant. Examples of TT rules are the Surface Water Treatment Rule (disinfection and filtration) and the Lead and Copper Rule (optimized corrosion control).

After determining an MCL or TT based on affordable technology for large systems, USEPA must complete an economic analysis to determine whether the benefits of that standard justify the costs. If not, USEPA may adjust the MCL for a particular class or group of systems to a level that "maximizes health risk reduction benefits at a cost that is justified by the benefits." USEPA may not adjust the MCL if the benefits justify the costs to large systems, and small systems are unlikely to receive variances.

States are authorized to grant variances from standards for systems serving up to 3,300 people if the system cannot afford to comply with a rule (through treatment, an alternative source of water, or other restructuring) and the system installs USEPA-approved variance technology. States can grant variances to systems serving 3,301 to 10,000 people with USEPA approval. SDWA does not allow small systems to have variances for microbial contaminants.

Under certain circumstances, exemptions from standards may be granted to allow extra time to seek other compliance options or financial assistance. After the exemption period expires, the public water supply (PWS) must be in compliance. The terms of variances and exemptions must ensure no unreasonable risk to public health.

Primary standards go into effect 3 years after they are finalized. If capital improvements are required, USEPA's administrator or a state may allow this period to be extended up to 2 years. The result of this decade-long process is that from the time a chemical is recognized as a potential hazard to the time a standard is enforced, community water supplies become contaminated, and a significant portion of the population can be exposed unknowingly to a harmful chemical.

Timing Is Everything

The delay in recognizing the impact of chemicals on human health is another layer of risk that is intrinsic to a "standard-based" pollution control policy. For example, such time delays were characteristic of hexavalent chromium,[6] which was the chemical of concern in "Erin Brockovich," and trichloroethylene, which was the chemical of concern in "A Civil Action."

Chromium has been known as a hazardous chemical since the 1850s, but not until the 1930s were there widespread reports of the toxicity of chromium to both aquatic life and humans (Fisher, 1938; Dugan, 1972; Chang, 1996; Klaasen, 1996; Krebs, 1998). By the1940s, the pollution of both surface water and groundwater by chromium was a frequent occurrence.

Because of this hazard, the USPHS in 1946 set a water quality standard of 0.05 ppm for hexavalent chromium. Even though chromium was a known toxic compound, after the first drinking water standards were set in 1925, it took another 21 years to add chromium to the list. Moreover, the current standard of 0.1 ppm is for total chromium and does not even include a standard for hexavalent chromium, which is suspected to be far more toxic. Unlike the federal government, the State of California has proposed a public health goal for total chromium in drinking water at 2.5 ppb and has initiated cancer studies on hexavalent chromium. The concern for hexavalent chromium is in no small part the result of a California drinking water survey in September 2001 that revealed 375 drinking water sources contained hexavalent chromium at

---

[6]A form of chromium that is characterized by its electrical charge or valance. Hexavalent chromium has a charge of +6 and is more toxic than trivalent chromium (+3). Hexavalent chromium in the environment is almost always manufactured while trivalent chromium occurs naturally and is a common constituent in vitamins. The analysis of chromium is reported as either total chromium (i.e., hexavalent + trivalent) or as hexavalent chromium.

levels greater than 5.0 ppb (Scharfenaker, 2001). Changes of this sort at both the federal and state level typify the capricious nature of setting water quality standards and cast doubt on the scientific validity of these modifications.

Trichloroethylene (TCE) is an organic compound that contains chlorine (Browning, 1953). TCE has been used as an organic solvent (i.e., it dissolves other organic compounds like oil) since the 1920s. Because of its widespread use in industry, most aspects of TCE toxicity were established in the 1930s. Beginning in the late 1940s, TCE was identified as an environmental pollutant. By 1975, TCE was identified as a possible carcinogen. Even with this knowledge, it took until 1987, or almost 40 years, for USEPA to establish a final maximum contaminant level in water of 0.005 ppm (DeZaune, 1997).

These two chemical examples, chromium and TCE, represent two of the most significant chemical pollutants of the past 75 years. Unfortunately, we still face the similar time delays as scientific data trickle in on only a very small number of the thousands of compounds that are released daily into our environment. In almost all cases, these chemicals do not have established federal drinking water standards.

Examples that typify the ongoing delays in USEPA's recognition of hazardous chemicals in the environment center on (1) methyl tertiary-butyl ether (MTBE),[7] (2) perfluoro-octanyl sulfonate (PFOS), (3) perchlorate, and (4) dioxin. In the late 1970s, MTBE was added to gasoline as an octane-enhancing compound. Before MTBE was introduced into gasoline, the Toxic Substance Control Act Interagency Testing Committee, an independent advisory committee to the USEPA, recommended that MTBE's toxicity be evaluated. Toxicity data (genetic, reproductive, and carcinogenic) developed in the early to mid-1990s led to a 1997 recommendation that MTBE exposure be limited to 0.01 to 0.02 ppm (NSTC, 1997). Thus, it took approximately 20 years after MTBE was added to gasoline for it to be identified as a chemical hazard by the USEPA. In spite of the toxic nature of this chemical, MTBE has yet to be added to the current primary drinking water standards. This is a serious omission since a January 2000 study by the American Water Works Association (Gullick and LeChevallier, 2000) reported that, "MTBE contamination of drinking water supplies is fairly widespread and may occur in almost any area where gasoline is used." What is more disturbing, however, is that this same article reports that existing water treatment methods do not remove MTBE. Since MTBE is currently unregulated, this means that MTBE polluted drinking water can be distributed directly to consumers without so much as a warning.

Perfluoro-octanyl sulfonate (PFOS) is the primary active ingredient of 3M Corporation's Scotchgard and is also incorporated into microwave

---

[7]Today, a common groundwater pollutant reported nationwide that usually requires costly remedial actions.

popcorn bags and fast-food wrappers. As of 2002, 3M Corporation will cease production[8] of PFOS because it has been discovered in the blood of humans and animals in pristine geographic areas of the world "where no apparent sources exist." This compound is now considered, approximately 40 years after its production began, a significant chemical hazard to humans and the environment based on its known toxicity and its extremely long persistence (Renner, 2001). This chemical, like MTBE, is also unregulated.

Perchlorate has been widely used since the late 1940s in rocket fuel and lubricating oils for the tanning and finishing of leather, as a fixer for fabrics and dyes, in electroplating, aluminum refining, rubber manufacture, and paint production. Perchlorate was not a chemical of environmental concern until 1997 when a new analytical method enabled scientists to detect it in water at concentrations as low as 4 ppb.[9] Using this new method, California water quality agencies found perchlorate in 140 public water supply wells. Because there was no federal drinking water standard for perchlorate (a chemical which, at latest count, has been identified in waters in 37 states), California ruled in 1997 that drinking water should not contain more than 18 ppb of perchlorate (USEPA, 1999).[10] In January 2002, California's Department of Health Services reduced the perchlorate action level to 4 ppb. Thus, some 50 years after the introduction of perchlorate, the State of California, but not the USEPA, finally recognized that perchlorate is a hazard in drinking water. It is a hazard because perchlorate is one of the many "endocrine-disrupting" chemicals that are not currently regulated by the USEPA, even though these compounds are believed to increase the risk of testicular, prostate, and breast cancer in humans (Roefer, 2000).

Dioxin, a chlorinated organic chemical identified as 2,3,7, 8-tetrachlorodibenzo-p-dioxin, received its initial notoriety in 1972 as the result of the first true environmental disaster in U.S. history. A chemical division of Syntex Corporation in Verona, Missouri gave a waste oil containing dioxin from the production of hexachlorophene (a common disinfectant) to a Mr. Russell Bliss. Mr. Bliss in turn sprayed the waste oil in horse arenas and on dirt roads throughout southeastern Missouri to control dust. The number of horses that died following the spraying were so numerous that they had to be mounded in piles and burned at several separate arenas. Mr. Bliss also sprayed the same waste oil on all the roads in Times Beach, Missouri. Once this was discovered, the entire town had to be evacuated and was eventually purchased by the federal government. The total cost of the Missouri dioxin cleanup

---

[8]PFOS production is less than 10,000 pounds per year.
[9]In water, ppb is expressed as micrograms of a chemical per liter of water or µg/L. For example, 1 ppb of sugar dissolved in water is the equivalent to mixing a packet of sugar (approximately 3 grams) into an Olympic-size pool.
[10]No one can agree on the appropriate standard. Arizona set a provisional health-based level at 31 ppb while Texas set an interim level at 22 ppb.

exceeded $1 billion. At that time, the Centers for Disease Control proclaimed dioxin to be the most toxic compound known to humans.

Since this incident and at about 10-year intervals, reports on the toxicity of dioxin have ranged from its being either extremely toxic or not really being a problem. As of the September 2000 Draft Dioxin Reassessment report, dioxin is now listed as a "known" human carcinogen. Regardless of its level of toxicity, the fact remains that dioxin is still allowed in drinking water[11] at an "acceptable level." Once again, it has taken the USEPA nearly 30 years to decide whether a chemical is toxic to humans. Such delays remain inherent to the USEPA programs that attempt to evaluate specific chemical hazards.

Delays by the USEPA, however, should not be surprising, as the National Research Council estimates that there are approximately 72,000 organic chemicals[12] in commerce within the United States, with nearly 2,000 new chemicals being added each year (NRC, 1999a). Given this number of chemicals, it would seem impossible that the USEPA could identify every chemical that poses a hazard to the public health. In fact, it is impossible. For example, under the Toxic Substances Control Act that was established in 1979, the USEPA has conducted an assessment program to determine which "new" chemicals present an unreasonable risk to human health or the environment. Since 1979, the USEPA has reviewed only approximately 32,000 new chemical substances. With so many chemicals left to evaluate, the USEPA chose to exclude all chemicals that are produced in amounts less than 10,000 pounds per year and all polymers[13] from further consideration. "The remaining 15,000 chemical subset has been identified as being the broad focus of the USEPA's existing chemical testing and assessment program, with the primary focus being on the 3000 high-production volume chemicals that are produced/imported at levels above 1 million pounds per year (NRC, 1999a)." Thus, the USEPA has limited the number of chemicals it will consider as having an "unreasonable risk to human health and the environment" without even considering toxicity. The error of such a selection process has already been demonstrated by the threat posed by perfluoro-octanyl sulfonate (as discussed previously). Because USEPA has adopted this approach to assessing which chemicals are a threat to human health, many chemicals that are truly hazardous will not be evaluated. As a result, the USEPA will never be able to collect the necessary toxicity data to determine which chemicals need to be regulated in our drinking water resources.

Once toxicity data are collected from animal studies, the USEPA must still identify those chemicals that should be regulated in drinking

---

[11]As of March 2001, only 71 public water supplies (out of 168,690) have been tested for dioxin.

[12]Either produced or imported by facilities in greater than 10,000 lb/year quantities.

[13]These are high-molecular-weight chemicals that are assumed to typically exhibit low toxicity and water solubility.

water. As of 1998, USEPA had completed its evaluation of approximately 400 compounds for potential regulation in drinking water. These top 400 chemicals were coalesced into one master list from existing lists of chemicals that were already incorporated into various regulations. Unfortunately, the top 400 were selected independently from the top 3,000 chemicals identified by USEPA's toxicity evaluation program. In other words, the USEPA failed to consider the inclusion of any of the "top 3000" chemicals into its initial list of chemicals to be considered for potential regulation in drinking water. This omission was pointed out by the National Research Council (NRC, 1999) as a major flaw in the USEPA program to select chemicals for addition to the drinking water standards, because not all chemicals of environmental concern were included for consideration.

This is a disturbing result. From a scientific and practical point of view, the USEPA never had either the time or resources to comprehensively evaluate and regulate the entire list of 72,000 chemicals. However, to select a minimum of five chemicals in 5 years for regulation is just as absurd. It is extremely difficult to accept the premise that of all the chemicals used in this country, only five additional chemicals are a hazard to human health. Yet, the USEPA expects us to believe that by regulating an additional five chemicals our drinking water will be safe. This is particularly egregious because the methods used by the USEPA for choosing these chemicals was flawed from the very beginning.

Consequently, many hazardous chemicals that should be regulated on the basis of their potential toxicity to humans will go unregulated and when these unregulated chemicals occur in drinking water, there will be no requirement to warn the consumer.[14] Furthermore, based on the USEPA's performance to date and its pace of evaluating and regulating chemicals in drinking water, we might all be dead by the time the USEPA ponders the fate of the chemicals on the CCL.

Such delays are inherent in current federal policy. For example, the Toxic Substances Control Act requires the USEPA demonstrate that a chemical is dangerous before it can take any action against that chemical (i.e., regulate that chemical). Because of this type of policy, the presence of unregulated chemicals in our drinking water represents a potential risk to human health. A perfect example of this problem is typified by chemicals identified as "endocrine-disrupting chemicals" (Trossell, 2001).

Endocrine-Disrupting Chemicals

The endocrine system regulates bodily functions through hormones released from the brain, thyroid, ovaries, testes, and other endocrine

---

[14]Community water systems are only required to notify consumers if regulated pollutants exceed their established criteria.

glands. Many chemicals are now being identified as substances that can interfere with normal hormone function. These endocrine-disrupting chemicals can affect reproduction and can increase a person's susceptibility to cancer and other diseases, even at low ppb levels of exposure. Therefore, many endocrine-disrupting chemicals do not have definable toxicologic threshold levels below which effects do not exist, in contrast to classic toxins that are assumed to have a monotonic (linear) dose response. Atrazine, the most commonly applied herbicide in the United States and a known endocrine disruptor, is a selective preemergent and postemergent herbicide used to control broadleaf and grassy weeds. Its primary uses are the control of weeds in corn and sugarcane fields and on residential lawns in Florida and the southeastern United States (USEPA, 2003). A study on the survivorship pattern of larval amphibians exposure to low concentrations of atrazine showed that atrazine exhibits a nonmonotonic dose-response curve similar to an inverted U (Figure 4-6). Survival at lower doses was lower than exposure at high doses. This pattern is characteristic of endocrine-disrupting compounds. Atrazine exhibits endocrine-disrupting effects in mammals by inhibiting androgen receptors and by inducing aromatase, the enzyme that converts androgen to estrogen in mammals, amphibians, and potentially reptiles.

Given the hazard that endocrine-disrupting chemicals pose, the USEPA has proposed an Endocrine Disruptor Screening Program that will gather the information necessary to identify those endocrine-disrupting chemicals that should be regulated. The USEPA's report to Congress in August 2000 (USEPA, 2000c), found that approximately 87,000[15] chemicals currently in "commerce" would have to be evaluated to define their potential risks. As a first step to the USEPA process, the USEPA proposed

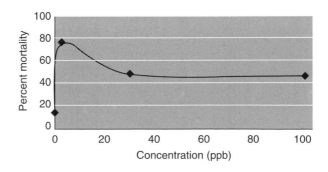

FIGURE 4-6. Nonmonotonic dose-response curve (survivorship pattern of larval amphibians) typical of endocrine disruptors.

---

[15]This number is significantly larger than the 72,000 chemicals reported by the National Research Council in 1999.

to focus on finding methods and procedures to detect and characterize the effects of pesticides, commercial chemicals, and environmental pollutants on endocrine activity. Their proposed schedule for selecting valid testing methods will take about 4 years. Obviously, actual testing using the approved methods will take much longer. Given this type of schedule, it will be decades before any specific endocrine-disrupting chemicals are identified by the USEPA as being a hazard to human health in general. Determining whether these chemicals should be allowed in drinking water will take even longer.

Thus, the vast majority of endocrine-disrupting chemicals will remain unregulated for a very long time. Although many of these compounds have yet to be identified, a general list of suspected endocrine-disrupting chemicals are presented in Appendix 4-3. Of these suspected endocrine-disrupting chemicals, 19 are regulated under the Primary Drinking Water Standards as being hazardous for other reasons (i.e., independent of any determination of their endocrine-disrupting characteristics). Because 93 chemicals are suspected of disrupting endocrine functions, another 74 chemicals are unregulated. Eight of these compounds are on USEPA's Contaminant Candidate List and may be regulated in the next 5 years. How can the remaining 74 chemicals not be a potential health risk? We may not know the answer to this question until we are well into the 21st century. Sadly, the American public is left to assume that if any of these unregulated chemicals occur in drinking water, they are not a threat to human health. This is an assumption that is hard to swallow.

In the meantime, an article by the American Water Works Association (Kolpin et al., 2002) recommends that "the utility [water provider to a community] should take measurements to establish whether compounds at issue [endocrine-disrupting chemicals] are present in the untreated water and how effective treatment is at reducing compound concentrations." Such actions alone will not solve the problem. However, if water utilities begin this process, at least the problem will become better defined.

As bad as this problem is, it is even more alarming that, for the most part, the current methods for evaluating toxicity are limited to individual chemical studies. Virtually no scientific data on the toxicity hazard posed by mixtures of both regulated and unregulated compounds currently exist. Furthermore, a whole class of chemical compounds is not even being tested by the USEPA. These compounds include the hormones, antibiotics, and synthetic organic compounds that are manufactured by the pharmaceutical industry.

Pharmaceutical Pollutants

In the early 1980s, seminars conducted by the American Association for the Advancement of Science, in cooperation with the USEPA, reported that mixtures of common pharmaceuticals (i.e., female hormones, tranquilizers, and diuretics) were widely and routinely detected in water supplies. The hazard posed by these compounds in low but chronic levels was not then known, nor is it known now (some of these compounds can also be endocrine disruptors). Additionally, the number of medications entering into and persisting in our water resources has increased. Currently, public water supplies do not monitor or report to the USEPA the occurrence of pharmaceuticals in drinking water. This raises questions as to the degree to which pharmaceuticals pollute our water resources, as well as their ultimate effect on the consumer.

Drugs are designed to have particular characteristics including lipophilicity and metabolism by-products. A large fraction of the drugs manufactured from 1992 to 1995 were lipophilic, dissolving in fat but not water (Halling-Sorensen et al., 1998). This means that chemicals that are excreted into the environment have the potential to bioconcentrate. Many drugs are also designed to be persistent, making their release into the environment extremely dangerous. The polar nature of the majority of drugs/metabolites leads to facile leaching from land disposal areas into groundwater or wet weather runoff into surface water. The remainder (largely those designed to pass the blood-brain barrier) have lipophilic character, rendering them prone to bioconcentration from consumption of water or bioaccumulation from the consumption of tissue.

When a human or animal is given a drug, the vast majority of the drug is not metabolized. Approximately 50 to 90 percent is excreted unchanged. The remaining 10 to 50 percent is excreted in the form of metabolites. Some of these metabolites may be more lipophilic and persistent than the original drug (ERF, 1998). Medications that pass through the human body and unused drugs that are dumped into sinks and toilets flow through sewers to publicly operated treatment works (POTW). Once in the POTW,[16] these compounds have been shown to survive the treatment process and ultimately end up being discharged to both surface and groundwater (Ternes, 1998). A 1998 report on drugs identified more than 20 pharmaceuticals in the effluent from POTWs (Hun, 1998). Two of these drugs were clofibrate (a cholesterol-lowering drug) and cyclophosphamide (a drug used in chemotherapy). Without monitoring for these chemicals, how do we determine their presence or absence and, if present, in what concentration? POTWs do not monitor for these compounds in their discharges.

In 1999, the United States Geological Survey (USGS) did begin a program to monitor pharmaceuticals, including prescription and nonprescription drugs, sex and steroidal hormones, and personal care products in surface and groundwater. Monitoring under the Toxic Substances

---

[16]The purpose of a POTW is to treat sewage to destroy bacterial hazards and reduce chemical pollutants before the waste is discharged back into the environment.

Hydrology Program was expanded to 95 chemicals in 2000 (Kolpin, 2002). These compounds, including some industrial, insecticides, and home care product chemicals, are listed in Appendix 4-4. The list includes 22 human and veterinary antibiotics, 13 prescription drugs, 5 nonprescription drugs, 39 industrial and household wastewater products (e.g., caffeine, flame retardants, personal care products), and 15 reproductive and steroidal hormones. The USGS study sampled 139 rivers and streams in 30 states for the list of 95 chemicals, of which only 14 have established water quality standards. The most frequently detected compounds included coprostanol (fecal steroid), cholesterol (plant and animal steroid), N-N-diethyltoluamide (insect repellant), caffeine (stimulant), triclosan (antimicrobial disinfectant), tri(2-chloroethyl)phosphate (fire retardant), and 4-nonylphenol (detergent metabolite). Concentrations of the detected compounds were generally less than 1 ppb. In half of the streams sampled, seven or more compounds were detected, and in one stream 38 chemicals were present. The researchers also stated that limited information is available with regard to the potential health effects to human and aquatic ecosystems from low-level, long-term exposure to these chemicals or combinations of these chemicals. One can only wonder as to the number of chemicals that will be monitored in the following years and what chemicals will be left out of the process. This type of research truly demonstrates that a problem exists, but offers no solution.

According to the *Journal of the American Water Works Association* (Regush, 2002), concern is growing over pharmaceuticals finding their way from medicine cabinets and sewage into the nation's water supplies, and no one is clear as to what the cumulative effect of this onslaught will be. As a result, it might be wise to avoid making facile generalizations about the safety of our drinking water. In the meantime, the ability to prevent prescription and nonprescription drugs, as well as sex and steroidal hormone pollution from POTWs, will require both better controls on discharges and additional wastewater treatment before discharge of effluents to the environment. Such controls, however, assume that the government will require the removal of these unregulated chemicals from wastewater. Since wastewater treatment costs would increase an estimated 10 percent (Kolpin, 2002), additional funds will be needed to limit this type of pollution. As a result, either the consumer will have to pay for the increased cost of treatment or the state and federal government must provide the funding. Until the public demands a change, inadequately treated wastewater will continue to be discharged into the country's water resources, polluting our drinking water with a vast array of chemicals.

These polluted waters often serve as the water resource for a downstream public water supply. Although most public water systems provide water treatment, the standard methods used do not necessarily remove all chemical pollutants from the water. When this deficiency is combined with the fact that public water systems do not even monitor for these compounds, consumers are exposed to an unknown pollution risk. This risk cycle needs to be broken.

A governmental policy that relies on water quality standards has left the public vulnerable to the hazards of both regulated and unregulated pollution. For example, a 1995 study by the United States Geological Survey, which is summarized in Exhibit 4-1, illustrates the extent of organic chemical pollution in the Mississippi River. Given the wastewater cycle of permitted sources of pollution (such as from POTWs), it is no wonder that cities such as New Orleans, located at the end of one of the longest wastewater/water cycles, the Mississippi River, have elevated cancer rates (NCI, 2000). This problem was most succinctly addressed in an article of *Journal of the American Water Works* (Maxwell, 2001): ". . . the boundaries between water and wastewater are already beginning to fade. For example, on some major rivers in the United States, water is used and reused up to 20 times as it travels to the sea so that the discharge water from one wastewater treatment plant comprising the raw water intake for a primary drinking water plant a few miles downstream." Advocates (Daughton and Ternes, 1999) that support the reuse of wastewater[17] contend that existing projects have "operated safely and reliably for nearly 40 years." This statement, however, refers only to biological pollution and not to chemical pollution, as these programs do not monitor for unregulated pollutants.

In addition to pollution arising from human use of pharmaceutical products, agricultural use of antibiotics is also a significant source of pollution. According to an August 2000 report from the Environmental Defense Fund to the USEPA, approximately 40 percent of all antibiotics in the United States are used in animal production. Furthermore, as much as 80 percent of the antibiotics administered orally pass through an animal unchanged. As a consequence, antibiotics pollute both surface and groundwater resources.

The pollution of water resources from both human and animal sources is well documented. Yet, no pharmaceuticals are currently being evaluated by the USEPA as a threat to drinking water. This condition continues to highlight the failure of federal water quality standards to effectively define chemical risks and protect human health within a reasonable time frame. In addition to the problems manifested by the USEPA efforts to define drinking water standards, the specific standards that are ultimately enacted at the state level are dependent on individual state regulatory agency decisions. Although states must, at a minimum, meet federal drinking water standards, state standards may be more strict. Thus, state regulatory decisions are also a concern, as is illustrated in Exhibit 4-2.

---

[17]Treated wastewater is usually injected into a groundwater aquifer where natural adsorption of pollutants is assumed to occur. Also it is assumed that the majority of clean water in the aquifer will dilute the pollution. These assumptions, however, are not always correct.

Exhibit 4-1. Mississippi River Report

A 1995 report by the United States Geological Survey (Circular 1133) presents data on "Organic Contamination of the Mississippi River from Municipal and Industrial Wastewater." According to this report, "The Mississippi River receives a variety of organic wastes, some of which are detrimental to human health and aquatic organisms. Urban areas, farms, factories, and individual households all contribute to contamination, by organic compounds, of the Mississippi River. This contamination is important because about 70 cities rely on the Mississippi River as a source of drinking water." This conclusion was based on a water sampling program that was conducted between Minneapolis-St. Paul, Minnesota, and New Orleans, Louisiana from 1987 to 1992.

These water analyses show that the Mississippi River is contaminated with the following groups of compounds:

- Methylene-blue substances from synthetic and natural anionic surfactants (e.g., detergents).
- Linear alkylbenzenesulfonates, a complex mixture of anionic surfactant compounds used in soap and detergent products.
- Nonionic surfactants such as nonylphenol and polyethylene glycol.
- Adsorbable halogen-containing organic compounds including solvents and pesticides.
- Polynuclear aromatic hydrocarbons from the combustion of fuels.
- Caffeine from beverages, food products, and medications.
- Ethylenediaminetetraacetic acid (EDTA), a widely used synthetic chemical for complexing metals.
- Volatile organic compounds including chlorinated solvents and aromatic hydrocarbons.
- Semivolatile organic compounds including priority pollutants such as trimethyltriazinetrione (a by-product of methylisocyanate) and trihaloalkylphosphates (a flame retardant).

This list differs considerably from the USEPA Primary Drinking Water Quality Standards presented in Table 1-9 and the chemicals listed in Exhibit 4-3. These differences further highlight the impossible task of defining which chemicals should be selected for monitoring, let alone which chemicals should be regulated.

Exhibit 4-2. *Standards in a Vacuum: Mixed Messages*

In California, which has long been in the vanguard of the environmental movement, a quiet debate is raging over what makes standards and who should be responsible for setting them. Two different agencies, the Office of Environmental Health and Hazard Assessment (OEHHA) and the Department of Health Services Division of Drinking Water and Environmental Management (DDWEM), are responsible for providing the public with maximum contaminant levels (MCLs) and public health goals (PHGs) for chemicals in drinking water. Why have multiple agencies create standards?

The DDWEM is the primary agency in the state responsible for setting and enforcing drinking water standards (MCLs). The mission statement for DDWEM states that the agency is responsible "for promoting and maintaining a physical, chemical, and biological environment which contributes positively to health, prevents illness, and assures protection of the public."

MCLs are enforceable regulatory standards under the California Safe Drinking Water Act of 1996 (Health and Safety Code Section 116365), and must be met by all public drinking water systems to which they apply. According to the California Health and Safety Code 116365(a) the Department of Health Services is to establish a contaminant's MCL "at a level as close as is technically and economically feasible to its public health goal (PHG), placing primary emphasis on the protection of public health (emphasis added)." DHS is therefore allowed to consider the technical feasibility of removing contaminants and the cost for removing contaminants (e.g., the very public debate about arsenic standards and the apparent reversal of the federal government of what is an acceptable standard). To date, the DHS has established primary MCLs for 78 chemicals and 6 radioactive contaminants.

In a twist of legislative genius, OEHHA as the agency whose mission is "to protect and enhance public health and the environment by an objective scientific evaluation of risks posed by hazardous substance," was given the added responsibility of defining what is a risk-free level for chemicals in drinking water. The same Health and Safety Code that requires DHS to establish a contaminant's MCL requires that OEHHA adopt PHGs based exclusively on public health considerations. PHGs are to be considered by DHS when establishing drinking water standards but are not enforceable standards. They are not supposed to "impose a regulatory burden on public water systems."

In a classic case of one hand not knowing what the other is doing, these two agencies have created conflicting standards. Although it may have been envisioned that DHS could reach out to the toxicologic experts at OEHHA for advice on the health implications of chemicals in the drinking water supply, the legislature has created a "Catch 22" for the agencies involved. Although DHS may create standards, it must seek OEHHA's input even though OEHHA's advice may be ignored when it is technologically infeasible (i.e., too expensive).

A major unaddressed issue regarding human health is the long-term effects of ingesting via potable waters very low, subtherapeutic doses of numerous pharmaceuticals many times a day for many decades. This concern especially relates to infants, fetuses, and people suffering from certain enzyme deficiencies (Daughton and Ternes, 1999).

Pharmaceuticals Detected in the Environment

The primary drug classes measured in the environment include some of the world's most readily used pharmaceuticals. They include (1) hormones/mimics, (2) antibiotics, (3) blood lipid regulators, (4) nonopioid analgesics/nonsteroidal anti-inflammatory drugs, (5) beta blockers/®₂-sympathomimetics, (6) antidepressants/obsessive-compulsive regulators, (7) antiepileptics, (8) antineoplastics, (9) impotence drugs, (10) tranquilizers, (11) retinoids, and (12) diagnostic contrast media (Daughton and Ternes, 1999). A brief description of these drug classes and their function is provided next.

Hormones

Hormones consist of amines, steroids, and polypeptides. Hormones are chemicals produced in one area of cells (or organ) which act at another place (cells or organs) of the body and are transported in the blood serum. The five functional hormone groups include: (1) homeostasis or maintenance of physiological conditions; (2) growth and development; (3) reproduction; (4) energy production, storage and use; and (5) behavior. Low-molecular-weight nonpeptidyl molecules can mimic hormones.

Antibiotics

Antibiotics include compounds that damage bacterial cell membranes (polymyxins), inhibit bacterial cell wall synthesis (penicillins), folic acid synthesis necessary for RNA and DNA replication (sulfonamides), DNA function (nalidixic acid), and protein synthesis (tetracyclines).

Blood Lipid Regulators (Hypolipidemic Drugs)

Hypolipidemic drugs are a class of drugs that lower the concentrations of lipoproteins, the agents that transport cholesterol and triglycerides in blood. The lipid-lowering drugs include statins, fibrates, bile-acid sequestrants, and nicotinic acid and acipimox (Xrefer, 2001). Statins (e.g., atorvastatin; cerivastatin; fluvastatin; pravastatin; simvastatin) act by inhibiting the enzyme hydroxy-3-methylglutaryl coenzyme A (HMG Co-A), which is involved in the synthesis of cholesterol (Xrefer, 2001).

Fibrates (e.g., bezafibrate; ciprofibrate; clofibrate; fenofibrate; gemfibrozil) lower plasma triglycerides and increase the breakdown of low-density lipoprotein cholesterol. Bile acid sequestrants (cholestyramine and colestipol) combine with bile acids and decrease the absorption of fats, thus increasing the amount of fat excreted in the feces and lowering plasma-cholesterol concentrations. Nicotinic acids (acipimox) are a member of the vitamin B complex that reduce concentrations of cholesterol and triglycerides in the blood and increases high-density lipoprotein cholesterol. It acts by inhibiting the breakdown of fats in fatty tissue.

Nonopioid Analgesics/Nonsteroidal Anti-Inflammatory Drugs

Analgesics are a group of drugs used for controlling pain. Nonopioid analgesics, mainly aspirin and paracetamol, provide effective relief of such pains as headache, toothache, and mild rheumatic pain (Xrefer, 2001). Nonsteroidal anti-inflammatory drugs (NSAIDs) are a large group of drugs used for pain relief, particularly in rheumatic disease. NSAIDs act by inhibiting the cyclo-oxygenase enzymes (COX-1 and COX-2) responsible for controlling the formation of prostaglandins, which are important mediators of inflammation (Xrefer, 2001). They include azapropazone, diflunisal, buprofen, ketoprofen, and naproxen. Adverse effects include gastric bleeding and ulceration.

Beta Blockers/®$_2$-Sympathomimetics

Beta blockers and ®$_2$-sympathomimetics act on the heart, reducing its force and speed of contraction, and on blood vessels, preventing vasodilation. They produce their effects by blocking the stimulation of beta-adrenergic receptors by noradrenaline in the sympathetic nervous system. Beta blockers are used to control abnormal heartbeats (arrhythmias), to treat angina, and to reduce high blood pressure (hypertension).

Antidepressants/Obsessive-Compulsive Regulators

Antidepressants act to relieve the symptoms of moderate to severe depression. Depression is thought to be related to abnormal function of the transmitters noradrenaline and/or serotonin (Xrefer, 2001). After these transmitters are released in the brain they stimulate brain cells and are taken up again by nerve endings, broken down, and hence inactivated. Some antidepressant drugs inhibit the reuptake mechanism, whereas others prevent breakdown of the transmitters; both actions increase the duration of action of these transmitters. The main classes of antidepressants are tricyclic antidepressants (TCAs), monoamine oxidase inhibitors (MAOIs), selective serotonin reuptake inhibitors (SSRIs), and lithium salts (Xrefer, 2001).

TCAs are a class of antidepressant drugs that block the reuptake of noradrenaline and serotonin into nerve endings, thus prolonging their action in the brain. TCAs can be divided roughly into two groups: those with sedative properties (amitriptyline hydrochloride, dothiepin hydrochloride, doxepin, and trimipramine), which tend to be of greater benefit to patients who are anxious and agitated, and those with only weakly sedative action (clomipramine hydrochloride, imipramine, lofepramine, nortriptyline, and protriptyline hydrochloride), which are more useful for treating lethargic and withdrawn patients (Xrefer, 2001). Related to the TCAs are the antidepressants maprotiline hydrochloride, mianserin hydrochloride, trazodone, and viloxazine hydrochloride.

MAOIs act by inhibiting monoamine oxidases, enzymes that break down noradrenaline and serotonin, and thus increase the activity of these neurotransmitters in the brain (Xrefer, 2001). MAOIs include isocarboxazid, phenelzine, tranylcypromine (which is the most hazardous), and moclobemide.

SSRIs inhibit the reuptake of serotonin (and possibly also of noradrenaline) and thus prolong its action in the brain. SSRIs include citalopram, fluoxetine, fluvoxamine maleate, paroxetine, and sertraline. Nefazodone hydrochloride and venlafaxine are related to the SSRIs (Xrefer, 2001).

Usually classified as antipsychotic drugs, lithium compounds are used primarily in the treatment of bipolar disorders, particularly the manic aspect. Lithium's manner of action in the body is unknown (Xrefer, 2001).

Antiepileptics

Antiepileptics or anticonvulsants prevent or reduce the severity and frequency of seizures in various types of epilepsy. Commonly used antiepileptic drugs include carbamazepine, lamotrigine, phenytoin, sodium valproate, topiramate, gabapentin, and vigabatrin (Xrefer, 2001).

Antineoplastic Agents

Antineoplastic agents act as nonspecific alkylating agents (i.e., specific receptors are not involved) and therefore have the potential to act as either acute or long-term stressors (mutagens/carcinogens/teratogens/ embryotoxins) in any organisms (Daughton and Ternes, 1999). Antineoplastic agents are a class of cytotoxic drugs that prevent cell replication by binding to DNA and preventing the separation of two DNA chains during cell division. The common alkylating drugs are cyclophosphamide, ifosfamide, chlorambucil, melphalan, busulfan, lomustine, carmustine, mustine, estramustine, treosulfan, and thiotepa (Xrefer, 2001).

Impotence Drugs

Impotence drugs enhance the erectile response to sexual stimulation and are used for the treatment of men who have difficulty in obtaining or maintaining an erection. During sexual stimulation, they act as a selective enzyme inhibitor, causing relaxation of smooth muscle and increasing blood flow to the erectile tissue of the penis (Xrefer, 2001). Sildenafil (Viagra) and alprostadil, prostaglandins that produces dilation of blood vessels, are the most commonly prescribed impotence drugs.

Tranquilizers

Tranquilizers are a generic label for any of several classes of drugs, all of which have one or more of the following properties: antianxiety, sedative, muscle relaxant, anticonvulsant, antiagitation (Xrefer, 2001). Antianxiety drugs or minor tranquilizers include the benzodiazepines such as chlordiazepoxide and diazepam; the muscle relaxant derivative meprobamate; sedatives such as the barbiturates; and buspirone, which was first introduced as an antipsychotic drug. Their primary action is to produce muscle relaxation and sedation through central nervous system action.

Antipsychotics or major tranquilizers are a label for several categories of drugs that are prescribed in cases of psychotic disorders. Included are the phenothiazines and related thioxanthines, the butyrophenones, the indolones, and the generally less effective rauwolfia alkaloids. Drugs in this group may have important effects on the autonomic nervous system, including epinephrine and norepinephrine blocking at sympathetic receptors and acetylcholine blocking at postganglionic parasympathetic receptors. They depress sensory input to the reticular formation and raise the general threshold for such stimuli in the brainstem.

Retinoids

Retinoids are low-molecular-weight lipophilic derivatives of vitamin A that can have profound effects on the development of various embryonic systems (Maden, 1996), especially amphibians in which retinoic acid receptors have been hypothesized to play a role in frog deformities. Retinoids have been used for a wide array of medical conditions including skin disorders (e.g., Accutane [isotretinoin] for acne), antiaging treatments (e.g., Retin-A [tretinoin] for skin wrinkles), and cancer (e.g., Vesanoid [tretinoin] for leukemia) (Daughton and Ternes, 1999).

Diagnostic Contrast Media

X-ray images of soft tissues are routinely captured by the use of contrast media. They include highly substituted and sterically hindered amidated iodinated aromatics such as diatrizoate and iopromide (Kalsch, 1999); which are used worldwide at annual rates exceeding 3,000 tons. Kalsch (1999) found these compounds to be quite resistant to transformation in sewage treatment works and river waters, although they have no bioaccumulation potential and low toxicity (Steger -Hartmann et al., 1998).

Personal Care Products

Personal care products are directed at altering odor, appearance, touch, or taste while not displaying significant biochemical activity. Most are used as active ingredients or preservatives in cosmetics, toiletries, or fragrances. They include fragrances, preservatives, disinfectants/antiseptics, sunscreen agents, and nutraceuticals/herbal remedies (Daughton and Ternes, 1999).

Given the number of pollutants detected in drinking water resources, including pharmaceutical compounds, that are known to have measurable impact when ingested, it is time to reevaluate the process by which standards are established.

Living with Risk

Because we accept federal standards, consumers of either public water supplies or bottled water are forced to live with the following unknown risks:

- What are the real health risks associated with consuming regulated chemical compounds at or below their established standards?
- What unregulated pollutants occur in drinking water and what are their concentrations and health risks?
- What are the actual concentrations of regulated compounds in drinking water?[18]

---

[18]Drinking water may be analyzed for regulated compounds but may not be analyzed at levels below concentrations that are deemed a safe and, thus, their concentrations are not reported.

- What effect do low levels of a multitude of pollutants have on the human body?

Given the current history of water pollution control in the United States, we will continue to suffer "legalized chemical pollution" until science demonstrates the actual nature of the harm or until there is a change in policy. Until that time, should we be expected to accept this risk[19] based on statistical assurances of safety? A blind acceptance of this risk is difficult to swallow because chemical species that occur in drinking water are known to be toxic at elevated concentrations, while their true hazard in lower concentrations remains unknown. In addition, few studies have ever been conducted to determine the actual concentration of these chemical pollutants in human blood and urine.

The Centers for Disease Control (CDC) reported the results of a 1999 study (CDC, 2001) that tested for the presence of 27 chemicals in the blood and urine of 5,000 individuals throughout the United States (Exhibit 4-3). Future CDC studies will expand this list of chemicals to 100.[20] The results of the initial study showed that elevated concentrations of mercury and phthalates (from plastics) were found in the human body and especially in women and children. In addition, a Public Broadcasting System report on chemical industry trade secrets by Bill Moyers in March 2001 described the presence of industrial chemicals in human blood. A blood sample taken from Mr. Moyers was analyzed by the Mt. Sinai School of Medicine for 150 industrial chemicals, and 84 chemicals were found to be present.

These reports clearly indicate that industrial chemicals permeate our bodies. The potential risk to human health is obvious. The only questions that remain are how many manufactured chemicals are actually in the human body, what is the range of concentrations for each chemical, and what are the health impacts? Such information will take a long time to accumulate. This information is critical to the use and application of standard-based pollution control. Yet given the fact that toxicity studies can take many years to complete for each chemical, analytical methods need to be validated and calibrated for each chemical, and recognizing that there are thousands of chemicals in the environment, this is a seemingly impossible task. Without this information, the public will always have to live with some unspecified level of risk.

---

[19]We accept other risks in today's environment. For example, living in geographic regions that have floods, earthquakes, tornadoes, and lightning; or participating in activities such as driving a car or flying in an airplane. These risks, however, have been well defined. It is much more practical to dislocate from an area with a known potential for natural disasters or not fly than to cease drinking water or breathing air.

[20]This number of compounds is insignificant to the number of chemicals manufactured and used in the United States.

Exhibit 4-3. The CDC Report

The CDC's 2001 National Report on Human Exposure to Environmental Chemicals found that virtually all humans have some "background" level of industrial chemicals in their bodies. The report states that ". . . the measurement of an environmental chemical(s) in a person's blood or urine does not by itself mean that the chemical causes disease. . . . For most of the other environmental chemicals [i.e., chemicals other than lead], we need more research to determine whether levels measured in the Report are of health concern."

The CDC knows that industrial chemicals can be found in human blood and urine, but does not know their impact on human health. To determine this impact will take decades of research. The research proposed by the CDC would appear to be an impossible task given the number of industrial chemicals used in the United States. As it is, the CDC report addressed only the following chemicals:

- Metals: lead, mercury, cadmium, cobalt, antimony, barium, beryllium, cesium, molybdenum, platinum, thallium, tungsten, and uranium.
- Organophosphate pesticides (by measuring six common metabolites that are representative of the following pesticides): chlorpyrifos, diazinon, fenthion, malathion, parathion, disulfoton, phosmet, phorate, temephos, and methyl parathion.
- Phthalate metabolites (phthalates are used in soap, shampoo, hair spray, nail polish, and plastics): mono-ethyl phthalate, mono-butyl phthalate, mone-2-ethylhexyl phthalate, mono-cyclohexyl phthalate, mono-n-octyl phthalate, mono-isononyl phthalate, and mono-benzyl phthalate.

Once again, this list of chemicals differs from those found in the USEPA Primary Drinking Water Quality Standards shown in Table 1-9, as well as the chemicals listed in Exhibits 4-1. Clearly, the current number of pollutants present in water exceeds the government's ability to even define the problem, no less determine the true health risk to humans.

Population, Pollution, Risk, and Precaution

The fact that the environment is damaged by human activities is a natural consequence of our existence. This knowledge is as fundamental as knowing the properties of water. Most individuals intuitively understand that human use and abuse of the earth's resources results in various impacts (i.e., dirty water, airborne odors, and damaged landscapes). Most individuals also understand that as the human population increases, the damage to the environment increases.

Paul Ehrlich's prophecy of environmental doom from increasing population as presented in his 1971 book, *Population Bomb*, is even more relevant today. Population in the United States continues to increase as the result of both natural birth/death rates and unchecked immigration (USBC, 2000). Clearly, as population and the gross domestic product of the United States increase, pollution increases. On the other hand, human beings' expanding use of natural resources coupled with technological advancements should not be viewed solely as creating harmful impacts. In contrast to the destructive aspects of expansion, there has been an ever-increasing production of food, as well as significant advancement in medical science. As a result, humans on average (at least in the developed nations) are living longer and better. The dichotomy presented by these facts will suggest to some that the continuing and expanding modification of the environment has little or no apparent affect on human health. Such a simplistic conclusion, however, does not take into account the diversity within the human species and habitat.

Because humans are living longer, we are also exposed to both acute[21] and chronic[22] levels of pollutants for longer periods. The ultimate effect of this exposure is not scientifically known but must be a concern to a society that has come to anticipate a prolonged life expectancy. This concern is more than justified because medical science has clearly determined that young individuals with varying degrees of sensitivity to illness and older individuals with their own degrees of sensitivity are generally more susceptible to health problems. In other words, humans on average may be living longer, but the impact of chemical pollution on those who live on or near the edge of the bell-shaped curve is unknown.

The ultimate effect on humans exposed to a specific chemical pollutant or combination of pollutants is not scientifically known. This lack of knowledge is addressed by scientists as a "calculated risk." When calculating a risk, one must rely on some degree of valid information to serve as a basis for the calculation and an estimate of the confidence in

---

[21]This is usually a one-time exposure to elevated levels of a pollutant. For example, an individual walking down a residential street inhales a high concentration of a pesticide that was just sprayed by a city or county agency to control mosquitoes.

[22]This is generally a fairly continuous exposure to low levels of a pollutant, for example, the ingestion of low concentrations of chemicals that are not removed from drinking water.

the resulting prediction. When considering the extremely limited amount of data on the exposure of humans to a specific toxic chemical or mixture of chemicals, there is currently no basis to assume that risk calculations can provide safeguards to the community. For example, a study by the United States Geological Survey (Gilliom et al., 1999) on pollution of the nation's water resources reports that, "the pervasive uncertainty of extrapolating results from laboratory animals to humans, estimating the risk associated with long-term consumption of drinking water that contains pesticides, even at levels below current regulatory standards, is speculative." In other words, the major problem with attempting to calculate the risk to humans is the lack of any realistic scientific database underlying the validity of the exercise. Regardless of these issues, the USEPA still relies on the risk assessment process to determine health risks to humans from specific chemicals in drinking water.

A seminal event in public health and water quality occurred in 1849 when John Snow, a founding member of the London Epidemiological Society, set out to discover the source of a large water-borne outbreak of cholera that had occurred in London. By studying the distribution of cases, he was able to determine that the cholera was most likely coming from a public well that was being polluted with human and animal waste. To control the epidemic, he simply took the handle off the pump. Thirty-four years later, *Vibrio cholerae*, the source of the epidemic, was identified. Snow was able to make the association between disease and the sources—and end the epidemic—without knowing all of the details.

Today with multiple sources of pollution, a massive number of manufactured chemicals in the environment, and the development of modern drinking water and wastewater treatment systems, the clear cause-and-effect relationship between pollutants, water quality, and public health has been obscured. Thus, the ability to implement public health policy, and in particular water quality policy, to protect the public health from exposure to toxic compounds is confounded by a fundamental lack of human exposure data. No simple method exists for predicting the true impact of consuming low levels of manufactured chemicals in drinking water. As a result, this problem cannot be solved by simply removing the "pump handle." Unlike John Snow, we now must spend millions of dollars and years of research deciding on the appropriate mode of action for ensuring sufficient water quality to protect the public health.

Because no real chemical exposure database exists for humans, environmental policy makers currently rely on an iterative process of risk management, risk communication, and risk assessments as a means of addressing the unknown consequences of consuming chemically polluted drinking water. To understand the purpose and limitations of this iterative process, it is important to understand the basic components of risk management, risk communication, and risk assessments.

Risk management is the process of weighing policy alternatives in light of the results of a risk assessment and implementing appropriate control options. The stated goal of risk management is to protect the public health by controlling risks as effectively as possible through the selection and implementation of appropriate measures. Risk communication is the exchange of information and opinion on risk among risk assessors, risk managers, and other interested parties, including the general public. Risk assessment is a structured process for determining the potential risks associated with any type of hazard—biological, chemical, or physical.

The Risk Assessment Process

Risk assessment has been defined (Asante-Duah, 1993) as "a systematic process for making estimates of all the significant risk factors that prevail over an entire range of failure modes and/or exposure scenarios due to the presence of some type of hazard. It is a qualitative or quantitative evaluation of consequences arising from some initiating hazard(s) that could lead to specific forms of system response(s), outcome(s), exposure(s), and consequence(s)." The risk assessment process should be the mechanism by which the best available scientific knowledge is used to establish case-specific responses that will ensure defensible decisions for managing hazardous situations in a cost-effective manner. However, this approach begs the question of whether this is the best mechanism.

Risk assessments guidelines were originally published in the National Academy of Sciences *Redbook*, which suggested that risk assessments should contain some or all of the following four steps (NRC, 1983):

- Hazard Identification
- Toxicity Assessment
- Exposure Assessment
- Risk Characterization

Hazard Identification

This process concerns the discussion of the toxicologic properties of a particular chemical. It is a qualitative evaluation that examines the applicable biological and chemical information to determine whether exposure to a chemical may pose a hazard or increase the incidence of a health condition or effect (e.g., cancer, birth defects) (USEPA, 1989). Human health effect studies are preferred over animal studies because of interspecies variation in response to chemical exposure; however, adequate human studies are generally not available for each chemical of concern. When this occurs, the results of animal studies are used to estimate the potential for human health effects from exposure to the same chemical.

For mixtures of chemicals, the toxicity may be evaluated in one of two ways. The first is to evaluate the mixture as a whole. This method is preferable, when adequate human and animal studies have been conducted, as it may account for various effects of chemical interaction, including antagonistic, synergistic, and potentiating effects, resulting from one or more of the chemicals (see the discussions earlier on the definitions of these terms). It is important to note that this type of information is usually developed from actual toxicity studies. The second method uses models to evaluate the toxicities of individual chemicals in a mixture when there is not adequate information to evaluate the mixture as a whole. In these cases, indicator chemicals (i.e., usually the most toxic or highest concentration in a mixture) are used to estimate the potential risk.

Toxicity Assessment

A toxicity assessment is the process of evaluating whether the possibility exists for an increase in the incidence of an adverse health effect (e.g., cancer, birth defect) resulting from human exposure to a substance. The assessment identifies the relationship between the dose of a substance and the likelihood of an adverse effect in the exposed population (Preuss and Ehrlich, 1987). Two methods of dose-response analysis are widely used to estimate effects in humans to low-exposure levels of chemicals.

The first is mathematical modeling of the dose-response relationship. The approach is often used to characterize the relationship between the dose of a defined carcinogenic chemical and incidence of cancer. Model outputs yield a cancer slope factor that represents an estimate of the ". . . largest possible linear slope (within the 95 percent confidence limit) at low extrapolated doses that is consistent with the data." (USEPA, 2005). A mathematical model, such as the nonphysiologically based Linearized Multi-Stage model, is used in conjunction with experimental data (when available) for this purpose. This model assumes that there is some risk associated with any dose of the chemical, even one molecule (i.e., the One-Hit Theory).

Chemicals that exhibit carcinogenicity are generally considered to have no exposure threshold (i.e., exposure to any amount of the chemical would result in some risk of cancer). Most modeling for quantitatively estimating the carcinogenic nature of chemicals at low doses to which people are exposed under environmental conditions is based on studies regarding human exposure to radiation (Paustenbach, 1989). While this assumption may be appropriate for radiation, many members of the scientific community believe that this model may not be suitable for all chemical carcinogens. Radiation is known to be genotoxic (i.e., it reacts directly with DNA) and an initiator of cancer. As a result, the dose is linearly related to the amount of radiation received at the target organ. On closer review, this approach appears somewhat ludicrous. Since there is no "threshold," one has to question the validity of a model that attempts to propose a method by which one can establish a "safe threshold level."

Nevertheless, this is the approach the USEPA uses to quantify risk which, by definition, is unquantifiable.

Chemical carcinogens fall into at least three major categories (Anderson, 1988): cytotoxicants (i.e., chemicals toxic to cells), initiators, and promoters (i.e., chemicals that promote the growth of cancer cells). The USEPA uses the Linearized Multi-Stage low-dose extrapolation model as the basis for estimating chemical-specific cancer risk at low doses. This model is recognized as a conservative approach to ensure the potential risk is not underestimated. Cancer slope factors derived from the Linearized Multi-Stage model are indices of carcinogenicity and are used in performing quantitative calculations to estimate carcinogenic risk.

Although there are some data on human exposures, most available information about dose-response relationships are based on data collected from animal studies and theoretical perceptions about what might occur in humans. The nature and strength of the evidence of the causation of cancer is an important aspect of the evaluation (Preuss and Ehrlich, 1987).

Carcinogenic classifications developed by the USEPA's Cancer Assessment Group classify candidate chemicals into one of the following groups, according to the weight of evidence for and against carcinogenicity from animal and epidemiological studies (USEPA, 2005):

- Group A - Human carcinogen (sufficient evidence of carcinogenicity in humans)
- Group B - Probable human carcinogen
- Group B1 - Limited evidence of carcinogenicity in humans
- Group B2 - Sufficient evidence of carcinogenicity in animals with inadequate evidence in humans
- Group C - Possible human carcinogen (limited evidence of carcinogenicity in animals; absence of human data)
- Group D - Not classifiable as to human carcinogenicity
- Group E - Evidence of noncarcinogenicity for humans (no evidence of carcinogenicity in adequate studies)

The second method of assessing the dose-response relationship is through the safety factor approach. This method is used to describe the relationship between the dose and the effects of defined noncarcinogenic chemical. The induced effect is assumed to have a threshold below which adverse health effects would not be seen. Exposure levels that do not result in adverse health effects in animals are extrapolated to human exposures using safety factors. According to USEPA (USEPA, 1992), a reference dose (RfD) or reference concentration (RfC) is a provisional estimate (with uncertainty spanning perhaps an order of magnitude) of the daily exposure to the human population (including sensitive subgroups) that is likely to be have no appreciable risk or deleterious effects during a portion of the lifetime, in the case of subchronic RfC or RfD, or during a lifetime, in the case of a chronic RfC or RfD.

Chemicals that exhibit adverse effects other than cancer or mutation-based developmental effects are believed to have a threshold (i.e., a dose below which no adverse health effect is expected occur). When extrapolating animal data to identify safe levels of human exposure, most researchers have focused on the use of a safety factor or uncertainty factor. The magnitude of the safety factor is, in turn, dependent on a number of quantitative and qualitative determinations based on the type, duration, and results of the animal research study. The concept of a threshold event is based on the assumption that there is a dose below which there is no effect. Health criteria levels are usually estimated from the no-observed adverse effect level (NOAEL) or the lowest observed adverse effect level (LOAEL) determined in chronic animal studies. The NOAEL is defined as the highest dose at which no adverse effects occur. The LOAEL is the lowest dose at which adverse effects begin to appear. NOEAL and LOAEL derived from animal studies are used by USEPA to establish oral and inhalation reference doses (RfDs) and reference concentrations (RfCs) for human exposures. The USEPA has used these approaches in establishing exposure route-specific reference doses (RfDs) for noncarcinogenic chemicals. An RfD is a daily dose level to which humans may be exposed throughout their lifetimes with no adverse health effect expected. An RfC is the concentration of a chemical in air to which humans may be exposed throughout their lifetimes with no adverse health effect expected. The highest degree of uncertainty identified with most risk assessments is normally associated with the extrapolation of results obtained from animals tested at high doses to those results that would be anticipated at low doses, which humans are more likely to encounter in the environment.

Exposure Assessment

An exposure assessment, as defined by the National Academy of Science (Preuss and Ehrlich, 1987), is the process of measuring or estimating the intensity, frequency, and duration of human exposure to an agent in the environment. The quantitative assessment of exposure, based on the chemical concentrations and the degree of absorption of each chemical, provides the basis for estimating chemical uptake (dose) and associated health risks (USEPA, 2005; CalEPA, 1994).

Potential exposure to chemicals in the environment is directly proportional to concentrations of the chemicals in environmental media (e.g., water, air) and the characteristics of exposure (e.g., frequency and duration). The characteristics of exposure are estimated using various exposure parameters. The concentrations of chemicals at specific exposure points will vary over space and time; however, a single estimate of an exposure point concentration may be used for risk assessment. This single value must be representative of the likely or average concentration to which a person would be exposed over the duration of the exposure. However, to be conservative and health-protective, point estimates based on concentrations greater than the average or most likely estimate of exposure are often used to avoid underestimating potential exposure.

Risk Characterization

Risk characterization is the description of the nature and magnitude of potential health risk, including attendant uncertainty. Risk characterization integrates the results of the exposure assessment and the toxicity assessment to estimate potential carcinogenic risks and noncarcinogenic health effects associated with exposure to chemicals. This integration provides quantitative estimates of either cancer risk or noncancer hazard indices that are compared to standards of acceptable risk.

With this transparent process, the quantitative determination of potential risks to a community is readily understood, improving decision-making processes for all parties involved. The importance of risk assessments in the decision-making process needs to be stated, especially when attempting to set standards for at-risk populations.

Importance of the Risk Assessment Process

The overall purpose of a risk assessment is to provide, as far as is feasible, sufficient information to risk managers to allow the best decision to be made concerning a potentially hazardous problem (Asante-Duah, 1993). Whyte and Burton (1980) defined the major objective of risk assessment as the development of risk management decisions that are more systematic, more comprehensive, more accountable, and more self-aware of what is involved than has often been the case. Tasks performed during a risk assessment are intended to help answer the questions "How safe is enough?" or "How clean is clean enough?" The risk assessment process is supposed to be a transparent method for characterizing the nature and likelihood of potential harm to the public. It is also supposed to help define the uncertainties and provide some level of comfort with the inferences that are made. And, finally, it is supposed to point out data gaps that can help prioritize research needs.

Most major federal environmental and food safety laws require the use of risk analysis to determine what constitutes a safe level of chemical exposure. Those laws include the Clean Air Act; the Federal Water Pollution Control Act of 1972; Comprehensive Environmental Response, Compensation, and Liability Act of 1980; Federal Insecticide Fungicide, and Rodenticide Act; Hazardous Materials Transportation Act; Occupational Safety and Health Act; Resources Conservation and Recovery Act; Superfund Amendments and Reauthorization Act of 1986; the Safe Drinking Water Act; Toxic Substances Control Act; and the Food, Drug, and Cosmetic Act.

In addition to their widespread use in the promulgation of drinking water quality standards, risk assessments are used to set standards for air quality and soil cleanup goals, and to protect food quality (Woteki, 1998).

When the Food and Drug Administration set out to revise the Recommended Dietary Allowances several years ago, it was concerned about how to define a tolerable upper level for the intake of nutrients. This question was of great importance to nutritionists at the Food and Drug Administration, because the use of dietary supplements, in conjunction with the practice of food fortification, may provide high levels of nutrients that might put consumers at risk. So whereas regulations for water are focused on "how little is safe enough," the standards regarding nutrients in foods often focus on "how much is safe enough." Another example of the value of risk assessment in food safety is the recently completed risk assessment by the U.S. Department of Agriculture, Food and Drug Administration, and the CDC scientists on *Salmonella enteritidis* (SE) in eggs and egg products. This was the first quantitative farm-to-table microbial risk assessment, and it is expected to serve as a prototype for future risk assessments. According to the Food and Drug Administration, data from the CDC indicated that SE is one of the most commonly reported causes of bacterial food-borne illness in the United States and has been increasing since 1976. Data from the risk assessment indicated that consumption of contaminated eggs results in an average of 661,633 human illnesses per year.

During the SE risk assessment, the Food and Drug Administration created a farm-to-table model that it can use to determine the effects of specific interventions on the incidence of illness. As part of this risk assessment, the team evaluated a number of possible interventions on the expected number of human illnesses, including shell egg cooling, diverting eggs from flocks with a high prevalence of SE-positive hens to breaker plants for pasteurization, and reducing the prevalence of SE-positive flocks. The outcome in the process was a decrease in the number of SE outbreaks. For other risk assessments, the determination may result in an acceptable standard of exposure for a community.

Risk Assessment Determinations

Risk assessments determine the probability that an adverse health effect may occur in a population exposed to a toxic agent. Risk assessments cannot determine whether any one individual will become ill after an exposure to an agent. Therefore, a risk assessment is a basic predictive device for risk managers to estimate potential accidental exposures, establish food tolerances, set environmental pollution and cleanup levels, define allowable workplace exposures, and evaluate the risk of chemical pollution from uncontrolled waste sites. For risk assessments to be used for these purposes, target risk levels need to be defined.

Target risk levels represent the tolerable limits to danger that society is prepared to accept as a consequence of potential benefits that could accrue. These levels may be represented by the *de minimis* or "acceptable" risk levels (Asante-Duah, 1993). Risk is *de minimis* if the incremental risk produced by the activity is sufficiently small so that there is no incentive to modify the activity (Whipple, 1987). Various demarcations of acceptable risk have been established by regulatory agencies. Cancer risks in excess of 1×10^{-5} (1 in 100,000) per chemical have been deemed unacceptable pursuant to the California Safe Drinking Water and Toxic Enforcement Act of 1986, otherwise known as Proposition 65 (California Health and Safety Code Sections 25249, 15 et seq.; 22 California Code of Regulations Section 12703(b)). The USEPA generally deems health risks to be significant if cancer risk exceeds the USEPA acceptable risk range of 1×10^{-6} to 1×10^{-4} (1 in 1,000,000 to 1 in 10,000) and/or the hazard index is greater than 1 (40 Code of Federal Regulations part 300.430(e)(2)(I)(A)(2)).

Because the USEPA and state environmental agencies consider a risk assessment as an important predictive tool, drinking water standards are strongly tied to the risk assessment process. This does not mean, however, that the process is necessarily protective of the public health. This failing is fundamentally associated with a sequential process that magnifies any error or uncertainties in the basic assumptions of the predictive models. Although the government attempts to "build in a safety factor" in setting standards, the truth remains an open issue: we really do not know the effect of a chemical, even at the lowest levels.

Validity of the Risk Assessment Process

As with any assessment process, the quality and quantity of the data collected for the assessment will affect the validity of the assessment. The validity of any risk assessment depends on the uncertainties inherent with each step in the risk assessment process. For example, the uncertainties associated with the area of hazard identification include (Whyte and Burton, 1980):

- The relative weights given to studies with differing results. For example, should positive results outweigh negative results if the studies that yield them are comparable?
- The relative weights given to results of different types of epidemiologic studies. Are the results of a prospective study more valid than those of a case-control study?
- Should experimental-animal data be used when the exposure routes in experimental animals and humans are different?
- How much weight should be placed on the results of various short-term tests?

The uncertainties associated with the area of toxicity assessment include these questions (Whyte and Burton, 1980):

- What dose-response models should be used to extrapolate from observed doses to relevant doses?

- How should different temporal exposure patterns in the study population and in the population for which risk estimates are required be accounted for?
- How should physiologic characteristics be factored into the dose-response relation?
- Should dose-response relations be extrapolated according to best estimates or according to upper confidence limits?
- What factors should be used for interspecies conversion of dose from animals to humans?

The uncertainties associated with the area of exposure assessment include these questions (Whyte and Burton, 1980):

- Should one extrapolate exposure measurements from a small segment of a population to the entire population?
- Should dietary habits and other variations in lifestyle, hobbies, and other human activity patterns be taken into account?
- Should point estimates or a distribution be used?
- How should exposures of special risk groups, such as pregnant women and young children, be estimated?

Uncertainties associated with the area of risk characterization include these questions (Whyte and Burton, 1980):

- Statistical uncertainties in estimating the extent of health effects. How are these uncertainties to be computed and presented?
- Which population groups should be the primary targets for protection and which provide the most meaningful expression of the health risk?

Given this extensive list of uncertainties that are associated with the risk assessment process, it is no wonder that the validity of the process is suspect. It may well be, however, that this is the best approach that science can furnish. But one must question whether mice and rats are really people.

Frustration over not having this type of chemical-specific data for human exposure prompted researchers at the Loma Linda Medical Center in California to begin a study designed to determine whether perchlorate in drinking water interferes with thyroid glands (Rogers, 2000). In this study, volunteers were paid $1,000 each to take pills containing perchlorate. Because this study is unique and privately funded, it is highly unlikely that the federal government will sponsor similar studies for other chemicals. This reality virtually ensures that valid human exposure/response data will not exist for a substantial number of chemicals that currently pollute our drinking water resources. Without this type of information, drinking water standards cannot be shown to protect human health.

This conclusion is supported by an article in the *Journal of the American Water Works Association* (Hoffbuhr, 2002), which quotes William Ruckelshaus, a former two-time administrator of the USEPA: "USEPA's laws often assume, indeed demand, a certainty of protection greater than science can provide given the current state of knowledge. The public thinks we know all the bad pollutants, precisely what adverse health or environmental effects they cause, how to measure them exactly, and control them absolutely. Of course, the public, and sometimes the laws are wrong." The author concludes that "He hit the nail right on the head. We expect too much of science and government when it comes to defining what is safe when dealing with trace amounts of chemicals."

This uncertainty, however, can also be used to argue the opposite opinion. For example, Bjørn Lomborg contends in *The Skeptical Environmentalist* (Lomborg, 2002) that although "we possess only extremely limited knowledge from studies involving humans, and by far the majority of our evaluations of carcinogenic pesticides are based on laboratory experiments on animals," these studies show that the risk is negligible. Lomborg estimates that the annual cancer mortality resulting from pesticide use in the United States is probably close to 20 in 560,000 and that this is an acceptable risk given the benefits of pesticides to agricultural production and the control of disease. His conclusions would be more valid if the toxicity studies were based on human data. Again, are mice and rats people? Considering that researchers lack both human toxicity data and knowledge of the effects of consuming a low-level mixture of chemical pollutants in drinking water, as well as the additive chemical exposures in food and air, it would seem prudent to limit, to the maximum possible, our exposure to chemicals in drinking water.

Considering that all of the issues associated with estimating risk and the policy conclusions are based on such risk predictions, what level of trust should the public have in the risk assessment process? Unfortunately, the public's trust has been damaged in the past by the miscommunication of risk information and such communication is as important as the actual quantification of the risk itself. Two examples of the miscommunication are the "confusion" associated with alar and dioxin.

In the case of alar, the Natural Resources Defense Council mounted a major public relations campaign in 1989 designed to force USEPA to speed up pesticide regulation (Friedman, 1996). National coverage of the alar controversy included risk estimates for which little or no explanation was given. Differences of opinion on the cancer potential of alar, how many apples were treated with alar, and the number of apples eaten by children resulted in significantly different risk estimates from the USEPA and the Natural Resources Defense Council. The confusion left the public panicked and resulted in ten cities including New York, Atlanta, Chicago, and Los Angeles, banning apples and apple products in school lunches. The apple industry estimated that it lost more than $100 million in apple sales. And most important, Americans' faith in the safety of the nation's food supply was shaken. Today, most scientists agree that the "concern" over alar was unwarranted.

In the second example, a series of articles in 1993 in the *New York Times*, which questioned the many standards used by USEPA for regulating toxic chemicals, highlighted the dioxin controversy. The environmental reporter for the *Times*, Keith Schneider (1991, 1993), wrote that "new research indicates that dioxin may not be so dangerous after all." He noted that many scientists and public health specialists said that "billions of dollars are wasted each year in battling problems that are no longer considered especially dangerous, leaving little money for others that cause far more harm." Here again miscommunication has resulted in misinformation. Inadequate or misleading risk communication by the media or scientific community contributes to the public's fear of environmental risks, such as alar and dioxin, or more recently arsenic and perchlorate. Although the media play an important role in influencing policy decisions and regulations, they also play an equally significant role in framing the discussion. Reporters must realize that even if they cannot place important risk assessment information into stories because of space or time limitations, they must understand this information to ask the right questions of sources and be sure they cover all of the important points.

When the questionable validity of risk assessments is combined with misleading communication to the public of the true risks, it is understandable that the risk assessment process creates an air of uncertainty. Given this uncertainty, it is not unreasonable to invoke the "precaution principle" (Appell, 2001) that states "when an activity raises threats of harm to human health or the environment, precautionary measures should be taken even if some cause and effect relationships are not fully established scientifically." The application of this principle is especially relevant to this problem, as many inorganic and synthetic organic compounds are known to be harmful to humans at elevated concentrations. However, the decision to use this principle can also have widespread social and economic repercussions.

Summary

Obviously, determining the extent to which precautionary measures are needed is the major problem in attempting to address any threat to human health. Although the application is complex, there are also solutions, some of which are also complex, that can meet this challenge. The precaution principle must be incorporated into the pollution policies of today, while reasonable solutions are still viable. Seeking alternative methods of risk management that address polluted drinking water is highly advisable.

References

Anderson, M.E., 1988, "Quantitative Risk Assessment and Industrial Hygiene," *American Industrial Hygiene Association Journal.*

Appell, David, 2001, "The New Uncertainty Principle," *Scientific American.*

Asante-Duah, K., 1993, *Hazardous Waste Risk Assessment*, CRC Lewis Publishers, Boca Raton, Florida.

Browning, E., 1953, *Toxicology of Industrial Organic Solvents*, Medical Research Council Industrial Health Research Board, London, Her Majesty's Stationery Office.

CalEPA, 1994, California Environmental Protection Agency, Supplemental Guidance for Human Health Risk Assessments and Preliminary Endangerment Assessment.

CDC, 2001, Centers for Disease Control, National Report on Human Exposure to Environmental Chemicals, Press release dated 3-21-2001.

Chang, L.W., 1996, *Toxicology of Metals*, CRC Lewis Publishers, Boca Raton, Florida.

Colvig, Timothy, 2001, "Evidence Issues: Getting Expert Opinions Past the Judicial Gatekeeper and into Evidence," In Sullivan, P.J, F.J. Agardy, and R.K. Traub, eds., *Practical Environmental Forensics*, John Wiley & Sons, New York.

Daughton, C.G. and T.A. Ternes, 1999, Pharmaceuticals and Personal Care Products in the Environment: Agents of Subtle Change? *Environmental Health Perspectives*, Vol. 107, No. 6, pp. 907–938.

DeZuane, John, 1997, *Handbook of Drinking Water Quality*, Van Nostrand Reinhold, New York.

Dugan, P.R., 1972, *Biochemical Ecology of Water Pollution*, Plenum/Rosetta, New York.

ERF, 1998, RACHEL Report On Drugs in the Water, Environmental Research Foundation, *http://www.igc.org/igc/en/hl/98090711157/hl9.html* (Sept 6, 1998).

Fisher, L.M., 1938, "Pollution Kills Fish," *Scientific American*, Vol. 160, pp. 144–146.

Friedman, Sharon, 1996, "The Media, Risk Assessment and Numbers: They Don't Add Up, *Risk: Health, Safety & Environment*, Franklin Pierce Law Center.

Gilliom, Robert J., Jack E. Barbash, Dana W. Kolpin and Steven J. Larson, 1999, "Testing Water Quality of Pesticide Pollution," *Environmental Science and Technology*, Vol. 33, No. 7.

Gullick, R.W. and M.W. LeChevallier, 2000, "Occurrence of MTBE in Drinking Water Sources," *Journal of the American Water Works Association*, Vol. 92, No. l, pp. 100–113.

Halling-Sorensen, B. et al., 1998, "Occurrence, Fate and Effects of Pharmaceutical Substances in the Environment: A Review," *Chemosphere*, Vol. 36, No. 2, pp. 357–393.

Hoffbuhr, Jack W., 2002, "Waterscape: An Executive Perspective, Risky Acronyms," *Journal of the American Water Works Association*, Vol. 94, April, p. 8.

Hun, Tara, 1998, "Water Quality: Studies Indicate Drugs in Water May Come from Effluent Discharges," *Water Environment & Technology*, Vol. 10, pp. 17–22.

IUPAC, 1993, *Glossary for Chemists of Terms Used in Toxicology.*

Kalsch, W., 1999, "Biodegradation of Iodinated X-ray Contrast Media Diatrizoate and Iopromide," *Science of the Total Environment*, Vol. 225, Nos. 1-2, pp. 143–153.

Klaassen, C.D., 1996, *Casarett & Doull's Toxicology*, McGraw Hill, New York.

Kolpin, Dana W. et al., 2002, *Pharmaceuticals, Hormones, and Other Organic Wastewater Contaminants in U.S. Streams, 1999-2000: A National Reconnaissance, Environmental Science and Technology*, Vol. 36, pp. 1202–1211.

Krebs, R.W., 1998, *The History and Use of Our Earth's Chemical Elements*, Greenwood Press, Westport, CT.

Lichtenstein, P. et al., 2000, "Environmental and Heritable Factors in the Causation of Cancer B Analyses of Cohorts of Twins from Sweden, Denmark, and Finland," *New England Journal of Medicine*, Vol. 343, No. 2, pp. 78–85.

Lomborg, Bjørn, 2002, *The Skeptical Environmentalist: Measuring the Real State of the World*, Cambridge University Press, Cambridge, United Kingdom.

Maden, M., 1996, "Retinoic Acid in Development and Regeneration," *Journal of Biosciences*, Vol. 21, No. 3, pp. 299–312.

Maxwell, Steve, 2001, "Ten Key Trends and Developments in the Water Industry," *Journal of the American Water Works Association*, Vol. 93, No. 4, p. 5.

NAS, 1983, *Risk Assessment in the Federal Government: Managing the Process*, National Academy Press, Washington, D.C.

NCI, 2000, *National Cancer Institute, Atlas of Cancer Mortality in the United States*, NIH Publication Nos. 99–4564.

NRC, 1999a, *Setting Priorities for Drinking Water Contaminants*, National Research Council, National Academy Press, Washington D.C.

NRC, 1999b, *Identifying Future Drinking Water Contaminants*, National Research Council, National Academy Press, Washington D.C.

NSTC, 1997, *Committee on Environmental and Natural Resources, Interagency Assessment of Oxygenated Fuels*, Executive Office of the President, National Science and Technology Council, Washington, D.C.

Paustenbach, D.J., ed, 1989, *The Risk Assessment of Environmental and Human Health Hazards: A Textbook of Case Studies*, John Wiley & Sons Publishing, New York.

Preuss, P.W. and A.M. Ehrlich, 1987, "The Environmental Protection Agency: Risk Assessment Guidelines," *Journal of Air Pollution Control Association*, Vol. 37.

Regush, Nicholas, 2002, "Questions on Protecting US Water Supplies," *Journal of the American Water Works Association*, Vol. 94.

Renner, Rebecca, 2001, "Scotchgard Scotched," *Scientific American*.

Roefer, Peggy, 2000, "Endocrine-Distrupting Chemicals in a Source Water," *Journal of the American Water Works Association*, Vol. 92, No. 8, pp. 52–58.

Rogers, Keith, 2000, "Drinking Water Study: Volunteers Ingesting Pollutant Perchlorate," *Las Vegas Review Journal*.

Scharfenaker, Mark A., 2001, "Chromium VI: A Review of Recent Developments," *American Water Works Association*, Vol. 93, No. 11, pp. 12–15.

Scharfenaker, Mark A., 2002, "Reg Watch, Water Suppliers Carefully Watching Liability Suits," *Journal of the American Water Works Association*, Vol. 94, pp. 28–42.

Schneider, Keith, 1993, "New View Calls Environmental Policy Misguided," *The New York Times*, March 21, 1993.

Schneider, Keith, 1991, "U.S. Backing Away from Saying Dioxin Is a Deadly Peril," *The New York Times*, August 15, 1991.

Steger-Hartmann, T., R. Länge and H. Schweinfurth, 1998, Environmental Behavior and Ecotoxicological Assessment, *Vorn Wasser*, Vol. 91, pp. 185–194.

Ternes, T.A., 1998, "Occurrence of Drugs in German Sewage Treatment Plants and Rivers," *Water Research*, Vol. 32, No. 11, pp. 3245–3260.

Trussell, Rhodes, 2001, "Endocrine Disruptors and the Water Industry," *Journal of the American Water Works Association*, Vol. 93, pp. 58–65.

USBC, 2000, U.S. Bureau of the Census, Population Division (March 30, 2000).

USEPA, 1989, Risk Assessment Guidance for Superfund: Volume I, *Human Health Evaluation Manual (Part A), Interim Final*, Office of Solid Waste and Emergency Response, EPA/540/1-89/002.

USEPA, 1992, *Supplemental Guidance to RAGS: Calculating the Concentration Term*, Office of Solid Waste and Emergency Response, EPA/9285/7/081.

USEPA, 1999, "Groundwater Cleanup: Overview of Operating Experience at 28 Sites," EPA 542-R-99-006, September 1999.

USEPA, 1999, Region 9 Perchlorate Update, United States Environmental Protection Agency, Region 9, 75 Hawthorne Street, San Francisco, California (June 1999).

USEPA, 2000a , *Providing Safe Drinking Water in America*, Office of Enforcement and Compliance Assurance, Washington, D.C., EPA 305-R-00-002 (April 2000).

USEPA, 2000b, National Drinking Water Contaminant Occurrence Database, Envirofacts Warehouse (April 28, 2000).

USEPA, 2000c, Endocrine Disruptor Screening Program, Report to Congress (August 2000).

USEPA, 2000, Liquid Assets 2000: American's Water Resources: A Turning Point, Office of Water, Washington, D.C., EPA-840-B00-001 (May, 2000).

USEPA, 2001, Technical Fact Sheet: Final Rule for Arsenic in Drinking Water, EPA 815-F-00-019.

USEPA, 2003, Atrazine Interim Reregistration Eligibility Decision (IRED) Q&As.

USEPA, 2005, Integrated Risk Information System.

USPHS, 1925, "Report of Advisory Committee on Official Water Standards," U.S. Public Health Service, *Public Health Reports*, Vol. 40, No. 15.

Whipple, De Minimus Risk, 1987, *Contemporary Issues in Risk Analysis*, Vol. 2, Plenum Press, New York.

Whyte, A.V. and I. Burton, eds., 1980, *Environmental Risk Assessment, SCOPE Report 15*, John Wiley & Sons, New York.

Williams, P.L et al., 2000, *Principles of Toxicology, Environmental and Industrial Applications*, 2nd ed., John Wiley & Sons, Inc., New York.

Woteki, C, 1998, Nutrition, Food Safety, And Risk Assessment-A Policy-Maker's Viewpoint, Remarks prepared for delivery by Dr. Catherine Woteki, Under Secretary for Food Safety, before Purdue University, West Lafayette, Indiana (June 1998).

Xrefer, 2001, http://www/xrefer.com

CHAPTER 5

Managing Risk and Drinking Water Quality

"How clean is clean enough *can only be answered in terms of how much we are willing to pay and how soon we seek success."*
Richard Nixon, Council on Environmental Quality, 1971

The general purpose of risk management is to identify and measure the potential impact of anticipated adverse events, thereby selecting a course or courses of action to minimize potential damages within existing economic constraints. Existing water quality programs around the world, common within the industrialized nations, typically allocate resources to study the toxicity of individual chemicals so that standards may be set and water treatment facilities upgraded, as required, to ensure that each chemical will not occur in drinking water at or above its established standard. Such an approach, however, allows for potential adverse effects resulting from the consumption of unregulated chemicals and chemical mixtures at low concentrations. This results from the historical evolution of regulations that accommodated the economic benefits associated with treatment costs that allowed low levels of pollution in drinking water. Clearly the current approach to managing health risks associated with polluted drinking water is flawed, but what other alternatives are available to both communities and individuals?

Obviously, in those countries that do not have a widely developed waste management and water treatment infrastructure, the greatest threat to human health is not having the availability of disinfected drinking water. This need has been once again highlighted by the catastrophic Asian tsunami of December 2004. The disinfection of drinking water is also an ongoing problem in industrialized nations. Such problems have been recently documented by Hrudey and Hrudey (2004) in their book on water-borne outbreaks of disease in affluent nations. The treatment of biohazards (pathogens) in drinking water, however, is well understood. The main problem is the proper implementation of known treatment technologies coupled with maintenance of water distribution systems. Thus, it is not our objective to further address this well-defined problem but rather focus on the risk management of chemicals, as opposed to pathogens, in drinking water resources.

Summarizing all of the information provided in the previous chapters, it is apparent that existing methods of protecting the public health from chemical pollution need to be reevaluated and updated in light of the current pace of industrial expansion, biotechnology, and the inability of life-science researchers to characterize the true risks to human health, "in real time." Therefore, we would be wise not to continually rely on "out of date" risk management practices as the means by which to ensure drinking water quality.

Historically, the control of chemical pollution has focused on the use of a region's natural resources to either assimilate or dilute pollution residues, with little or no economic cost to the discharger. As damages to the environment became obvious, even to lay people, polluters were required to reduce, but not necessarily eliminate, levels of contamination introduced into the environment. Clearly economics played a role (i.e., little or no perceived health risk balanced with the benefits of economic growth). In today's environment, the human body, rather than receiving waters, has taken over the role of "assimilation" or "dilution" of the chemical residues resulting from industrial development, saving industry from the burden of additional pollution control costs.

Given this condition, is there really a health risk that needs to be addressed or managed beyond existing environmental control programs? Based on past historical approaches to managing the risks of hazardous chemicals in the environment, there is sufficient justification for doubting the ability of existing programs in timely fashion to assess, respond, and manage risks. Thus, seeking alternative methods of risk management addressing polluted water is highly advisable.

Learning from the Past and Present

The inability of both science and society to assess risk in timely fashion can be illustrated by evaluating how industrialized societies have typically dealt with pollution of surface and groundwater. For example, in the United States, numerous examples of surface water pollution by unregulated[1] chemicals have been reported since the beginning of the 20th century (Wilson and Calvert, 1913; Kraft, 1927). Likewise, evidence of chemical pollution from unregulated compounds has also been reported for groundwater in the 1930s, 1940s, and 1950s (Brown, 1935;

---

[1] A chemical without an established water quality criteria. This also includes chemical mixtures as a group pollutant.

Pickett, 1947; Muehlberger, 1950). Decades before Rachael Carson[2] came on the scene, the common unregulated weed killer, 2,4-D, was found to have polluted wells in Montebello, California. Similarly, TCE, which until recently was an unregulated compound, was identified in groundwater as far back as 1949. Concern over groundwater pollution in the 1940s and 1950s prompted the underground waste disposal task group of the American Water Works Association (1953) to state that "the proper time to control underground pollution is before it occurs." The American Water Works Association was only one of many who advocated this approach. Despite the warnings in the early 1950s, methods by which to prevent pollution of groundwater remained essentially unchanged.

Approximately 55 years later, historic pollution from unregulated chemicals is now being remediated in both soil[3] and groundwater at a great expense. Furthermore, the USEPA (1999) has now recognized the failure of currently available technology to completely clean up groundwater (i.e., once polluted groundwater resources may never be fully restored). By classic neglect, resulting in the presence of unregulated pollutants in water resources, industries as well as government agencies have been forced to expend billions of dollars for remedial actions that are not capable of totally eliminating the hazard. The net result is that our water resources, in many areas of the United States, will remain polluted for decades or possibly centuries.

Because risk management is based on regulating specific chemicals, complicated and overlapping environmental control programs have been put in place. For example, in the United States, there are 32 federal executive agencies in 10 cabinet departments, including the Executive Office of the President, that are active within 25 separate programs[4] to manage and protect water quality.[5] All of these programs are linked to the concept of water quality standards. As a result of this linkage, the USEPA adopted a quantitative method (i.e., the risk assessment process) for assigning a level of hazard to specific chemical standards.

The USEPA risk assessment process is based on the presumption that, as implemented, drinking water does not pose a threat to the public health of the American public. These statistical models have been used for several decades by the USEPA to justify the selection of standards by which to control the release of specified levels of a chemical pollutant into our water resources, while simultaneously assuring the public that regulated pollution is "protective of human health and the environment."

---

[2]Author of the 1962 book *Silent Spring*, which eventually resulted in governmental policies to ban the use of DDT.

[3]Chemicals in soil are a potential source of pollution to groundwater. Therefore, before groundwater remediation occurs, soil source removal is usually evaluated and implemented.

[4]Includes 200 sets of federal rules, regulations, and laws.

[5]United States of America's submission to the 5th Session of the Commission on Sustainable Development, April 1997.

Faith in this method, however, should be tempered by the following statement on the reliability of a risk assessment as ". . . at best a dubious exercise. In most cases it involves assumptions built upon other assumptions, the effect comparable to a house built on sand" (Brown, 1987). Reliability of these methods is also suspect because no human toxicity data are available for use in this calculation. As with all models, the axiom "garbage in, garbage out" truly applies to risk assessment methods.

The inability of regulatory agencies to make timely decisions on identifying chemical risks combined with risk assessment methods that cannot be calibrated or validated for humans suggests that past and current risk management methods lack credibility. This problem is exacerbated by the fact that agencies, within the United States and worldwide, that are responsible for water quality regulate different sets of compounds (see Exhibit 2-8). The credibility of current risk management methods is even more questionable given the approach of the 1958 Delaney Amendment. As part of the Delaney Amendment to the Food, Drug, and Cosmetic Act, it was specified that "no" chemical additive shown to cause cancer in humans or animals could be added to food or cosmetics. This amendment recognized the potential hazard to public health. Yet current regulations, including the 1996 Food Quality Protection Act, allow known toxic and carcinogenic pesticides in food when it was clearly recognized as a hazard.

In a study by California Public Interest Research Group (CPIRG) in 1999, the following common unregulated pesticides have been found in California drinking water resources: diuron, bromacile, aldicarb, methyl bromide, diazinon, chlorthal, carbaryl, methoachlor, EPTC, norflurazon, chloropyrifos, cyanazine, trifluralin, dieldrin, and metrobuzin. Two of these pesticides, dieldrin and metrobuzin, were originally on the USEPA Contaminant Candidate List in 1998 but were removed from this list in 2004. For the purpose of comparison with water quality standards, those pesticides that are allowed in food at specified "tolerance" levels are given in Appendix 2-8.

That humans are exposed to a mixture of chemical pollutants in both food and water suggests that total chemical exposures must be considered when evaluating their potential impacts on human health. It is not the intent of this book to address the risk of chemicals in food or the combined risk in food and water; however, this issue is briefly discussed in Exhibit 5-1. This issue aside, there are unregulated pesticides in drinking water resources (see Appendix 2-8) that are obviously not considered toxic in water but are deemed a hazard in food.

There may be various reasons for this dichotomy. The USEPA may assume that many of these chemicals have limited production volume, distribution, or solubility, so they would not pose a significant water pollution risk to the consumer. However, they may pose a significant risk to the local residents in those areas where these chemicals are used. General presumptions about the distribution and fate of pesticides by the USEPA do not provide assurance that unregulated pesticides are not

Exhibit 5-1.   Chemicals in Food and Water

If one combines the chemical exposure from both food and water, the level of exposure is obviously increased. However, is this increase a greater risk? According to Lomborg (2001), the risk is increased but is primarily associated with the food we eat. "The average American consumes 295 pounds of fruit and 416 pounds of vegetables a year. A rough calculation shows that he therefore consumes about 24 mg of pesticide each year. Even if one drinks 2 liters of water a day for a whole year from a well with a pesticide concentration exactly at the EU limit (which would be a maximally pessimistic scenario), one would absorb about 300 times less pesticide from the water than from fruit and vegetables." Such a general comparison does not negate the need to limit chemical exposure in drinking water, nor does it address site-specific exposure or the potential synergistic effects of nonpesticide chemical mixtures. However, this finding creates another problem.

If foods produced without the use of pesticides (i.e., organic) are assumed to be truly free from pesticide residues (i.e., no unintended air pollution of a crop or cross-contamination during processing), then individuals who want to eliminate their exposure to manufactured chemicals should only consume organic foods and drink pure water. Such a decision, however, has a social cost because people who cannot afford this choice will continue to be exposed to higher levels of chemical pollutants.

entering our water resources and are not present in drinking water supplies.[6] Clearly, unless a chemical is regulated as a hazardous substance, it is considered *nontoxic* or an "acceptable risk" by omission. This type of risk management process (i.e., use of identified water quality standards) lacks the basic credibility necessary to ensure the protection of human health. Yet, why do water quality programs that are mandated to protect the public rely on water quality standards? Not surprisingly, the answer is economics.

Risk and Economics

Economics have always been a major factor in determining the level at which a specific water quality standard is set. The most recent example

---

[6]There are no monitoring data that would rule out the occurrence of these chemicals in the drinking water.

occurred in March 2001 when the Bush administration put on hold a planned reduction in the arsenic standard for drinking water from 50 ppb to 10 ppb. The administration then partially reversed itself and mandated that the 10 ppb standard be met no later than January 2006. The reason offered by the Bush administration for delaying implementation of the 10 ppb standard was because of the need to make sure that this new standard was valid and affordable. In our view, this justification was unsupportable given that epidemiologic data suggest that the standard should be zero (USEPA, 2001a). Indeed, the government's maximum contaminant level goal is currently set at zero. The real and unspoken reason for the delay was its potential cost to the U.S. economy. It has been previously estimated (Pontius, 2002) that the annual cost of meeting this new standard would range from $6,494 to $1,340,716 per year[7] for each public water system. For example, when the 10 ppb arsenic standard is put into effect,[8] the following cities would have to upgrade their water treatment systems (USEPA, 2001b): Albuquerque, New Mexico; Chino Hills, California; Lakewood, California; Landcaster, California; Midland, Texas; Moore, Oklahoma; Norman, Oklahoma; Rio Rancho, New Mexico; Scottsdale, Arizona; and Victoria, Texas. In all, approximately 4,000 of the 74,000 systems regulated by this new arsenic level will have to install additional treatment to comply with the new standard (USEPA, 2000). Another reason given for the delay was that a reduced arsenic[9] standard would impact water treatment costs of mining operations in the United States.

If society is expected to accept drinking water that contains pollutants at "safe" levels, which it has for the last 75 years, two critical criteria must be addressed and met. First, there should be a minimum of risk associated with the consumption of drinking water containing unregulated pollutants or pollutants at or below accepted standards. This assumption of "minimum risk" remains to be proven. Second, pollutant limitations that are established for our domestic water need to be strictly enforced by both state environmental agencies and the USEPA. The enforcement function, however, has major flaws. In August 2001, the USEPA's inspector general released a report on the enforcement effectiveness of the existing water pollution programs. This report concluded that:

- At least 40 percent of the nation's waters do not meet the standards states have set for them. Polluted runoff, both regulated and unregulated, is causing the majority of the nation's water quality problems.

---

[7] The reason for this cost range is the size of the public water system. The larger the system the greater the cost.

[8] The Bush administration accepted the 10 ppb arsenic standard in October 2001 (a 7-month delay).

[9] Arsenic is a pollutant in acid mine drainage from coal and metal ores with sulfides.

- All 50 states found that their ability to identify water quality problems was constrained by a lack of nonpoint water quality data to determine the cause or source of water pollution. This lack of information has prevented states from setting nonpoint discharge standards. Because no standards have been set, no definitive policy decisions have been made to control runoff from sources such as animal feed lots and agricultural lands.
- Environmental protection of water resources is principally enforced by state environmental agencies. The USEPA reports that when a state agency identifies a company that has violated a clean water regulation, they often fine the company too little and/or may never collect the fine. States frequently delay enforcement actions up to a year after a violation. Some states report that more than half of the facilities that violated a clean water regulation in 1999 continued to have the same violation in 2000.
- To fix the problem, the federal government needs to send more money to the states to enforce environmental programs. However, even with this financial support, some states are reluctant to enforce programs that may impact "small businesses and economically vital industries."

Because effective enforcement is one of the most important components of protecting the nation's drinking water, the deficiencies noted by the USEPA's inspector general casts serious doubts on the ability of some state agencies to enforce even existing federal water quality programs. Furthermore, the lack of monitoring data for a substantial number of unregulated chemicals found in both nonpoint and point sources of pollution is a significant flaw in the current enforcement program.[10] Without such data on the true distribution of unregulated chemicals in water resources, how can the USEPA determine if a water quality problem exists? This dilemma is a classic Catch-22 and is discussed in greater detail in Exhibit 5-2.

Even before more funds[11] are spent to enforce existing programs, an extensive and costly nationwide nonpoint water quality monitoring program must be implemented. Otherwise, enforcement cannot be effective. If our basic water resources do not "come up to standard," then one must question the technical implementation of any program aimed at minimizing pollutants in our water supplies. Sadly, the ability to implement such a broad scale program is close to being both economically and technically impossible under the current environmental control programs.

---

[10]There is no comprehensive federal program to evaluate the occurrence and magnitude of unregulated pollutants in drinking water sources around the United States (i.e., only a very small number of compounds are currently monitored in the larger community water systems).

[11]Given the 2004 projected federal deficit, there is no reason to believe that any funds will be available for increased enforcement any time soon. This is particularly true when given the massive funding that is required for the "War on Terrorism."

Exhibit 5-2. Regulatory Policy and Monitoring for Unregulated Pollutants in Wastewater

A city located in the Napa-Sonoma wine country of California recently proposed to divert a portion of its treated wastewater effluent for irrigation of local vineyards. Homeowners who obtained their drinking water from wells in the vicinity of the growers immediately sued the city in an attempt to block the implementation of the proposal. As environmental experts, we were retained by the homeowners to determine if the wastewater posed a pollution threat to the local groundwater.

Because the region is dominated by vineyards, the chemicals of concern in wastewater were pesticides specifically used for grapes in the Napa-Sonoma area (i.e., dimethoate, diphacinone, fenarimol, mancozeb, myclobutanil, oxyfluorfen, and propargite). Of these pesticides, the city only monitored "occasionally" for diphacinone. In addition to the pesticides, the concern was raised that some unknown pharmaceuticals could pass through the wastewater treatment system and be discharged along with the wastewater effluent into the groundwater. Thus, the city was requested to monitor for (1) those pesticides that are specific to the Napa-Sonoma region and (2) those pollutants that may be an indicator that organic chemicals are passing through the city's wastewater treatment systems.[12] The selected indicator compounds included a wide range of phthalates, phenols, and alkylbenzenesulfonates common to household products, caffeine, and selected pharmaceuticals (i.e., aspirin, ibuprofen, estrogen, clofibrate erythromycin, and tetracycline).

If any of the indicator compounds were found in the effluent, then the treatment system might be allowing an unknown amount of various compounds to be discharged with the effluent. This finding would imply that the existing treatment system was not adequately designed to remove these pollutants. Thus, there would be no guarantee that the effluent was not a hazard to public health.

The city responded to the request by stating that there was no need to analyze for these chemicals in their wastewater because "criteria must exist to evaluate the toxicity of the chemicals." In other words, because no state or federal standards had been set for these chemicals (i.e., they are unregulated), there was no hazard nor need to monitor for them. This attitude clearly illustrates that after

---

[12]Because it would be impossible to analyze for all the potential pharmaceuticals that could pass through a wastewater treatment system, it was necessary to select a set of compounds that could be used as an indicator that the problem exists. These compounds were selected based on their occurrence in other municipal wastewater effluents (see Chapters 1 and 3).

years of accepting "pollution-based standards" water quality agencies are still willing to allow toxic or hazardous chemicals in water simply because it is "politically" acceptable to pollute if the state or federal government has not set a standard.

Furthermore, the federal government has historically negotiated with both local public agencies and industry to reach a consensus on standards. Because a consensus is based on both science and economics, standards heralded to "protect human health and the environment" will always lack full scientific credibility. In addition to the influence of economics on risk management decisions, the basic scientific validity of water quality standards is questionable (see the discussion in Exhibit 5-3). All of these conditions strongly suggest that an alternative approach to risk management and drinking water quality needs to be seriously considered.

Exhibit 5-3. Why Water Quality Standards Are Not a Credible Method for Protecting Human Health

Water quality standards are not a credible risk management method for the following reasons:

The vast majority of water quality standards are *not* based on actual human exposure data to specific chemicals. Damage to human health has not been, and is unlikely to ever be, calibrated to a specific chemical concentration in drinking water,[13] nor can these standards be validated in the real world.

1. It is currently impossible to collect human exposure data because the human environment is too complex, and toxicity studies for the most part will not use humans in place of laboratory animals.
2. The time period between the introduction of a chemical into the environment and when it is recognized as being toxic to humans or animals often spans decades. This problem will only get worse as we fall further and further behind in our evaluation of these chemicals.
3. The overwhelming number of chemicals that have already been introduced into the environment have not, as yet, even begun to be evaluated to see if they are a hazard to humans.

(continued)

---

[13]Volunteers in a medical study conducted by Loma Linda Medical Center in California were being paid $1,000 each to take pills containing perchlorate in experiments aimed at establishing a drinking water standard for perchlorate (*Las Vegas Review-Journal*, November 2000).

Exhibit 5-3. *(continued)*

4. The rate at which new chemicals are introduced into the environment is significantly greater than the rate at which they are studied and recognized as being toxic to animals.
5. There is virtually no data on the toxicity of chemical mixtures on animals, let alone humans.
6. Given the current pace of evaluating chemicals for inclusion to the Primary Drinking Water Standards, it is impossible for the USEPA to identify and regulate the current number of chemicals already in the environment, let alone the approximately 2,000 new chemicals introduced each year.
7. No consensus exists within the various federal programs as to which chemicals pose a threat in food and which should be regulated in drinking water. Nor is there a justifiable scientific basis for selecting those chemicals for which drinking water standards are set. By their omission, unregulated chemicals are therefore defined as nontoxic or as being an acceptable risk.
8. The methods used to select chemicals for potential inclusion to the Primary Drinking Water Standards are flawed.
9. Even if a chemical is identified as being toxic to humans, economic considerations may preclude its addition to the Primary Drinking Water Standards or influence the level at which the standard is set.
10. When standards are exceeded, the offending pollutant is not required to be removed from the drinking water supply. Consumers are only warned not to drink the water. For many individuals, this may not be a viable option.
11. When unregulated chemicals are identified in drinking water, the offending community water supply has to either notify its customers of the chemical's presence or remove it.
12. Given the potentially large number of chemicals that can be found in drinking water, it is impossible to monitor for specific unsuspected compounds.

Given these characteristics, water quality standards should not be used as a guide to a risk management program that is supposed to protect human health and the environment.

An Alternative Approach

Because the fundamental economics of industrial development in the United States are supported by governmental programs that allow chemical pollution of our waters, there is very little chance that a zero pollution discharge policy will become a reality. Nor are there as yet any practical regulatory solutions for controlling pharmaceutical and wide-spread nonpoint source pollution. Also, given the geographic and hydrologic diversity of water resources that are drinking water sources, source protection programs, although beneficial, are no panacea. All of these circumstances virtually guarantee that under the present approach, pollutants will always be in our drinking water. From a scientific and engineering viewpoint, the only way to achieve drinking water quality that poses the minimum health risk to humans is to ensure that chemical pollutants are reduced in drinking water sources to the lowest levels possible. This means implementing a technology-based risk management program that is not inconsistent with current USEPA programs that already allow limited treatment techniques (TT)-based standards (see Table 1-9).

Given the diversity of public and private drinking water supplies globally, the implementation of technology-based risk management programs will require (1) tailored solutions based on appropriate technology considering local circumstances,[14] (2) an expansion of certification testing to ensure water treatment technologies operate at maximum removal efficiencies, and (3) the approval of appropriate analytical methods, such as high-performance liquid chromatography (HPLC), to provide chemical finger printing monitoring capabilities. Utilizing a TT-based risk management program will not be without economic consequence considering that the USEPA[15] has estimated that approximately $1 trillion would be required just to meet existing infrastructure and water quality objectives, let alone to TT upgrades to maximum pollution removal.

A TT-based risk management drinking water quality program will actively protect the public health by removing both regulated and unregulated chemical pollutants in our drinking water. There is also a need for a TT-based drinking water quality program because the current standard-based risk management program is now open to lawsuits. A recent article in the *Journal of the American Water Works Association* states (Scharfenaker, 2002) that "of particular concern to water suppliers is the litigation potential linked to the expanding body of research indicating possible reproductive developmental risks from relatively short exposures by pregnant women to elevated levels of certain disinfection by-products, which have been and continue to be regulated based on a potential cancer risk associated with chronic exposure." In this same article, the Deputy Director of Government Affairs for the American Water Works Association remarked that "compliance with a standard should protect a drinking water utility from liability for any damages that might be associated with a substance or contaminant in drinking

---

[14]At a minimum, all water systems should have advanced oxidation, membrane filtration, and a compatible posttreatment disinfectant regimen.
[15]Federal Water Review of September/October of 2000.

water. . . . The whole point of a standard is reducing risks to levels that are achievable and socially acceptable." If a TT-based program was in place, this concern would not even be an issue.

Implementing a TT-based risk management approach to water quality protection could not occur rapidly. Such a major change would require time to evaluate, engineer, and plan for the proper implementation of the program. Given the likely inertia of both governments and existing water utilities to resist such a change, the consumer will have to evaluate the risks associated with drinking water (which may and often does contain pollutants) and take the necessary steps to reduce exposure independent of governmental or water utility actions.

Consumer-Based Protection

Consumer confidence in drinking water can be based on a number of factors. Each of these factors, either individually or jointly, can influence consumer confidence in determining if added protection is necessary for drinking water sources. Each of these factors, as discussed next, can influence the consumer's potential exposure to chemicals in their drinking water. Therefore, guidelines[16] are provided as a means of allowing each consumer to evaluate their potential exposure to drinking polluted water. These guidelines address a consumer's main source of drinking water (i.e., home and/or office) as well as bottled water. Also, each consumer will have to do some minor research to evaluate his or her own unique "water" environment.

Main Source of Drinking Water

The first and most important factor is the degree to which a consumer's water source may or may not be polluted. This includes water provided by a utility or from a consumer's own groundwater well or surface water resource. The following source water characteristics have a direct bearing on the quality of drinking water:

1. Source waters that originate from protected watersheds have the least probability of containing manufactured chemicals. This is true for both groundwater and surface water resources. In those very special cases where the point of groundwater extraction (e.g., an artesian wellhead or spring) is actually located in a wilderness[17] area and all of the groundwater in the aquifer that feeds the well or

---

[16]The guidelines and recommendations are based on the information provided in the previous chapters.

[17]A tract of land, or a region, uncultivated and uninhabited by human beings, whether a forest or barren plain. This means that there can be no roads or other manufactured structures that give motorized access to a region.

spring is only recharged from a wilderness area, this source water will have the least probability of containing manufactured pollutants. A surface water reservoir that receives its only source of water from a wilderness area provides the next best source of drinking water (i.e., assuming little or no air pollution) relative to manufactured chemical pollutants.

Given the fact that man-made features and habitations can cross aquifer boundaries without our knowledge, the ability to accurately establish wilderness source areas is not simple. Therefore, the next best source of drinking water is from National Parks and designated wilderness areas that have the least man-made impacts but no mining or agriculture. These environments are illustrated in Figure 5-1. It should be noted, however, that the process of transporting the source water to the user can impart chemical pollutants such as petroleum hydrocarbons from machinery and distribution systems.

2. Groundwater sources that originate in any mixed land use (e.g., a combination of urban, agricultural, mining, forest, grassland, or protected area) have a much greater probability of containing manufactured pollutants, depending on specific pollution characteristics. In the area

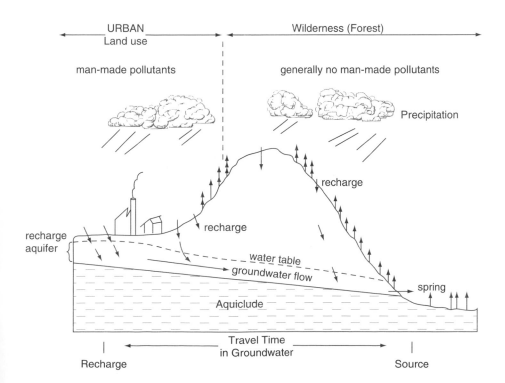

FIGURE 5-1. Pollution and groundwater recharge.

where the source water originates, the following information could be used to evaluate the potential exposure. For example, in the source area, who discharges to surface waters and what is being discharged? Are there any known sources of groundwater pollution? What percentage of the area is used for cultivated agriculture, animal agriculture, or mining, and is wastewater injected into groundwater aquifers?

In a mixed land use environment, a groundwater source has a much lower probability of containing man-made pollutants if (a) the area that recharges the aquifer does not contain sources of pollution; (b) the travel time from the recharge area to the extraction well is long (decades to centuries) so that aquifer materials have a greater opportunity to remove pollutants; (c) the aquifer materials are sedimentary (i.e., silts, sand, clay, and gravel mixtures) as opposed to fractured rocks;[18] (d) the aquifer is separated from the earth's surface by at least one impervious geologic strata (i.e., an aquiclude); (e) wastewater is not injected into the aquifer; (f) the groundwater is not locally recharged by surface water resources and; (g) an extraction well is not located down-gradient from waste disposal systems (i.e., septic tanks and leach fields, cesspools, and sewers). These environments are illustrated in Figure 5-2.

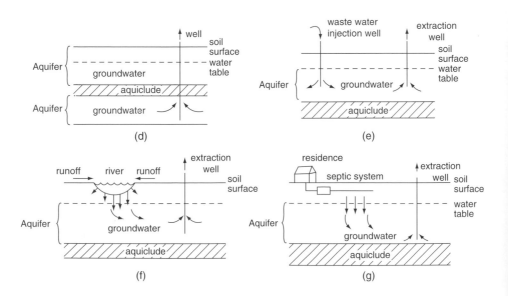

FIGURE 5-2. Pollution and groundwater environment examples.

---

[18]In general, fractured carbonate rocks (limestone and dolomite), metamorphic rocks (e.g., quartzite and slate), and igneous rocks (e.g., granite and basalt) are less likely to remove pollutants.

3. Surface water sources that originate in any mixed land use (e.g., a combination of urban, agricultural, mining, forest, grassland, and protected areas) have a high probability of containing manufactured pollutants either from permitted NPDES discharges, surface runoff from both urban and agricultural land uses or combinations thereof. This is particularly true for river systems (e.g., Ohio, Missouri, Mississippi) where pollutants are cycled by the river as a result of publicly owned treatment works, and industries discharging waste-waters into rivers also serve as water sources used for drinking water (with or without subsequent treatment). This cycle occurs a hundred times over along the length of major rivers.

4. There are also some important aspects associated with water utilities and water quality that individual (i.e., business or residence) water users having their own water supply do not usually encounter. First, some water utilities blend a combination of groundwater and surface waters. Second, individual water users normally do not disinfect their water with halogens (i.e., chlorine or bromine). Thus, they do not have the problem of halogenated disinfection by-products.

5. Water utilities that use halogens (i.e., chlorine or bromine) to disinfect water will generate halogenated organic disinfection by-products unless the amount of dissolved organic matter in the source water is very low. Generally, groundwater will contain less dissolved organic matter than surface water. Therefore, halogenated surface water will tend to have more halogenated disinfection by-products than groundwater sources.[19] However, one should not discount the formation of disinfection by-products resulting from the interaction of free chlorine (residual) with organic matter resident in distribution systems. This problem can be alleviated to some degree by the water utility switching to chloramine as the treatment chemical. Obviously, those water utilities (such as those in many European countries) that disinfect their finished water with ozone or ultraviolet light (UV) do not have the same problem.[20]

6. In addition to traces of untreated or incompletely treated chemicals leaving the treatment works, and the possibility of disinfection by-products, distribution systems may contain lead and copper metals that can be leached into the water.

Given these characteristics of source waters, what should a consumer do to determine the potential quality of drinking water? This depends on whether the water source is from a utility or from a private source.

---

[19]In the United States, water utilities are not required to disinfect groundwater.

[20]The downside to not using halogenated chemicals, however, is that once the water is in the distribution system, there is no residual disinfectant in the distribution system.

Consumers with Water Utility Products

A consumer should either access the Internet or contact the water utility directly to obtain the water utility's annual water quality report for every year that an annual report was produced. If any of these reports show detectable concentrations of any regulated chemical or unregulated chemical, the consumer should seriously consider installing a point-of-entry or point-of-use water treatment system (see discussions of these types of systems in Chapter 3). If there is no history of any chemical pollution, the consumer is still urged to go one step further and try to determine the nature of the source water and whether advanced water treatment technologies are used. The consumer should recognize that the utility is not required to test for constituents that may be present but are not on the government (federal or state) list of chemicals that require analyses.

Even if the annual water quality report(s) indicates the nature of the water source (i.e., surface, groundwater, or a mix of both), the consumer should contact the water utility and inquire as to the origin of the source (surface or groundwater) and request a candid appraisal of what source protection programs that are in place and if there are any known sources of pollution in the area that may influence the water product.[21] If representatives cannot provide this information or if there are no source protection programs in place, the consumer should consider installing a point-of-entry or point-of-use water treatment system. Any consumer who receives water from a utility that obtains any portion of its source water from surface water[22] should consider installing a point-of-entry or point-of-use water treatment system. It should be noted, however, that if the water utility uses advanced water treatment systems (e.g., granulated activated carbon, oxidation or reverse osmosis technologies), in-home water treatment is not necessary.

In the absence of advanced water treatment, source waters from a wilderness or protected national park would be ideal; however, such sources are rare. Thus, ideal source water is more realistically groundwater that has the following characteristics: (1) the source aquifer is overlain by one or more regionally extensive aquitards; (2) the recharge area is protected; (3) no wastewater is injected in the aquifer; (4) the aquifer does not receive recharge from a polluted river, creek, or lake; (5) the aquifer is composed of sands, silts, clays, and gravels; and (6) the travel time from the recharge area to the point of extraction lasts for decades.[23] Groundwater resources that do not meet these criteria may still be

---

[21]Questions to the water utility can be focused on the pollution characteristics described previously.
[22]Unless it originates from a wilderness watershed.
[23]This type of technical information, if available, can be obtain from the water utility, a state department that regulates environmental quality, or contact the geology department at a regional university or college (most professors are more than willing to discuss the subject).

relatively pollution free. Based on the available information, it is up to consumers to determine whether they are comfortable with the environmental characteristics of their source water. If not, then they should consider installing a point-of-entry or point-of-use water treatment system.

Consumers with Their Own Source Water

Home owners or businesses that have their own groundwater well should have the supply tested for all of the regulated drinking water pollutants and pesticides. Analytical laboratories across the United States can test for all of these chemicals for approximately $3,000 per sample. Testing for other unregulated chemicals and specific pesticides could at least double the cost depending on the chemicals selected. Because most unregulated chemicals are most likely associated with agricultural practices and human waste disposal,[24] it would be possible just to monitor for additional pesticides (i.e., other than those in Table 1-9) and indicators of pollution by human waste disposal systems (i.e., septic systems, cesspools, and sewers). A broad range of pesticides can be determined by analyzing for the organochlorine pesticides (EPA Method 8081A) and the organophosphate pesticides (EPA Method 8141A).[25] Indicators of human waste disposal could include tests for total coliforms[26] and caffeine.[27] If the results of the tests detect any of the regulated chemicals or indicators of human waste disposal, the user should consider treating the source with a point-of-entry water treatment system that uses activated carbon and UV disinfection (costs range from $2,000 to $3,000). Any consumer who uses a surface water supply should consider forgoing the tests and just install a point-of-entry water treatment system, with the cost of the system typically within the range of the cost for analytical testing.

Bottled Water as an Alternative

No bottled water company provides all of the information on the bottle label for the consumer to (1) evaluate whether that product has the potential for containing manufactured pollutants, (2) determine the concentration of the major cations and anions in the product, (3) determine the concentration of trace elements and radioactivity in the product, (4) establish whether the product is bottled on site, (5) determine whether the product was disinfected, and (6) if a plastic bottled is used, determine the concentration of those chemicals that typically leach from plastic.

---

[24]As discussed previously, these wastes can and do contain a wide range of chemicals from personal care products and pharmaceuticals.
[25]The cost of these analyses would be approximately $300.
[26]Total coliform levels should be below current standards.
[27]From the standpoint of environmental monitoring, caffeine analysis is a nonstandard method. Thus, laboratories that are willing to perform this type of analysis may charge a premium for there services. A total coliform test runs from $50 to $100.

As a consequence, if the consumer is genuinely interested in getting the "complete story," the consumer is forced to contact the manufacturer or review the manufacturer's published information, to the degree available, on web sites. Because virtually 99 percent of bottled water companies do not have this type of information available on a web site or from consumer services, it should be required by law to be either on the label or on the company's web site.

Without comprehensive information, the consumer has no real ability to select a bottled water based on its environmental characteristics. Many bottled water companies attest that their water product comes from a protected source and is, therefore, pure or free of manufactured pollutants. Yet, these same companies do not provide actual chemical monitoring data or source protection evidence on their web sites (if available) that would verify these marketing claims. Unfortunately, consumers can only compare the available data and information on each bottled water before selecting a product that they feel minimizes health risks.

Based on environmental characteristics alone, consumers should, at a minimum, consider not purchasing a bottled water if (1) it does not list an identifiable source, (2) it is not disinfected, or (3) it is in a plastic bottle.[28]

Summary

For the foreseeable future, the protection of water quality and human health is better served by consumer choices and independent decisions by water utilities to apply advanced treatment processes as opposed to governmental programs.[29] Furthermore, without water industry or governmental intervention, there will be no change in the current risk management of our water quality. Clearly, the implementation of a TT-based risk management approach will depend on the potential actions by the water industry and governmental agencies. Generally, the water industry has demonstrated a greater concern and has implemented treatment programs in advance of increased government regulations due solely to satisfying the concerns of its customer base.

Potential Action by the Water Industry

It is important to realize that not all community water supplies and private sources of drinking water require additional treatment to remove

---

[28]Unless the bottled water manufacturer publishes actual analytical data on their products, for which analysis was completed for those constituents that are known to leach from specific plastic products.

[29]Although this observation is specific to the industrial nations, it is even more relevant for the underdeveloped countries of the world that lack established environmental programs.

chemical pollutants. For example, water resources that originate in pristine watersheds require little or no additional treatment to remove chemicals, simply because the pollutant content is *de minimis*. For example, San Francisco[30] and Seattle receive their water supply from protected watersheds. Similarly, deep groundwater resources that originate from pristine watersheds that are not recharged with partially treated wastewater or by polluted rivers and streams, or are not polluted by human activities on the earth's surface (e.g., agriculture, leaking gasoline tanks, waste sites), may not require additional treatment.[31] These conditions need to be better defined so that we focus on those water resources that do require additional treatment to properly address the issue of chemical pollutants. This type of study is consistent with mandated source water protection programs.

Water resources that are used as a source of drinking water and may require additional treatment to remove manufactured chemicals include:

- Surface waters that pass through land masses that include mining, agricultural, and urban development.
- Surface water and groundwater that receive municipal or industrial wastewater discharges.
- Surface waters that are recharged by polluted groundwater.
- Groundwater under the influence of polluted surface waters (i.e., polluted river water that percolates into the groundwater).

Obviously, the level of pollution in each resource category can vary widely depending on the environmental conditions. Thus, water systems that obtain their water from any of these potentially polluted resources need to critically evaluate sources of chemical pollution that may contribute to the water resource and determine how either to control the sources of pollution or to implement the necessary treatment technologies to guarantee their customers a source of drinking water that is as pollution free as technically possible.

The first step in providing drinking water with a minimum concentration of chemical pollutants is to assess the ability of water utilities to achieve this level of quality. One of the major hurdles facing utilities is that only a very small fraction of the treated and distributed water is actually consumed as "drinking water." For example, approximately 90 percent of treated water is used by industry and commercial business, small agricultural businesses, landscape irrigation, firefighting, and a multitude of other uses outside the home (e.g., washing cars, walls, and

---

[30]However, the city of San Francisco needs to upgrade the pipeline from the watershed to the city at an estimated cost of more than $4 billion just to maintain the integrity of this transmission system.

[31]This conclusion also assumes that there are not elevated concentrations of trace metals, such as arsenic, that can occur naturally in water resources.

sidewalks; swimming pools; fountains; fish ponds; and lawn irrigation). These nondrinking water uses obviously do not have to be treated to current standards, let alone to a zero level of pollution. Yet, a water utility has no choice because it distributes treated water without distinction as to use. As a result, water utilities who are already treating huge quantities of water far in excess of drinking water requirements would have to treat that same quantity to a much higher degree of purity.

An alternative would be the development of dual use systems. In such an approach, a separate drinking water supply would be made available to home owners and businesses, and a lesser quality water system would be provided for agriculture, landscaping, firefighting, and wastewater carriage. This approach would be a direct and much more cost-effective method for obtaining high-quality drinking water, as a much smaller volume would require advanced treatment technologies. The drawback to such a system, however, would be the cost of installing a dual distribution system. This separate system would require a massive expenditure of both federal and state funds to build new and parallel treatment plants and distribution systems. Clearly, privately owned water systems, of which there are many, could not afford such an undertaking without massive subsidies. Indeed, for the whole nation to adopt this approach would be an impossible task in real time and within current budgets. However, this approach can be effectively planned and implemented in new communities, industrial parks, and subdivisions. When these engineering and economic constraints are considered along with the need to address posttreatment pollution in the distribution system, many water utilities could not justify the production of a higher quality water to their consumers. Fortunately, there are other alternatives.

Professor Walter Weber of Michigan State University (Landers, 2000) has proposed the use of advanced water treatment technologies in a "satellite mode." These highly advanced treatment systems (i.e., coupling reverse osmosis with carbon filtration) would be used at the neighborhood level to improve the quality of the water coming from the central water treatment plant. The objective of using such a system would be to provide the highest quality water to a limited but specialized consumer base (e.g., housing subdivisions, apartment complexes, or commercial districts). This approach has merit because a satellite system installed at a "point of need" would require a much shorter length of "parallel" distribution systems. According to Weber, the use of satellite systems is necessary since ". . . potable water is of questionable quality . . . we're going to be facing the reality that water supply is, in fact, wastewater." Weber further points out that "although water treatment technologies continue to advance, they are too expensive to treat large quantities of water to potable[32] levels." Professor

---

[32]This comment is significant considering that Weber's statement does not even address water treatment to a zero pollution level.

Daniel A. Okun of the University of North Carolina has recommended duel water systems for more than 40 years, recognizing the advantages cited previously.

Another approach to reducing the volume of water requiring treatment to achieve minimum pollution water is to make even greater use of recycled water for those uses that do not require potable water. Although this approach does not affect the quality of our drinking water, it does reduce the volume of water that must be treated to a near zero pollution level. In the western United States, recycled wastewater has been used for decades for watering lawns, gardens, and golf courses.[33] In addition to these uses, a new trend toward the direct use of recycled wastewater for flushing toilets in commercial buildings has started. The Irvine Ranch Water District in Southern California began delivering recycled water to high-rise office buildings in Irvine's Jamboree Center complex in 1991.[34] More recently, the East Bay Municipal Utility District announced a similar use of recycled water in a 20-story office building in Oakland, California.[35] The building is fitted with a dual plumbing system to use treated wastewater for toilet flushing. Unlike the land application of wastewater, dual use systems in high-rise commercial buildings pose no threat to groundwater. Both of these approaches reduce the cost of producing high-quality water because smaller volumes actually require treatment.

Ultimately, numerous alternatives exist for removing chemical pollutants in our drinking water. In some cases, water systems are already approaching or are at minimum chemical pollution in their product, and all that needs to be done is to slightly increase the efficiency levels to further reduce pollutant levels. In other cases, major technical changes will have to be made to minimize the level of chemical pollutants. However, the basis of any policy must be predicated on the application of the best available technologies to attain a water product with minimum levels of chemical pollutants.

The water industry, like most other industries, is more often controlled and limited by economic rather than technical constraints. Concerns include meeting federal and state drinking water standards, delivering sufficient water to meet their costumer base, maintaining a satisfactory product quality level, and staying within a budget. If these concerns are met, there is usually no incentive to provide an even higher quality product to its customers unless the community served is willing to pay the bill for this enhanced purity and added protection.

---

[33]As discussed previously, the land application of wastewater always presents the possibility of groundwater pollution.
[34]Association of California Water Agencies—Water Facts, "Water Recycling" (Winter 2000 to 2001).
[35]*San Francisco Chronicle,* Tuesday, July 31, 2001.

Implementing advanced treatment technologies should add only 15 to 25 percent to a water utility's budget.[36] This situation is not an unlikely scenario. For example, as a result of the widespread pollution of drinking water by pesticides, some water utilities in the "Corn Belt" region of the United States have already upgraded water treatment facilities (Cohen and Wiles, 1997) at a significant cost to both the community and consumers.

With the implementation of the Stage 2 Disinfection Rule, those community water systems that exceed the new standards will have to reduce the amount of dissolved organic carbon in the raw water, switch to chloramine or to nonchlorine/bromine disinfection systems (e.g., use ultraviolet light or ozone, as practiced in Europe) or install advanced treatment technologies. In those cases where some degree of advanced treatment technologies are implemented to comply with the Stage 2 Rule, the cost of the additional improvements needed to reach minimum pollution in drinking water should be marginal. Thus, for some water utilities, the leap to providing a truly pollution-free drinking water to their customers may not be that great. Ultimately, are water utilities capable of implementing the necessary technologies to provide drinking water that is as close to pollution free as possible without state or federal funding support? The answer is yes, but what incentives are required to implement the currently available technology on a broad scale?

Another problem facing water utilities in their attempt to improve water quality through new treatment technologies is their aging distribution systems. Because of the potential threat of a terrorist polluting a water resource, community water systems will also need funding to address security issues. The federal government is currently spending approximately $3 billion a year to repair our water supply infrastructure. The American Water Works Association reported in April 2001 (Scharfenaker, 2001a) that an anticipated expenditure for drinking water infrastructure would cost approximately $151 billion over the next 20 years. Of this amount, approximately 25 percent of these funds would be used to upgrade treatment systems. To further confuse the cost projections, the American Water Works Association in July 2001 (Scharfenaker, 2001b) estimated drinking water infrastructure costs at $150.9 billion. Furthermore, the *Journal of the American Water Works Association* (Scharfenaker, 2001a) reported that the EPA projected expenditures of $250 billion over 30 years to upgrade tens of thousands of miles of aging drinking water system pipes, with no cost estimates for water treatment technology upgrades.

The American Water Works Association prepared an independent cost analysis that estimated a per capita expenditure of an additional

---

[36]Based on communications with private and municipal water utility managers in California.

$100 per customer[37] for each of the next 20 years. Based on this American Water Works Association assessment (Means et al., 2002), approximately 68 percent of this estimated expenditure, or $102.6 billion, is needed immediately. Finally, the USEPA has estimated that $1 trillion will be required between 2010 and 2020 to meet current infrastructure objectives (Federal Water Review, 2000). According to the 2000 Census, there are 105,480,101 households in the United States. Based on the USEPA estimate, the federal government would need to spend $9,480 per household over a 10-year period to improve infrastructure. A more recent publication[38] projects a $50 billion annual expenditure to "build, operate and maintain needed drinking water facilities over the next 20 years." The report projects an average per capita share (assuming a U.S. population of 285 million people) of $175/person/year. These costs do not even guarantee the consumer that chemical pollutants will be absent "at the tap." These high expenditure projections are required just to maintain the status quo. It should be emphasized that these cost projections have been developed and proposed by the water supply industry and the USEPA. They represent the combined wisdom of industry and government.

Given the difficulty water utilities face in (1) obtaining funds from federal and state governments, and (2) raising the water bill rates of its consumers, it will be a challenge to provide high quality drinking water "across the board." However, the water industry is capable of meeting this challenge. Because of the extensive investment in community water systems, it makes sense that these utilities should be the preferred and least costly distributor of drinking water.[39] To facilitate this transformation of returning the consumer to community-provided water, it is recommended that the water industry perform the following tasks:

- Educate local government and the communities they serve to the potential health risks and the solutions that can be adopted to meet community-specific requirements. Through education, communities can make the necessary economic choices. By knowing the alternatives, many communities may be willing to pay a higher water bill to reap the potential health benefits.
- Communities can also allow for graduated water bills so that the lower income families will not bear undue cost increases.
- Begin the development of long-term plans to rebuild water infrastructure into a dual use and/or satellite water supply system so that minimum pollution water can be delivered to customers at lower costs.

---

[37]Based on the AWWA model, this analysis only applies to consumers of the 20 large and medium-size systems in the study.

[38]Office of the Inspector General for Audit, Western Division, San Francisco, California, Report No. 2001-P-00013 (August 2001).

[39]To put the issue into dollars and cents, the typical cost of treated domestic water in California is about 50 cents per cubic meter, while bottled water in California costs, on average, $995 per cubic meter.

- Turning once again to the terrorist threat, it should be noted that it would be much more cost effective to design access security into new distribution pipelines than to attempt to retrofit existing distribution access. Such "designed in" security would significantly reduce the ability of terrorists to intentionally poison drinking water.
- Because maximum treatment will, in some cases, also remove natural minerals, the water industry should begin to evaluate methods of replacing minerals in treated drinking water (i.e., similar to the several bottled water products that mineralize their product after treatment).
- A TT-based water quality system will also require a new method of monitoring water treatment performance. Because many of the chemical pollutants that can occur in our water supplies cannot be realistically measured only by monitoring for specific chemicals or indicator chemicals, the water industry should be the leader in the development of fingerprinting for water resources both before and after treatment.

Of all the tasks just discussed, the development of fingerprinting methods is critical, because the vast number of organic compounds that can be dissolved in water is beyond the ability of chemical-specific monitoring (i.e., quantification of individual compounds) to adequately characterize a water resource. This means that nonspecific chemical fingerprinting methods should be used to characterize the complete range of organic compounds that can be found in a given water resource or wastewater.[40] Given this need, the only commercially available technology with adequate sensitivity is HPLC coupled with a mass spectrometer (MS) or HPLC/MS. For example, pesticide studies conducted by the U.S. Geological Survey (Lee and Strahan, 2003) showed that the estimated mean method detection limit (MDL) for all of the compounds and their degradation products ranged from 0.004 to 0.051 µg/L. A list of the MDLs for each pesticide is given in Table 5-1. An example of a total ion

TABLE 5-1
Method Detection Limits for Selected Pesticides Using HPLC/MS

Compound	Mean Concentration (µg/L)	Estimated Method Detection Limit (µg/L)
Acetochlor	0.027	0.021
Alachlor	0.021	0.019
Dimethenamid	0.024	0.018
Flufenacet	0.023	0.011
Metolachlor	0.023	0.004
Propachlor	0.022	0.008

---

[40]Fingerprinting methods can be used to determine if groups of compounds are effectively removed by water treatment processes.

chromatogram of a 1.0 µg/L standard in a buffered reagent-water sample is given in Figure 5-3. It should be noted that in this sample, the herbicide 2,4-D is used as an internal standard.

This chromatogram illustrates the ability for HPLC/MS to provide a chemical fingerprint. Even with a chemical fingerprint of a wastewater or drinking water source, however, it is important to spike samples before treatment with known chemical surrogates (i.e., like the 2,4-D internal standard). Selected chemical surrogates (i.e., compounds representative of various organic chemical classes) can help define the chemical characteristics of a fingerprint and serve as an indicator of treatment efficiency. HPLC is not the only monitoring technology that could be used for fingerprinting polluted water. Raman spectroscopy may also be appropriate for online monitoring applications.

Although there are no commercially available Raman spectroscopy units with the current sensitivity of HPLC, the technology has been used to characterized dissolved organic compounds in groundwater in the part per billion range. Raman spectroscopy provides chemical information about molecular vibrations that can be used for sample identification and quantization. The technique involves shining a laser, at a selected wavelength, on a sample and detecting the scattered light. The majority of the reflected light will be the same frequency as the laser; however, a very small amount of light interacts with the molecules of a compound so that the original wavelength emitted by the laser is shifted. This shifted light produces a Raman spectra or chemical fingerprint. For example, studies on monitoring cyanide in wastewater using surface-enhanced Raman spectroscopy found that cyanide in water could be detected down to 10 µg/L (Premasiri et al., 2001). An example of the cyanide spectra is given in Figure 5-4.

In addition to all of these traditional actions, water utilities can also provide key services to residences/businesses relative to point-of-entry water treatment systems. The water utility could offer to install (for a fee) a standardized point-of-use-system to those persons who would like to have such a system. Furthermore, for all those individuals who have point-of-use systems in the region of service, the water utility could provide both routine maintenance and monitoring services for a monthly or yearly fee to ensure that these systems are functioning properly.

Clearly, the water industry should lead the way toward a TT-based water quality program, as the most feasible solutions are grounded in the communities that they serve. However, federal and state governments can also provide valuable actions.

Potential Governmental Actions

Federal and state governments should not abandon the use of water quality criteria to protect the beneficial uses of water (i.e., aquatic environment, agriculture, industry, and recreation) and to control the discharge

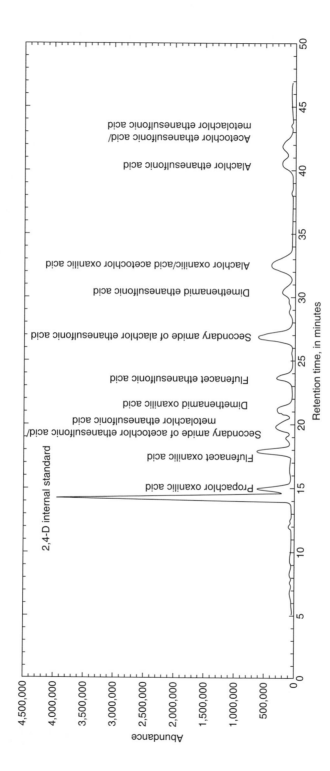

FIGURE 5-3. High-performance liquid chromatography.

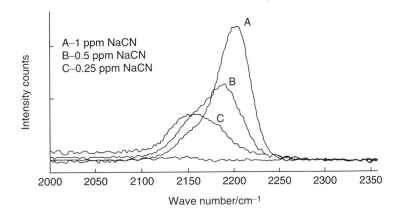

FIGURE 5-4. Raman Spectra (sodium cyanide).

of pollutants to the environment. It is critical, however, that federal and state governments abandon the concept of the drinking water standard as the means by which to protect the public health. The only true and complete protection is to have drinking water based on TT-based standards of performance.

For the American public to have access to water that is as free of pollution as possible, the federal government must make a fundamental change in the approach by which "quality water" is provided to the consumer. Some of these changes will take more time than others. These should include (1) support to water utilities in their effort to provide higher quality water, (2) assist in the upgrading of system infrastructure, (3) support for small in-home and workplace water treatment systems, (4) require the bottled water and beverage industry to improve their guarantees of water purity, and (5) implement appropriate regulatory controls that are supportive of TT-based drinking water quality.

Support for Community Water Systems

The federal government needs to ensure that community water systems that serve transient populations (e.g., national parks, rest areas along interstate highways, federal lands, and reservations) are upgraded to provide the same quality drinking water as currently required of regulated water systems. In some cases, the federal government may need to provide funding to community water systems that are unable to upgrade their facilities via rate increases. In addition to monetary support, the federal government should assist the water industry in developing chemical monitoring methods to fingerprint chemical pollution both before and after treatment to ensure that water treatment systems are functioning properly. Such procedures would drastically reduce monitoring costs and simplify regulatory oversight and related governmental costs.

Water Infrastructure

As discussed previously, the federal government needs to assist local communities with the planning and funding of dual and satellite water systems to replace old distribution systems as they are repaired and/or replaced. This involvement will obviously be long term. As a result, the USEPA should be mandated to begin long-term program planning, fund research grants to universities to evaluate engineering alternatives, complete cost-benefit analyses, conduct material evaluations of materials to be used in distribution systems to ensure that they are "nonpollution contributors," implement pilot scale projects for dual and satellite systems, develop and integrate *in situ* methods of chemical monitoring for real-time treatment and control systems, and evaluate the effectiveness of treatment systems to remove specific chemicals and chemical compound classes.

Support for Small In-Home and Workplace Water Treatment Systems

Until the water infrastructure in the United States can be fully, or even partially, replaced or community systems voluntarily upgraded, the federal[41] government should sponsor programs to promote the use of in-home and workplace water treatment systems as required. This step would be one of the quickest ways to help the American public attain drinking water that contains the minimum concentration of chemical pollution, until community-based upgrades are installed. Such help is particularly important for that portion of the population relying on private sources of drinking water. Indeed, the federal government has already approved the use of point-of-use systems to meet the new arsenic standard.

An important part of this solution should be the provision of low interest government loans and a tax credit program for the purchase of small water treatment systems. This approach has been practiced for many years with regard to home owner efforts to reduce energy consumption. Tax credit and rebate programs already exist for insulation, energy saving utilities, and solar heating. Federal and state governments should also begin to provide small treatment systems for public buildings, medical facilities, and federally sponsored low-income housing.

---

[41]Although a federal program would be preferred, state or local governments could implement the same program.

Installation of such systems in key federal buildings that might be the target of a terrorist attack is particularly important.

At a minimum, federal or state governments should offer support to those households and business that must rely on private sources of drinking water that may be of questionable quality. Support should also be extended to those individuals who obtain their drinking water from community water systems that do not have a pristine source water and have not been upgraded to remove the maximum amount of chemical pollutants. The response to manufactured chemicals should be no different than the government's approach to point-of-use systems for arsenic removal.

Federal and state governments could also be more aggressive in safeguarding individuals who consume water currently not protected by best available treatment technologies by requiring that (1) all new home and building construction in these unprotected areas incorporate a water treatment system that removes the maximum concentration of manfactured chemical pollutants, (2) a minimum pollution drinking water treatment system be installed whenever a private residence or building is remodeled (i.e., to the degree that a permit is required), and (3) a minimum pollution drinking water treatment system be installed whenever a private residence or building is sold before the sale. This type of program is consistent with other building code requirements that are usually implemented at the city or county level.

This approach is certainly not new, and there are many examples in which building codes have over time achieved the desired objectives. For example, this method has already been applied to smoke detectors, sprinkler systems, electrical wiring and plumbing, low-flow toilets, energy-saving utilities, reduced window areas, exterior wall insulation, seismic strengthening, flood plain zoning, and even the limitation in new construction of wood burning fireplaces. Providing minimum pollution drinking water through the mechanism of building codes will not guarantee a speedy implementation, but it will certainly move the nation more rapidly to a minimum pollution objective.

A program of this scope also requires the federal government to provide greater control of the testing, certification, and availability of water purity data from standardized small treatment systems. This level of control is currently not available, although the National Sanitation Foundation does test and evaluate individual components of treatment systems. The USEPA should also fund university research on small system treatment design that will optimize chemical and biological water purity, while assuring easy maintenance and low operating costs.

State and federally sponsored programs that assist unprotected consumers in acquiring in-home/workplace treatment systems will necessarily drive down the unit cost of these systems simply through mass production. Such capital and maintenance cost reduction will benefit all consumers and will further enable the public to install affordable water treatment technologies.

Bottled and Beverage Water

The federal government should require that all bottled water be properly labeled to reflect the environmental characteristics of the product, not the "real" and sometimes "invented" sources of bottled water. Current labels such as "artesian, spring, or mineral water" are environmentally meaningless. The following types of bottled water are possible:[42]

- Pollutant-free water: Water treated to remove all solids and soluble manufactured organic compounds to as close to zero as possible using the best available technology. This product should be certified "pure" using chemical fingerprinting methods. No such product is currently available.
- Treated water: Water from community water systems or water from a community water system that has received additional treatment but not certified as a pollutant-free water.
- Naturally pure water: Bottled water that is verified to have originated from either a natural spring that occurs in mountainous regions of the world with minimal human impact (i.e., wilderness and national park designations or high alpine regions untouched by mining or towns) or from verified protected deep aquifers whose water source originates from a wilderness area, or from deep thermal sources that have groundwater flow transport times in hundreds of years.
- Potable water: Bottled water that meets current drinking water quality criteria with no guarantee that the product does not contain manufactured pollutants.

Such a classification will help keep the consumer from making uninformed decisions.

Proposed New Regulatory Programs

With the implementation of a TT-based water quality program, regulatory monitoring and notice programs can be greatly simplified, as individual chemical components in drinking water will no longer be an issue. Monitoring compliance can rely on chemical finger printing methods to assess treatment performance. This approach will not only reduce the need for costly governmental oversight, but will also reduce monitoring costs for community water systems. As a result, regulatory agencies will need to modify compliance monitoring programs to augment and support TT-based performance criteria.

A TT-based treatment program is expected to generate a somewhat greater quantity of wastes resulting from the removal of greater amounts

---

[42]All source water would also be purified (e.g., UV light treatment).

of chemicals. It should be both technically and economically feasible for most communities to dispose of these wastes back to the untreated water source.[43] This disposal method is especially true for those areas that meet existing beneficial use water quality standards for agriculture, industrial, recreation, and wildlife uses. The allowance of such discharges should be part of any TT-based drinking water quality program.

An Alternative Risk Management Program

Today our drinking water is not "certifiably free" from manufactured chemical pollutants and is not protected from either further chemical pollution or possible terrorist acts (i.e., contamination). Protection of drinking water cannot be assured under existing programs, as they are founded on outdated methods that lack scientific credibility. Water resources can be protected to a much greater degree if federal and state governments considered the following:

- Do not rely exclusively on drinking water standards to protect public health.
- Require the use of the best available treatment technologies to remove chemical pollutants from drinking water that does not originate from pristine water resources.
- Implement programs that shift the primary responsibility of attaining maximum quality drinking water to community water systems, with the cost of upgrading water treatment passed on to the consumer.
- When appropriate, provide economic assistance to community water systems to implement the best available water technologies.
- Implement programs that provide maximum quality drinking water to areas of the country that do not have access to quality sources of water. This goal could be accomplished by requiring that (1) all new building construction include a point-of-entry treatment system for drinking water, (2) a point-of-entry water treatment system be installed whenever a private residence or building is remodeled (i.e., to the degree that a permit is required), and (3) a point-of-entry water treatment system be installed before a private residence or building is sold.
- Provide the impetus for reduction in the cost of point-of-use and point-of-entry systems by initiating a 5-year program to install these systems in federal facilities, buildings, and federally sponsored housing projects that currently do not receive the highest quality drinking water.

---

[43]In those areas where the removal of solids will not have a material influence on the beneficial use classification of the source water, the solids should be allowed to be discharged back into the source water.

- Implement long-term programs to assist local communities with planning and funding of dual and satellite water systems as an integral part of a program to repair and replace aging distribution systems. The dual use and satellite systems must be designed to provide pollution-free water and be protected from terrorist incursion.
- Require that bottled water and beverages that use water as an additive be properly labeled to accurately reflect the environmental characteristics of their contents.

To a large extent, the degree to which our government agencies implement these policies will depend on how much the public cares. Protection of drinking water from all forms of pollution must be a public goal given the potential hazards to the public health. For example:

- We know that today we are all consuming water containing both regulated and unregulated pollutants.
- We know that present government programs cannot guarantee pollution-free water.
- We know that water supplies are not totally safe from terrorist actions.
- We know that the nation's entire water treatment and distribution system needs major overall, upgrading, and replacement.
- We know that projections for system upgrade costs range upwards of a trillion dollars.
- We know that there will never be a comprehensive drinking water monitoring program that adequately addresses all unregulated pollutants.

Unfortunately, the general public may not be as informed or appreciative of these issues as those of us in the scientific, engineering, and regulatory community. Fortunately, for each one of these hazards there are defined and achievable solutions:

- We know that the technology exists to enable community water systems to produce and distribute a much higher quality water to their customers.
- We know that reliable point-of-entry and point-of-use home water purification systems are available commercially.
- We know that dual-use systems have been implemented and have the capability of delivering high-quality water at a lower cost.
- We know that the economics of furnishing high-quality water to the majority of our population are within the boundaries of reality.
- We know that community drinking water systems can be designed to eliminate or significantly reduce the threat of terrorist acts.

Finally, we know that an enlightened and informed public can make decisions that are in their best interest. All that remains is that these potential actions be considered and debated by consumers, water utilities, and governmental agencies.

References

American Water Works Association, 1953, "Findings and Recommendations of Underground Waste Disposal: Task Group Report," *Journal of the American Water Works Association*, Vol. 45, No. 19, pp. 1295–1296.

Lomborg, B., 2001, *The Skeptical Environmentalist*, Cambridge University Press, Cambridge, United Kingdom.

Brown, Michael, H., 1987, *The Toxic Cloud, The Poisoning of America's Air*, Harper & Row, New York.

Brown, W., 1935, "Industrial Pollution of Ground Waters," *Water Works Engineering*, Vol. 88, No. 4, pp. 171–177.

Cohen, Brian and Richard Wiles, 1997 "Tough to Swallow, How Pesticide Companies Profit from Poisoning America's Tap Water," *Environmental Working Group*.

Federal Water Review, 2000, "WIN Calls For Federal Infrastructure Assistance," Association of Metropolitan Water Agencies, Washington, D.C.

Hrudey, Steve E. and Elizabeth J. Hrudey, 2004, *Safe Drinking Water, Lessons from Recent Outbreaks in Affluent Nations*, IWA Publishing, London.

Kraft, R., 1927, "Locating the Chemical Plant," *Chemical and Metallurgical Engineering*, Vol. 34, No. 11, pp. 678–679.

Landers, Jay, 2000, "Treatment Changes Needed to Ensure Sustainable Water Supply," *Water Environment & Technology*, Vol. 12, pp. 12–13.

Lee, E.A. and A.P. Strahan, 2003, "Methods of Analysis by the U.S. Geological Survey Organic Geochemistry Research Group—Determination of Acetamide Herbicides and Their Degradation Products in Water Using Online Solid-phase Extraction and Liquid Chromatography/Mass Spectrometry," Open-File Report 03-173.

Means, Edward G. et al., 2002, "The Coming Crisis: Water Institutions and Infrastructure," *Journal of the American Water Works Association*, Vol. 94, pp. 34–38.

Muehlberger, C., 1950, "Possible Hazards form Chemical Contamination in Water Supplies," *Journal of the American Water Works Association*, Vol. 42, No. 11, pp. 1027–1034.

Pickett, A., 1947, "Protection of Underground Water from Sewage and Industrial Wastes," *Sewage Works Journal*, Vol. 19, No. 3, pp. 464–472.

Pontius, Frederick W., 2002, "Regulatory Compliance Planning to Ensure Water Supply Safety," *Journal of the American Water Works Association*, Vol. 94, pp. 12–14.

Premasiri, W.R., R.H. Clark, S. Londhe and M.E. Wombie, 2001, "Determination of Cyanide in Waste Water by Low-Resolution Surface Enhanced Raman Spectroscopy on Sol-Gel Substrates," *Journal of Raman Spectroscopy*, Vol. 32, pp. 919–922.

Scharfenaker, Mark, 2001a, "Second National Need Survey Pegs Drinking Water Infrastructure Costs at $150.9 Billion," *Journal of the American Water Works Association*, Vol. 93, pp. 26–39.

Scharfenaker, Mark, 2001b, "Mythic Monster Aids First Scientific Assessment of Nationwide Water Infrastructure Needs," Discussion of the AWWA Report "Dawn of the Replacement Era: Reinvesting in Drinking Water Infrastructure," *Journal of the American Water Works Association*, Vol. 93.

Scharfenaker, Mark A., 2002, Reg Watch, Water Suppliers Carefully Watching Liability Suits, *Journal of the American Water Works Association*, Vol. 94, pp. 28–40.

U.S. Environmental Protection Agency, 1999, Groundwater Cleanup: Overview of Operating Experience at 28 Sites, EPA 542-R-99-006, September 1999.

Wilson, H. and H. Calvert, 1913, *Trade Waste Waters: Their Nature and Disposal*, Lippincott, Philadelphia.

USEPA, 2000, "Liquid Assets 2000: American's Water Resources: A Turning Point," Office of Water, Washington, D.C., EPA-840-B00-001 (May, 2000).

USEPA, 2001a, "Technical Fact Sheet: Final Rule for Arsenic in Drinking Water," EPA 815-F-00-019.

Gleick, Peter H. et al., *The World's Water—The Biennial Report on Freshwater Resources*, 2004–2005, Island Press, Washington, D.C.

APPENDIX 1-1

Average Elemental Abundance in the Earth's Crust

Element	Concentration (mg)[1]	Percentage	Total (%)
Oxygen	464,000	46.16	
Silica	282,000	28.05	
Aluminum	82,000	8.16	
Iron	56,000	5.57	
Calcium	41,000	4.08	
Sodium	24,000	2.39	
Magnesium	23,000	2.39	
Potassium	21,000	2.01	
			98.81 (top 8 elements)
Titanium	5,700	0.56	
Hydrogen	1,400	0.14	
Phosphorous	1,050	0.10	
Manganese	950	0.095	
Fluorine	625	0.062	
Barium	425	0.042	
Strontium	375	0.037	
Sulfur	260	0.025	
Carbon	200	0.020	
Zirconium	165	0.016	
Vanadium	135	0.013	
Chlorine	130	0.012	
			99.93 (top 20 elements)
Chromium	100		
Rubidium	90		
Nickel	75		
Zinc	70		
Cerium	67		

(continued)

[1]Average elemental concentration in 1,000 grams of rock.

Element	Concentration (mg)[1]	Percentage	Total (%)
Copper	55		
Yttrium	33		
Neodymium	28		
Lanthanum	25		
Cobalt	25		
Scandium	22		
Lithium	20		
Nitrogen	20		
Niobium	20		
Gallium	15		
Lead	12.5		
Boron	10		
Thorium	9.6		
Samarium	7.3		
Gadolinium	7.3		
Praseodymium	6.5		
Dysprosium	5.2		
Ytterbium	3		
Hafnium	3		
Cesium	3		
Beryllium	2.8		
Erbium	2.8		
Uranium	2.7		
Bromine	2.5		
Tin	2		
Arsenic	1.8		
Germanium	1.5		
Molybdenum	1.5		
Tungsten	1.5		
Holmium	1.5		
Europium	1.2		
Terbium	1.1		
Lutetium	0.8		
Iodine	0.5		
Thallium	0.45		
Thulium	0.25		
Cadmium	0.2		
Antimony	0.2		
Bismuth	0.17		
Indium	0.1		
Mercury	0.08		
Silver	0.07		
Selenium	0.05		
Gold	<0.05		
Platinum	<0.05		
Tellurium	<0.05		
Rhenium	<0.05		

Adapted from Krauskopf, 1967.

APPENDIX 1-2

Chemical Compounds with Established Water Quality Criteria—1952

Abietic acid
Acetates
Acetic acid
Acetone
Acetonitrile
Acetylene
Acids (hydrochloric, nitric sulfuric, phosphoric, acetic, propionic, lactic, benzoic, boric, chromic, cresylic, cyclohexanecarboxylic, formic, oleic, gallic, humic, hydrofluoric, linoleic, naphthalic, salicylic, selenious, oxalic, picric, sulfurous, tannic, tartaric)
Acridine
Albumin,
Alcohols (amyl, ethyl, methyl, butyl, octyl, phytosterol)
Alkalinity (sodium hydroxide, ammonium hydroxide, sodium carbonate)
Alum (aluminum sulfate, aluminum ammonium sulfate)
Aluminum chloride, nitrate, oxide and potassium sulfate
Ammonia
Ammonium carbonate, chloride, ferrocyanide, nitrate, picrate, sulfate, sulfide, thiocyanate
Aniline
Antimonyl potassium tartrate
Antimony trioxide, trichloride
Arsenic
Arsenic trioxide
Barium
Barium chloride, nitrate
Benoclor
Benzaldehyde
Benzene
Benzene hexachloride
Beryllium

Boron
Bromine
Cadmium
Cadmium chloride, nitrate, sulfate
Calcium carbonate, chloride, hydroxide, hypochlorite, nitrate, sulfate
Camphor
Carbonates
Carbon disulfide
Chemopodium oil
Chloramines
Chloradane
Chlorides
Chlorinated hydrocarbons
Chlorine
Chloroform
Chlorophenols
Chromium
Chromates
Cloroben
Cobalt
Cobalt chloride, nitrate
Coniline
Copper
Copper ferrous sulfate, ammonium chloride, chloride, nitrate, sulfate
Creatine
Cresols
Cyanides
Cyclohexane and cyclohexene
Dichlorobenzene
DDT
Dinitrophenol
Ethylamine
Ethylene
Ferric chloride, oxide, potassium sulfate, sulfate, cyanides, carbonate
Ferrous chloride, sulfate
Fluorides
Formaldehyde
Gammexane
Glycerin
Glycerol
Guaiacol
Heptane
Hydrazine
Hydrocarbons
Hydrogen sulfide
Insecticides
Iodine
Lead
Lead acetate, arsenate, chloride, nitrate, sulfate
Lithium
Lithium chloride

Lindane
Magnesium
Magnesium acetate, bicarbonate, chloride, nitrate, oxide, sulfate
Manganese
Manganese chloride, nitrate
Mercaptans
Mercuric chloride
Mercuro-organics
Mercury
Metaphosphates
Methane
Methanethiol
Molybdenum
Naphtha
Naphthalene
Nickel
Nickel acetate, chloride, nitrate
Nicotine
Nigrosine
Nitramine
Nitrates
Pyridine
Pyrophosphates
Quinine
Quinoline
Radioactivity
Radium
Resins
Ricin
Rotenone
Selenium
Silver
Soaps
Sodium
Sodium arsenate, arsenite, benzoate, bisulfate, borate, bromate, bromide,
    carbonate, chlorate, chloride, cholate, chromate, citrate, cyanate, cyanide,
    dichromate, ferrocyanide, fluoride, formate, hydroxide, iodate, iodide, nitrate,
    nitrile, nitroferricyanide, oxalate, perborate, phosphate, selenite, sulfate,
    sulfite, tetraborate, thiocyanate, thiosulfate
Stannic and stannous salts
Strontium
Strontium chloride, nitrate
Strychnine
Sulfates
Oil, petroleum
Oxalates
Paradichlorobenzene
Pentachlorophenol
Pentachlorophenates
Pentane
Pentene

Pentone
Petroleum benzine
Phenanthrene
Phenol
Phosphates
Phosphorus
Phytosterol
Picrates
Picrotoxin
Potassium
Potassium bicarbonate, carbonate, chloride, chromate, cyanate, cyanide, dichro-
    mate, ferricyanide, ferrocyanide, hydroxide, iodide, nitrate, permanganate,
    sulfate, sulfide, thiocyanate, xanthogenate
Sulfides
Sulfites
Sulfur
Sulfur dioxide
Synthetic detergents
Tars
Tartrates
Tetryl
Thallium
Thiophene
Thiophenol
Tin
Toluene
Toxaphene
Trinitrophenol
Trinitrotoluene
Tryptophane
Tungsten
Turpentine
Uranium
Urea
Vanadium
Xylenes
Zinc
Zinc acetate, chloride, nitrate, oxide, sulfate

APPENDIX 1-3

USEPA National Recommended Water Quality Criteria for Freshwater and Human Consumption of Water + Organism: 2002

Pollutant	Criteria Maximum Concentration (μg/L)	Human Consumption Water + Organism (μg/L)
Antimony		5.6
Arsenic	340	0.018
Cadmium	2.0	
Chromium (III)	570	
Chromium (VI)	16	
Copper	13	
Lead	65	
Mercury	1.4	
Nickel	470	610
Selenium		170
Silver	3.2	
Thallium		1.7
Zinc	120	7,400
Cyanide	22	700
TCDD		0.000000005
Acrolein		190
Acrylonitrile		0.051
Benzene		2.2
Bromoform	4.3	
Carbon tetrachloride		0.23
Chlorobenzene	680	
Chlorodibromomethane	0.40	
Chloroform		5.7
Dichlorobromomethane	0.55	

(continued)

Pollutant	Criteria Maximum Concentration (µg/L)	Human Consumption Water + Organism (µg/L)
1,2-Dichloroethane		0.38
1,1-Dichloroethylene		0.057
1,2-Dichloropropane		0.50
1,3-Dichloropropene		10
Ethylbenzene	3,100	
Methyl bromide		47
Methylene chloride		4.6
1,1,2,2-Tetrachloroethane		0.17
Tetrachloroethylene		0.69
Toluene		6,800
1,2-Trans-dichloroethylene		700
1,1,2-Trichloroethane		0.59
Trichloroethylene		2.5
Vinyl chloride		2.0
2-Chlorophenol		81
2,4-Dichlorophenol		77
2,4-Dimethylphenol		380
2-Methyl-4,6-dinitrophenol		13
2,4-Dinitrophenol		69
Pentachlorophenol	19	0.27
Phenol	21,000	
2,4,6-Trichlorophenol		1.4
Acenaphthene	670	
Anthracene		8,300
Benzidine		0.000086
Benzo(a)anthracene		0.0038
Benzo(a)pyrene		0.0038
Benzo(b)fluoranthene	0.0038	
Benzo(k)fluoranthene	0.0038	
Bis(2-chloroethyl)ether		0.030
Bis(2-chloroisopropyl)ether		1,400
Bis(2-Ethylhexyl)phthalate		1.2
Butylbenzyl phthalate	1,500	
2-Chloronaphthalene		1,000
Chrysene		0.0038
Dibenzo(a,h)anthracene		0.0038
1,2-Dichlorobenzene		2,700
1,3-Dichlorobenzene		320
1,4-Dichlorobenzene		400
3,3'-Dichlorobenzidine		0.021
Diethyl phthalate		17,000
Dimethyl phthalate		270,000
Di-n-butyl phthalate		2,000
1,2-Diphenylhydrazine		0.036
Fluoranthene	130	
Fluorene	1,100	
Hexachlorobenzene		0.00028
Hexachlorobutadiene		0.44

Pollutant	Criteria Maximum Concentration (µg/L)	Human Consumption Water + Organism (µg/L)
Hexachlorocyclopentadiene		240
Hexachloroethane	1.4	
Ideno(1,2,3-cd)pyrene	0.0038	
Isophorone	35	
Nitrobenzene		17
N-Nitrosodimethylamine		0.00069
N-Nitroso-din-propylamine	0.0050	
N-Nitrosodiphenylamine		3.3
Pyrene		830
1,2,4-Trichlorobenzene		260
Aldrin	3.0	0.000049
alpha-BHC		0.0026
beta-BHC		0.0091
gamma-BHC (Lindane)	0.95	0.019
Chlordane	2.4	0.00080
4,4'-DDT	1.1	0.00022
4,4'-DDE		0.00022
4,4'-DDD		0.00031
Dieldrin	0.24	0.000052
alpha-Endosulfan	0.22	62
beta-Endosulfan	0.22	62
Endosulfan sulfate		62
Endrin	0.086	0.76
Endrin aldehyde		0.29
Heptachlor	0.52	0.000079
Heptachlor epoxide	0.52	0.000039
Polychlorinated biphenyls		0.000064
Toxaphene	0.73	0.00028
Aluminum	750	
Chloride	860,000	
Chlorine	19	
Chloropyrifos	0.083	
Ether, Bis(chloromethyl)		0.00010
Hexachlorocyclo-hexane		0.0123
Manganese		50
Methoxychlor		100
Nitrates		10,000
Parathion	0.065	
1,2,4,5-Tetarachlorobenzene		0.97
Tributyltin	0.46	
2,4,5-Trichlorophenol		1,800

APPENDIX 2-1

Dow Industrial Chemicals, Solvents and Dyes in 1938

Acetanilid
Acetic anhydride
Acetylene tetrabromide
Aniline oil
Anthranilic acid
Barium bromide
Benzoyl chloride
Bis phenol-A
Bromine
Bromoform
Cadmium bromide
Carbon bisulfide
Carbon tetrachloride
Caustic soda
Ciba blue
Ciba red
Ciba scarlet
Ciba violet
Chloracetyl chloride
Chloroform
Dichloracetic acid
Diethylaniline
Diethyl benzene
Diethylene glycol
Dimethylaniline

Diphenyl
Diphenyloxide
Dowanone blue
Dowanone yellow
Ethyl benzene
Ethyl bromide
Ethyl chloride
Ethylene chlorobromide
Ethylene dibromide
Ethylene dichloride
Ethylene glycol
Ethylene oxide
Ethyl monobromacetate
Ethyl monochloracetate
Ferric chloride
hexachlorethane
Hydrabromic acid
Indigo
Isopropyl benzene
Magnesium bromide
Methyl monochloracetate
Midland vat blue
Monobrombenzene
Monochloracetic acid
Monochlorobenzene

Sources: Dow Industrial Chemicals and Dyes, The Dow Chemical Company, Midland, Michigan, 1938, and Dow Industrial Solvents, The Dow Chemical Company, Midland, Michigan, 1938.

Muriatic acid
Orthodichlorbenzene
Orthachlor paranitraniline
Orthochlorphenol
Paradibrombenzene
Paradow
Paraphenetidin
Paraphenylphenol
Para tertiary butyl phenol
Perchlorethylene
Propylene dichloride
Phenol
Phenyl acetate
Phenyl glycine
Phenyl hydrazine
Phenyl methyl pyrazolone
Phthelimide

Potassium bromide
Salicylaldehyde
Salicylic acid
Sodium acetate
Sodium bromate
Sodium hydrosulfide
Sodium sulfide
Sulfur chloride
Sulfur monochloride
1,1,2,2-Tetrachlorethane
Trichlorbezene
1,1,2-Trichlorethane
Trichlorethylene
Trimethylene bromide
Triphenyl phosphate
Zinc bromide

APPENDIX 2-2

USEPA List of Priority Pollutants

Chemical Compound	Compound Type[3]
Acenapthene	Base-Neutral Extractable
Acenapthylene	Base-Neutral Extractable
Acrolein	Volatiles
Acrylonitrile	Volatiles
Aldrin	Pesticides
Anthracene	Base-Neutral Extractable
Antimony	Inorganics
Arsenic	Inorganics
Asbestos	Inorganics
Beryllium	Inorganics
Benzene	Volatiles
Benzidine	Base-Neutral Extractable
Benzo(a)anthracene	Base-Neutral Extractable
Benzo(b)fluoranthene	Base-Neutral Extractable
Benzo(k)fluoranthene	Base-Neutral Extractable
Benzo(ghi)perylene	Base-Neutral Extractable
Benzo(a)pyrene	Base-Neutral Extractable
alpha-BHC	Pesticides
beta-BHC	Pesticides
delta-BHC	Pesticides
gamma-BHC (Lindane)	Pesticides
Bis(2-chloroethoxy)methane	Base-Neutral Extractable
Bis(2-chloroethyl)ether	Base-Neutral Extractable
Bis(2-chloroisopropyl)ether	Base-Neutral Extractable
Bis(2-ethylhexyl)phthalate	Base-Neutral Extractable

(continued)

[3]The compound type relates to application of appropriate laboratory analytical methods.

Chemical Compound	Compound Type
Bis(chloromethyl)ether	Volatiles
Bromodichloromethane	Volatiles
Bromoform	Volatiles
Bromomethane	Volatiles
4-Bromophyenyl phenyl ether	Base-Neutral Extractable
Butyl benzyl phthalate	Base-Neutral Extractable
Cadmium	Inorganics
Carbon tetrachloride	Volatiles
Chlordane	Pesticides
Chlorobenzene	Volatiles
Chloroethane	Volatiles
2-Chloroethyl vinyl ether	Volatiles
Chloroform	Volatiles
p-Chloro-m-cresol	Acid Extractable
Chloromethane	Volatiles
2-Chloronapthalene	Base-Neutral Extractable
2-Chlorophenol	Acid Extractable
4-Chlorophenyl phenyl ether	Base-Neutral Extractable
Chromium	Inorganics
Chrysene	Base-Neutral Extractable
Copper	Inorganics
Cyanide	Inorganics
4,4'-DDD	Pesticides
4,4'-DDE	Pesticides
4,4'-DDT	Pesticides
Dibenzo(a,b)anthracene	Base-Neutral Extractable
Dibromochloromethane	Volatiles
Di-n-butyl phthalate	Base-Neutral Extractable
1,2-Dichlorobenzene	Base-Neutral Extractable
1,3-Dichlorobenzene	Base-Neutral Extractable
1,4-Dichlorobenzene	Base-Neutral Extractable
3,3'-Dichlorobenzidine	Base-Neutral Extractable
Dichlorodifluoromethane	Volatiles
1,1-Dichloroethane	Volatiles
1,2-Dichloroethane	Volatiles
1,1-Dichloroethylene	Volatiles
trans-1,2-Dichloroethylene	Volatiles
2,4-Dichlorophenol	Acid Extractable
1,2-Dichloropropane	Volatiles
cis-1,3-Dichloropropene	Volatiles
trans-1,3-Dichloropropene	Volatiles
Dieldrin	Pesticides
Diethyl phthalate	Base-Neutral Extractable
Dimethyl phthalate	Base-Neutral Extractable
2,4-Dimethylphenol	Acid Extractable
4,6-Dinitro-o-cresol	Acid Extractable
2,4-Dinitrophenol	Acid Extractable
2,4-Dinitrotoluene	Base-Neutral Extractable
2,6-Dinitrotoluene	Base-Neutral Extractable

Chemical Compound	Compound Type
Di-n-octyl phthalate	Base-Neutral Extractable
alpha Endosulfan	Pesticides
beta Endosulfan	Pesticides
Endosulfan sulfate	Pesticides
Endrin	Pesticides
Endrin aldehyde	Pesticides
Ethylbenzene	Volatiles
Fluoranthene	Base-Neutral Extractable
Fluorene	Base-Neutral Extractable
Heptachlor	Pesticides
Heptachlor epoxide	Pesticides
Hexachlorobenzene	Base-Neutral Extractable
Hexachlorobutadiene	Base-Neutral Extractable
Hexachlorocyclopentadiene	Base-Neutral Extractable
Hexachloroethane	Base-Neutral Extractable
Indeno(1,2,3-c,d)pyrene	Base-Neutral Extractable
Isophorone	Base-Neutral Extractable
Mercury	Inorganics
Methylene chloride	Volatiles
Napthalene	Base-Neutral Extractable
Nickel	Inorganics
Nitrobenzene	Base-Neutral Extractable
2-Nitrophenol	Acid Extractable
4-Nitrophenol	Acid Extractable
N-Nitrosodimethylamine	Base-Neutral Extractable
N-Nitrosodiphenylamine	Base-Neutral Extractable
N-Nitroso-din-propylamine	Base-Neutral Extractable
Pentachlorophenol	Acid Extractable
Phenathrene	Base-Neutral Extractable
Phenol	Acid Extractable
Polychlorinated biphenyl's (PCB)	Base-Neutral Extractable
Pyrene	Base-Neutral Extractable
Selenium	Inorganics
Silver	Inorganics
1,1,2,2-Tetrachloroethane	Volatiles
1,1,2,2-Tetrachloroethene	Volatiles
Thallium	Inorganics
Toluene	Volatiles
Toxaphene	Base-Neutral Extractable
1,2,4-Trichlorobenzene	Volatiles
2,3,7,8-Tetrachlorodibenzo-p-dioxin	Base-Neutral Extractable
1,1,1-Trichloroethane	Volatiles
1,1,2-Trichloroethane	Volatiles
Trichloroethylene	Volatiles
Trichlorofluoromethane	Volatiles
2,4,6-Trichlorophenol	Acid Extractable
Total phenols	Acid Extractable
Vinyl chloride	Volatiles
Zinc	Inorganics

APPENDIX 2-3

Summary of Surface Water Data

The following data represent the frequency (as a percentage) that a specific pesticide was detected in a water sample.

Compound	Basin Land Use		
	Agricultural	Urban	Mixed
Acetochlor			9.80
Acifluorfen	0.43		2.74
Alachlor	36.36	13.46	39.02
Aldicarb	0.32		
Aldicarb sulfoxide	0.11		
Atrazine	77.20	86.24	88.62
Atrazine, deethyl	52.80	50.46	62.20
Azinphos-methyl	2.61	0.93	1.22
Benfluralin	0.50	3.36	1.63
Bentazon	4.58	1.29	8.68
Bromacil	0.74	0.63	
Bromoxynil	0.64		
Butylate	7.70	1.83	8.16
Carbaryl	10.69	45.26	21.14
Carbofuran	11.99	2.75	9.35
Chlorothalonil	0.32		
Chlorpyrifos	15.60	40.67	19.59
Cyanazine	27.67	8.26	45.93
2,4-D	11.62	13.50	9.13
Dacthal mono-acid	0.43		
2,4-DB	0.32	0.32	
DCPA	22.20	29.36	30.20
DDE	6.30	1.22	4.90
Diazinon	16.90	74.85	45.31
Dicamba	0.53	0.32	1.83
Dichlobenil	0.11	1.59	0.90

(continued)

Compound	Basin Land Use		
	Agricultural	Urban	Mixed
Dichlorprop	0.43	0.96	0.46
Dieldrin	6.90	3.67	3.27
2,4-Diethylaniline	4.50	0.61	4.90
Dinoseb	0.11		
Disulfoton	0.50	0.92	
Diuron	7.96	13.02	9.46
DNOC		0.32	
EPTC	25.13	3.98	22.86
Ethalfluralin	3.30		0.41
Ethoprop	3.40	1.83	3.25
Fenuron	0.11		0.45
Fluometuron	2.86		
Fonofos	5.80	2.14	10.20
HCH, alpha	0.40		0.41
HCH, gamma	1.90	0.92	2.86
Linuron	3.40	1.53	2.04
Malathion	5.60	19.57	8.16
MCPA	1.81	4.82	
Methiocarb	0.11		0.45
Methomyl	0.96	0.32	
Methyl parathion	0.90	0.31	0.41
Metolachlor	73.23	58.59	81.30
Metribuzin	13.70	6.73	14.29
Molinate	4.90	0.92	2.04
Napropamide	7.59	1.83	3.66
Neburon		0.32	
Norflurazon	0.64	0.32	
Oryzalin	0.74	3.81	0.45
Parathion	0.20		0.31
Pebulate	2.10		4.08
Pendimethalin	11.00	19.57	8.57
Permethrin, cis	0.30		0.82
Phorate	0.10		0.41
Prometon	34.97	83.79	61.79
Pronamide	2.20	0.31	2.85
Propachlor	3.00	1.83	3.66
Propanil	1.90	2.75	
Propargite	5.30	0.31	3.27
Propham	0.11	0.63	
Propoxur	0.22	0.33	
Simazine	61.74	87.77	82.93
Tebuthiuron	15.88	31.19	34.15
Terbacil	4.64	1.23	3.28
Terbufos	0.20		0.41
Thiobencarb	3.10	1.22	1.22
Triallate	8.50	0.31	6.94
Triclopyr	1.17	2.25	
Trifluralin	17.50	6.42	15.92

APPENDIX 2-4

Summary of Shallow Groundwater Data

The following data represent the frequency (as a percentage) that a specific pesticide was detected in a water sample.

Compound	Basin Land Use	
	Agricultural	Urban
Acetochlor	0.25	
Alachlor	3.14	0.33
Aldicarb sulfoxide	0.34	
Atrazine	43.72	24.25
Atrazine, deethyl	42.27	20.27
Azinphos-methyl	0.44	
Benfluralin	0.11	0.33
Bentazon	1.01	0.35
Bromacil	1.23	2.42
Butylate	0.32	
Carbaryl	0.32	1.33
Carbofuran	0.76	0.66
Chlorpyrifos	0.54	
Cyanazine	1.73	1.33
2,4-D	0.45	1.04
2,4-DB	0.11	
Dacthal mono-acid	0.11	
DBCP	1.35	
DCPA	1.41	
DDE	3.68	1.99
Diazinon	0.54	1.66
Dicamba	0.11	0.35
Dichlobenil	0.33	0.35
1,2-Dichloropropane	1.75	0.33
Dichlorprop	0.11	

(continued)

| Compound | Basin Land Use | |
	Agricultural	Urban
Dieldrin	0.97	5.65
2,4-Diethylaniline	1.19	
Dinoseb	0.34	
Diuron	2.34	2.77
EDB	0.41	
EPTC	1.62	
Ethalfluralin	0.33	
Ethoprop	0.11	
Fluometuron	0.45	
HCH, alpha	0.11	
Linuron	0.11	
Malathion	0.32	
Methyl parathion	0.22	
Metolachlor	18.37	9.97
Metribuzin	3.46	1.66
Norflurazon	0.11	
Oryzalin	0.69	
Pebulate	0.32	
Pendimethalin	0.22	
Permethrin, cis	0.22	
Picloram	0.11	
Prometon	13.40	27.24
Pronamide	0.22	
Propachlor	0.32	
Propanil	0.76	0.33
Propargite	0.11	
Propoxur		0.35
Simazine	22.38	15.28
Tebuthiuron	1.73	6.31
Terbacil	0.87	0.34
Terbufos	0.11	0.33
Triallate	0.32	0.33
1,2,3-Trichloropropane	1.08	
Trifluralin	0.43	0.66

APPENDIX 2-5

Organic Chemicals Found in Landfill Leachate and Gas

Chemical Class	Pollutant
Alcohols	Methanol
	Ethanol
	Propanol
	Butanol
	2-Methyl propanol
	Pentanol
	2-Ethyl hexanol
Alkanes	Propane
	Butanes
	Pentanes
	Hexanes
	Heptanes
	Octanes
	Nonanes
	Decanes
	Undecanes
	Dodecanes
Alkenes	Propene
	Butenes
	Pentadienes
	Pentenes
	Hexenes
	Heptenes
	Octenes
	Nonadienes
	Nonenes
	Decadienes
	Decenes
	Undecadienes
	Undecenes

(continued)

Chemical Class	Pollutant
Amines	Dimethylamine
Aromatic	Benzene
	Toluene
	Styrene
	Xylenes
	Ethylbenzene
	Methyl styrene
	Dimethyl styrenes
	Naphthalene
Carboxylic Acids	Ethanoic acid
	Butanoic acid
Cycloalkanes	Cyclohexane
	Methyl cyclopentane
	Methyl cyclohexane
	Diemethyl cyclohexane
	Trimethyl cyclohexane
	Propyl cyclohexanes
Cycloalkenes	Limonene
	Terpenes
Esters	Methyl ethanoate
	Ethyl ethanoate
	Methyl propanoate
	Methyl butanoate
	Methyl 2-methyl propanoate
	Ethyl propanoate
	Propyl ethanoate
	2-Propyl ethanoate
	Butyl methanoate
	Methyl pentanoate
	Methyl 2-methyl butanoate
	Ethyl butanoate
	Ethyl 2-methyl propanoate
	Propyl propanoate
	Butyl ethanoate
	2-Butyl ethanoate
	Ethyl pentanoate
	Ethyl 2-methyl butanoate
	Propyl butanoate
	Propyl propanoate
	Butyl propanoate
	Methyl hexanoate
Ethers	Diethyl ether
Halogenated	Chloromethane
	Chlorofluoromethane
	Dichloromethane
	Chlorodifluoromethane
	Dichlorodifluoromethane
	Trichloromethane
	Dichlorodifluoromethane

Chemical Class	Pollutant
	Trichlorofluoromethane
	Chloroethane
	1,1-Dichloroethane
	Vinyl chloride
	1,1,1-Trichloroethane
	1,2-Dichloroethylene
	1,1-Dichloroethylene
	Trichloroethylene
	Tetrachloroethylene
	1,1,1-Trifluorochloroethane
	1,2-Dichlorotetrafluoroethane
	Chlorotrifluoroethylene
	1,1,1-Trichlorotrifluoroethane
	1,1,2-Trifluorotrichloroethane
	1,1,1,2-Tetrafluorochloroethane
	Dichlorobenzenes
Organosulfur	Carbonyl sulfide
	Carbon disulfide
	Methanethiol
	Propanethiols
	Dimethyl sulfide
	Dimethyl disulfide
	Thiophene
	Hydrogen sulfide
Oxygenated	Acetone
	2-Butanone
	Tetrahydrofuran
	Furan
	Methyl furans
	Dimethyl furans
	2(2-hydroxypropoxy)propanol

APPENDIX 2-6

Unregulated Pollutants Discharged to or Identified in Water Resources

Acetone
Acetochlor
Acifluorfen
Alcohol ethoxylate
Aldicarb
Aldicarb sulfoxide
Alkylbenzenesulfonates
Alkyl diamine
Alkyl lead
Aluminum
Atrazine, deethyl
Azinphos-methyl
Benfluralin
Bentazon
Benzaldehyde
1,2-Benzene dicarboxylic acid
Benzene, 1-methyl-4-2(methyl propyl)
Benzoic acid
Benzophenone
Benzothiazole, 2,2-(methylthio)
Benzyl alcohol
Bromacil
Bromoxynil
Butyl methyl ketone
Butyl 2-methylpropyl ester
Butylate
Caffeine
Carbamazepine
Carbofuran
Cholestanol
4-Chloroaniline
3-Chloro-2-butanol

p-Chloro-m-cresol
2-Chloroethyl vinyl ether
4-Chlorophenol
Chlorothalonil
Chlorpropamide
Chlorpyrifos
Chromium, hexavalent
Cyanazine
1,3,5-Cycloheptatriene
Cyclohexanone
Cyclohexanone,
    4-(1,1-dimethyethyl)
Cyclopentanol 1,2-dimethyl-
    3-(methylethenyl)
Dacthal mono-acid
2,4-DB
Decahydro naphthalene
Decanal
Diacetate, 1,2-ethanediol
Diazinon
Dibenzofuran
Dicamba
Dichlobenil
1,2-Dichloropentane
Dichlorprop
Dicyclohexylamine
2,4-Diethylaniline
N,N-Diethyl-3-methyl benzamide
2,5-Dimethyl 3-hexanol
2,2-Dimethyl 3-pentanol
Disulfoton
Diuron

DNOC
EPTC
Ethalfluralin
Ethanol, 2-butoxy-phosphate
Ethanone 1-(2-naphthalenyl)
Ethoprop
Ethyl citrate
Ethylenediaminetetraacetic acid
(EDTA)
Fenuron
Fluometuron
Fonofos
HCH, alpha
HCH, gamma
Heptanal
2-Heptanone, 3 hydroxy-3 methyl
Hexadecanoic acid
Hexanal
3 Hexanol
Isobutyl methyl ketone
Linuron
Malathion
MCPA
Meprobomate
Methiocarb
Methomyl
3-Methoxy-3methyl-hexane
2-Methoxy-1-propanol
Methylene-blue
Methyl ethyl ketone
2-Methylnaphthalene
Methyl parathion
2-Methylphenol
4-Methylphenol
Methyl tertiary-butyl ether (MTBE)
Metolachlor
Metribuzin
Molinate
Mirex
Neburon
Nickel
2-Nitroaniline
3-Nitroaniline
4-Nitroaniline
Nonanal
Nonylphenol
Norflurazon

Octachlorostyrene
Octadiene, 4,5 dimethyl-3,6
dimethyl
Octandecanoic acid
Octanal
Oryzalin
Parathion
Pebulate
Pendimethalin
Pentobarbital
Perfluoro-octanyl sulfonate (PFOS)
cis-Permethrin
Phenol
Phenol 2,4(bis(1,1-dimethylethyl))
Phenol 4,4(1,2-diethyl-1,2-
ethanediyl)bisphenol nonyl
Phensuximide
1-Phenyl ethanone
Phorate
Piperonyl butoxide
Polyethylene glycol
Propanic acid 2 methyl-2,
2-dimethyl-1-(2hr...)
1-Propanol, 2-(2-hydroxypropoxy)
2-Propanone, 1-(1-cyclohexen-
1-yl)
Propargite
Propham
Propoxur
Sodium dodecylbenzenesulfonate
Sulfonamide
Tebuthiuron
Terbacil
Terbufos
Tetradecanal
Tetradecanoic acid
Thallium
Thiobencarb
m-Toluate
Triallate
Trichlopyr
Trichlorofluoromethane
2,4,5-Trichlorophenol
Trihaloalkylphosphates
Trifluralin
Trimethyltriazinetrione
Vinyl acetate

APPENDIX 2-7

Chemicals Known to the State of California to Cause Cancer or Reproductive Toxicity (April 20, 2001)

Chemicals Known to Cause Cancer

A-alpha-C (2-Amino-9H-pyrido [2,3-b]indole)
Acetaldehyde
Acetamide
Acetochlor
2-Acetylaminofluorene
Acifluorfen
Acrylamide
Acrylonitrile
Actinomycin D
Adriamycin (Doxorubicin hydrochloride)
AF-2;[2-(2-furyl)-3-(5-nitro-2-furyl)]acrylamide
Aflatoxins
Alachlor
Aldrin
2-Aminoanthraquinone
p-Aminoazobenzene
ortho-Aminoazotoluene
4-Aminobiphenyl (4-aminodiphenyl)
1-Amino-2,4-dibromoanthraquinone
3-Amino-9-ethylcarbazole hydrochloride
2-Aminofluorene
1-Amino-2-methylanthraquinone
2-Amino-5-(5-nitro-2-furyi)-1, 3,4-thiadiazole
4-Amino-2-nitrophenol
Amitrole

Analgesic mixtures containing phenacetin
Aniline
Aniline hydrochloride
ortho-Anisidine
ortho-Anisidine hydrochloride
Antimony oxide (Antimony trioxide)
Aramite
Arsenic (inorganic arsenic compounds)
Asbestos
Auramine
Azacitidine
Azaserine
Azathioprine
Azobenzene
Benz[a]anthracene
Benzene
Benzidine [and its salts]
Benzidine-based dyes
Benzo[b]fluoranthene
Benzo[o]fluoranthene
Benzo[k]fluoranthene
Benzofuran
Benzo[a]pyrene
Benzotrichloride
Benzyl chloride
Benzyl violet 4B
Beryllium and beryllium compounds
2,2-Bis(bromomethyl)-1,3-propanediol

257

Bis(2-chloroethyl)ether
N,N-Bis(2-chloroethyl)-2-naphthyl-
   amine (Chlornapazine)
Bischloroethyl nitrosourea (BCNU)
   (Carmustine)
Bis(chloromethyl)ether
Bis(2-chloro-1-methylethyl)ether,
   technical grade
Bitumens
Bracken fern
Bromodichloromethane
Bromoethane
Bromoform
1,3-Butadiene
IIA-Butanediol dimethanesulfonate
   (Busulfan)
Butylated hydroxyanisole
beta-Butyrolactone
Cacodylic acid
Cadmium and cadmium compounds
Caffeic acid
Captafol
Captan
Carbazole
Carbon tetrachloride
Carbon-black extracts
Ceramic fibers (airborne particles of
   respirable size)
Certain combined chemotherapy for
   lymphomas
Chlorambucil
Chloramphenicol
Chlordane
Chlordecone (Kepone)
Chlordimeform
Chlorendic acid
Chlorinated paraffins
p-Chloroaniline
p-Chloroaniline hydrochloride
Chloroethane (Ethyl chloride)
1-(2-Chloroethyl)-3-cyclohexyl-
   1-nitrosourea (CCNU) (Lomustine)
1-(2-Chloroethyl)-3-(4-methylcyclo-
   hexyl)-1-nitrosourea
   (Methyl-CCNU)
Chloroform
Chloromethyl methyl ether
   (technical grade)
3-Chloro-2-methylpropene
1-Chloro-4-nitrobenzene
4-Chloro-ortho-phenylenediamine

p-Chloro-o-toluidine
p-Chloro-o-toluidine, and its strong
   acid
5-Chloro-o-toluidine and its strong
   acid salts
Chloroprene
Chlorothalonil
Chlorotrianisene
Chlorozotocin
Chromium (hexavalent compounds)
Chrysene
C.I. Acid Red 114
C.I. Basic Red 9 monohydrochloride
C.I. Direct Blue 15
C.I. Direct Blue 218
C.I. Solvent Yellow 14
Ciclosporin (Cyclosporin A;
   Cyclosporine)
Cidofovir
Cinnamyl anthranilate
Cisplatin
Citrus Red No. 2
Clofibrate
Cobalt metal powder
Cobalt [11] oxide
Cobalt sulfate heptahydrate
Conjugated estrogens
Creosotes
para-Cresidine
Cupferron
Cycasin
Cyclophosphamide (anhydrous)
Cyclophosphamide (hydrated)
Cytembena
D&C Orange No. 17
D&C Red No. 8
D&C Red No. 9
D&C Red No. 19
Dacarbazine
Daminozide
Dantron (Chrysazin;
   1,8-Dihydroxyanthraquinone)
Daunomycin
DDD (Dichlorodiphenyl-
   dichloroethane)
DDE (Dichlorodiphenyl-
   dichloroethylene)
DDT (Dichlorodiphenyl-
   trichloroethane)
DDVP (Dichlorvos)
N,N'-Diacetylbenzidine

2,4-Diaminoanisole
2,4-Diaminoanisole sulfate
4,4'-Diaminodiphenyl ether
(4,4'-Oxydianiline)
2,4-Diaminotoluene
Diaminotoluene (mixed)
Dibenz[a,h]acridine
Dibenz[a,j]acridine
Dibenz[a,h]anthracene
7H-Dibenzo[c,g]carbazole
Dibenzo[a,e]pyrene
Dibenzo[a,h]pyrene
Dibenzo[a,i]pyrene
Dibenzo[a,l]pyrene
1,2-Dibromo-3-chloropropane (DBCP)
2,3-Dibromo-1-propanol
Dichloroacetic acid
p-Dichlorobenzene
3,3'-Dichlorobenzidine
3,3'-Dichlorobenzidine
dihydrochloride
1,4-Dichloro-2-butene
3,3'-Dichloro-4,4'-diaminodiphenyl
ether
M-Dichloroethane
Dichloromethane (Methylene
chloride)
1,2-Dichloropropane
1,3-Dichloropropene
Dieldrin
Dienestrol
Diepoxybutane
Diesel engine exhaust
Di(2-ethylhexyl)phthalate
1,2-Diethylhydrazine
Diethyl sulfate
Diethylstilbestrol (DES)
Diglycidyl resorcinol ether (DGRE)
Dihydrosafrole
Diisopropyl sulfate
3,3'-Dimethoxybenzidine (ortho-
Dianisidine)
3,3'-Dimethoxybenzidine dihydrochloride (ortho-Dianisidine
dihydrochloride)
Dimethyl sulfate
4-Dimethylaminoazobenzene
trans-2-[(Dimethylamino)methylim-
ino]-5-[2-(5-nitro-2-furyl)vinyl]-1,3,
4-oxadiazole
7,12-Dimethylbenz(a)anthracene

3,3'-Dimethylbenzidine (ortho-
Tolidine)
3,3'-Dimethylbenzidine
Dimethylcarbamoyl chloride
1,1-Dimethylhydrazine (UDMH)
1,2-Dimethylhydrazine
Dimethylvinylchloride
3,7-Dinitrofluoranthene
3,9-Dinitrofluoranthene
1,6-Dinitropyrene
1,8-Dinitropyrene
Dinitrotoluene mixture, 2,442,6
2,4-Dinitrotoluene
2,6-Dinitrotoluene
Di-n-propyl isocinchomeronate (MGK
Repellent 326)
1,4-Dioxane
Diphenylhydantoin (Phenytoin)
Diphenylhydantoin (Phenytoin),
sodium salt
Direct Black 38 (technical grade)
Direct Blue 6 (technical grade)
Direct Brown 95 (technical grade)
Disperse Blue 1
Epichlorohydrin
Erionite
Estradiol 17B
Estragole
Estrone
Estropipate
Ethinylestradiol
Ethoprop
Ethyl acrylate
Ethyl methanesulfonate
Ethyl-4,4'-dichlorobenzilate
Ethylene dibromide
Ethylene dichloride
(1,2-Dichloroethane)
Ethylene oxide
Ethylene thiourea
Ethyleneimine
Fenoxycarb
Folpet
Formaldehyde (gas)
2-(2-Formylhydrazino)-4-(5-nitro-
2-furyl)thiazole
Furan
Furazolidone
Furmecyclox
Fusarin C
Ganciclovir sodium

Gemfibrozil
Glu-P-1 (2-Amino-6-methyldipyrido
[1,2 a:3′,2′-d]imidazole)
Glu-P-2 (2-Aminodipyrido[1,2-a:3′,
2′-d]imidazole)
Glycidaldehyde
Glycidol
Griseofulvin
Gyromitrin (Acetaldehyde
methylformylhydrazone)
HC Blue 1
Heptachlor
Heptachlor epoxide
Hexachlorobenzene
Hexachlorocyclohexane (technical
grade)
Hexachlorodibenzodioxin
Hexachloroethane
Hexamethylphosphoramide
Hydrazine
Hydrazine sulfate
Hydrazobenzene (1,2-
Diphenylhydrazine)
Indeno (1,2,3-cd)pyrene
Indium phosphide
IQ (2-Amino-3-methylimidazo[4,5-
f]quinoline)
Iprodione
Iron dextran complex
Isobutyl nitrite
Isoprene
Isosafrole
Isoxaflutole
Lactofen
Lasiocarpine
Lead acetate
Lead and lead compounds
Lead phosphate
Lead subacetate
Lindane and other
hexachlorocyclohexane isomers
Lynestrenol
Mancozeb
Maneb
Me-A-alpha-C (2-Amino-3-methyl-
9H-pyrido[2,3-b]indoie)
Medroxyprogesterone acetate
MeIQ(2-Amino-3,4-dimethylimi-
dazo[4,5-f]quinoline)
MelQx(2-Amino-3,8-dimethylimi-
dazo[4,5-f]quinoxaline)

Melphalan
Merphalan
Mestranol
Metham sodium
8-Methoxypsoralen with ultraviolet
A therapy
5-Methoxypsoralen with ultraviolet
A therapy
2-Methylaziridine (Propyleneimine)
Methylazoxymethanol
Methylazoxymethanol acetate
Methyl carbarmate
3-Methylcholanthrene
5-Methylchrysene
4,4′-Methylene bis(2-chloroaniline)
4,4′-Methylene bis(N,N-dimethyl)
benzenamine
4,4′-Methylene bis(2-methylaniline)
4,4′-Methylenedianiline
4,4′-Methylenedianiline
dihydrochloride
Methyl hydrazine and its salts
Methyl iodide
Methyl mercury compounds
Methyl methanesulfonate
2-Methyl-l-nitroanthraquinone (of
uncertain purity)
N-Methyl-N′-nitro-N-nitrosoguani-
dine
N-Methylolacrylamide
Methylthiouracil
Metiram
Metronidazole
Michler's ketone
Mirex
Mitomycin C
Monocrotaline
5-(Morpholinomethyl)-3-[(5-nitro-
furfurylidene)-amino]-
2-oxazolidinone
Mustard Gas
MX (3-chloro-4-dichloromethyl-
5-hydroxy-2(5H)-furanone)
Nafenopin
Nalidixic acid
1-Naphthylamine
2-Naphthylamine
Nickel and certain nickel compounds
Nickel carbonyl
Nickel refinery dust from the
pyrometallurgical process

Nickel subsulfide
Niridazole
Nitrilotriacetic acid
Nitrilotriacetic acid, trisodium salt
  monohydrate
5-Nitroacenaphthene
5-Nitro-o-anisidine
o-Nitroanisole
Nitrobenzene
4-Nitrobiphenyl
6-Nitrochrysene
Nitrofen (technical grade)
2-Nitrofluorene
Nitrofurazone
1-[(5-Nitrofurfurylidene)-amino]-
  2-imidazolidinone
N-[4-(5-Nitro-2-furyl)-
  2-thiazolyl]acetamide
Nitrogen mustard
  (Mechlorethamine)
Nitrogen mustard hydrochloride
  (Mechlorethamine hydrochloride)
Nitrogen mustard N-oxide
Nitrogen mustard N-oxide
  hydrochloride
Nitromethane
2-Nitropropane
1-Nitropyrene
4-Nitropyrene
N-Nitrosodi-n-butylamine
N-Nitrosodiethanolamine
N-Nitrosodiethylamine
N-Nitrosodimethylamine
p-Nitrosodiphenylamine
N-Nitrosodiphenylamine
N-Nitrosodi-n-propylamine
N-Nitroso-N-ethylurea
3-(N-Nitrosomethylamino)propioni-
  trile
4-(N-Nitrosomethylamino)-1-(3-
  pyridyl)l-butanone
N-Nitrosomethylethylamine
N-Nitroso-N-methylurea
N-Nitroso-N-methylurethane
N-Nitrosomethylvinylamine
N-Nitrosomorpholine
N-Nitrosonornicotine
N-Nitrosopiperidine
N-Nitrosopyrrolidine
N-Nitrososarcosine
o-Nitrotoluene

Norethisterone (Norethindrone)
Norethynodrel
Ochratoxin A
Oil Orange SS
Oral contraceptives, combined
Oral contraceptives, sequential
Oxadiazon
Oxazepam
Oxymetholone
Oxythioquinox
Panfuran S
Pentachlorophenol
Phenacetin
Phenazopyridine
Phenazopyridine hydrochloride
Phenesterin
Phenobarbital
Phenolphthalein
Phenoxybenzamine
Phenoxybenzamine hydrochloride
o-Phenylenediamine and its salts
Phenyl glycidyl ether
Phenylhydrazine and its salts
o-Phenylphenate, sodium
o-Phenylphenol
PhiP(2-Amino-1-methyl-6-phenylimi-
  dazol[4,5-b]pyridine)
Polybrominated biphenyls
Polychlorinated biphenyls
  Polychlorinated biphenyls
Polychlorinated dibenzo-p-dioxins
Polychlorinated dibenzofurans
Polygeenan
Ponceau MX
Ponceau 3R
Potassium bromate
Primidone
Procarbazine
Procarbazine hydrochloride
Procymidone
Progesterone
Pronamide
Propachlor
1,3-Propane sultone
Propargite
beta-Propiolactone
Propylene oxide
Propylthiouracil
Quinoline and its strong acid
  salts
Radionuclides

Reserpine
Residual (heavy) fuel oils
Safrole
Salicylazosulfapyridine
Selenium sulfide
Soots, tars, and mineral oils
Spironolactone
Stanozolol
Sterigmatocystin
Streptozotocin (streptozocin)
Styrene oxide
Sulfallate
Tamoxifen and its salts
Terrazole
Testosterone and its esters
2,3,7,8-Tetrachlorodibenzo-para-
dioxin (TCDD)
1,1,2,2-Tetrachloroethane
Tetrachloroethylene
(Perchloroethylene)
p-a,a,a-Tetrachlorotoluene
Tetrafluoroethylene
Tetranitromethane
Thioacetamide
4,4'-Thiodianiline
Thiodicarb
Thiourea
Thorium dioxide
Toluene diisocyanate
ortho-Toluidine
ortho-Toluidine hydrochloride
Toxaphene (Polychlorinated cam-
phenes)

Treosulfan
Trichlormethine (Trimustine
hydrochloride)
Trichloroethylene
2,4,6-Trichlorophenol
1,2,3-Trichloropropane
Trimethyl phosphate
2,4,5-Trimethylaniline and its strong
acid salts
Triphenyltin hydroxide
Tris(aziridinyl)-para-benzoquinone
(triaziquone)
Tris(l-aziridinyl)phosphine sulfide
(thiotepa)
Tris(2-chloroethyl)phosphate
Tris(2,3-dibromopropyl)phosphate
Trp-P-1 (Tryptophan-P-1)
Trp-P-2 (Tryptophan-P-2)
Trypan blue (commercial grade)
Unleaded gasoline
Uracil mustard
Urethane (Ethyl carbamate)
Vinclozolin
Vinyl bromide
Vinyl chloride
4-Vinylcyclohexene
4-Vinyl-l-cyclohexene diepoxide
(Vinyl cyclohexenedioxid
Vinyl fluoride
Vinyl trichloride (1,1,
2-trichloroethane)
2,6-Xylidine (2,6-Dimethylaniline)
Zileuton

Chemicals Known to Cause Reproductive Toxicity

Developmental Toxicity

Acetazolamide
Acetohydroxamic acid
Actinomycin D
All-trans retinoic acid
Alprazolam
Altretamine
Amantadine hydrochloride
Amikacin sulfate
Aminoglutethimide
Aminoglycosides
Aminopterin
Amiodarone hydrochloride

Amitraz
Amoxapine
Angiotensin converting enzyme (ACE)
inhibitors
Anisindione
Arsenic (inorganic oxides)
Aspirin
Atenolol
Auranofin
Azathioprine
Barbiturates
Beclomethasone dipropionate

Benomyl
Benzene
Benzphetamine hydrochloride
Benzodiazepines
Bischloroethyl nitrosourea (BCNU)
  (Carmustine)
Bromacil lithium salt
Bromoxynil
Bromoxynil octanoate
Butabarbital sodium
1,4-Butanediol dimethanesulfonate
  (Busulfan)
Cadmium
Carbamazepine
Carbon disulfide
Carboplatin
Chenodiol
Chinomethionat (oxythioquinox)
Chlorambucil
Chlorcyclizine hydrochloride
Chlordecone (Kepone)
Chlordiazepoxide
Chlordiazepoxide hydrochloride
1-(2-Chloroethyl)-3-cyclohexyl-
  l-nitrosourea (CCNU)
  (Lomustine)
Chlorsulfuron
Cidofovir
Cladribine
Clarithromycin
Clobetasol propionate
Clomiphene citrate
Clorazepate dipotassium
Cocaine
Codeine phosphate
Colchicine
Conjugated estrogens
Cyanazine
Cycloate
Cycloheximide
Cyclophosphamide (anhydrous)
Cyclophosphamide (hydrated)
Cyhexatin
Cytarabine
Dacarbazine
Danazol
Daunorubicin hydrochloride
2,4-D butyric acid
o,p,-DDT
p,p,-DDT
2,4-DP (dichloroprop)

Demeclocycline hydrochloride
  (internal use)
Diazepam
Diazoxide
Dichlorophene
Dichlorphenamide
Diclofop methyl
Dicumarol
Diethylstilbestrol (DES)
Diflunisal
Dihydroergotamine mesylate
Diltiazem hydrochloride
Dinocap
Dinoseb
Diphenylhydantoin (Phenytoin)
Disodium cyanodithioimidocarbonate
Doxorubicin hydrochloride
Doxycycline (internal use)
Doxycycline calcium (internal use)
Doxycycline hyclate (internal use)
Doxycycline monohydrate
  (internal use)
Endrin
Ergotamine tartrate
Estropipate
Ethionamide
Ethyl alcohol in alcoholic beverages
Ethyl dipropylthiocarbamate
Ethylene dibromide
Ethylene glycol monoethyl ether
Ethylene glycol monomethyl ether
Ethylene glycol monoethyl ether
  acetate
Ethylene glycol monomethyl ether
  acetate
Ethylene thiourea
Etodolac
Etoposide
Etretinate
Fenoxaprop ethyl
Filgrastim
Fluazifop butyl
Flunisolide
Fluorouracil
Fluoxymesterone
Flurazeparn hydrochloride
Flurbiprofen
Flutamide
Fluticasone propionate
Fluvalinate
Ganciclovir sodium

Goserelin acetate
Halazepam
Halobetasol propionate
Haloperidol
Halothane
Heptachlor
Hexachlorobenzene
Histrelin acetate
Hydramethylnon
Hydroxyurea
Idarubicin hydrochloride
Ifosfamide
Iodine-1 31
Isotretinoin
Lead
Leuprolide acetate
Levodopa
Linuron
Lithium carbonate
Lithium citrate
Lorazepam
Lovastatin
Mebendazole
Medroxyprogesterone acetate
Megestrol acetate
Melphalan
Menotropins
Meprobamate
Mercaptopurine
Mercury and mercury compounds
Methacycline hydrochloride
Metham sodium
Methazole
Methimazole
Methotrexate
Methotrexate sodium
Methyl bromide as a structural
    fumigant
Methyl chloride
Methyl mercury
Methyltestosterone
Metiram
Midazolam hydrochloride
Minocycline hydrochloride
    (internal use)
Misoprostol
Mitoxantrone hydrochloride
Myclobutanil
Nabam
Nafarelin acetate
Neomycin sulfate (internal use)

Netilmicin sulfate
Nickel carbonyl
Nifedipine
Nitrapyrin
Nitrogen mustard (Mechlorethamine)
Nitrogen mustard hydrochloride
    (Mechlorethamine hydrochloride)
Norethisterone (Norethindrone)
Norethisterone acetate
    (Norethindrone acetate)
Norethisterone
    (Norethindrone)/Ethinyl estradiol
Norethisterone
    (Norethindrone)/Mestranol
Norgestrel
Oxadiazon
Oxazepam
Oxymetholone
Oxytetracycline (internal use)
Oxytetracycline hydrochloride
    (internal use)
Paclitaxel
Paramethadione
Penicillamine
Pentobarbital sodium
Pentostatin
Phenacemide
Phenprocoumon
Pimozide
Pipobroman
Plicamycin
Polybrominated biphenyls
Polychlorinated biphenyls
Potassium dimethyldithiocarbamate
Pravastatin sodium
Prednisolone sodium phosphate
Procarbazine hydrochloride
Propargite
Propylthiouracil
Pyrimethamine
Quazepam
Resmethrin Retinol/retinyl esters,
    when in daily dosages in excess of
    10,000 1 U, or 3,000 retinol
    equivalents.
Ribavirin
Rifampin
Secobarbital sodium
Sermorelin acetate
Sodium dimethyldithiocarbamate
Streptomycin sulfate

Streptozocin (streptozotocin)
Sulindac
Tamoxifen citrate
Temazepam
Teniposide
Terbacil
Testosterone cypionate
Testosterone enanthate
2,3,7,8-Tetrachlorodibenzo-para-
   dioxin (TCDD)
Tetracycline (internal use)
Tetracyclines (internal use)
Tetracycline hydrochloride
   (internal use)
Thalidomide
Thioguanine
Tobramycin sulfate
Toluene

Triadimefon
Triazolam
Tributyltin methacrylate
Trientine hydrochloride
Triforine
Trilostane
Trimethadione
Trimetrexate glucuronate
Uracil mustard
Urethane
Urofollitropin
Valproate (Valproic acid)
Vinblastine sulfate
Vinclozolin
Vincristine sulfate
Warfarin
Zileuton

Female Reproductive Toxicity

Aminopterin
Amiodarone hydrochloride
Anabolic steroids
Aspirin
Carbon disulfide
Chlorsulfuron
Cidofovir
Clobetasol propionate
Cocaine
Cyclophosphamide (anhydrous)
Cyclophosphamide (hydrated)
o,p,-DDT
p,p,-DDT
Diflunisal Dinitrotoluene (technical
   grade)
Ethylene oxide
Etodolac
Flunisolide
Flurbiprofen
Gemfibrozil
Goserelin acetate
Haloperidol
Lead
Leuprolide acetate
Levonorgestrel implants
Nifedipine
Oxydemeton methyl
Paclitaxel
Pimozide

Rifampin
Streptozocin (streptozotocin)
Sulindac
Thiophanate methyl
Triadimefon
Uracil mustard
Zileuton
Male Reproductive Toxicity
Altretamine
Amiodarone hydrochloride
Anabolic steroids
Benomyl
Benzene
Cadmium
Carbon disulfide
Chlorsulfuron
Cidofovir
Colchicine
Cyclohexanol
Cyclophosphamide (anhydrous)
Cyclophosphamide (hydrated)
2,4-D butyric acid
o,p,-DDT
p,p,-DDT
1,2-Dibromo-3-chloropropane (DBCP)
m-Dinitrobenzene
o-Dinitrobenzene
p-Dinitrobenzene
2,4-Dinitrotoluene

2,6-Dinitrotoluene
Dinitrotoluene (technical grade)
Dinoseb
Doxorubicin hydrochloride
Epichlorohydrin
Ethylene dibromide
Ethylene glycol monoethyl ether
Ethylene glycol monomethyl ether
Ethylene glycol monoethyl ether
  acetate
Ethylene glycol monomethyl ether
  acetate
Ganciclovir sodium
Gemfibrozil
Goserelin acetate
Hexamethylphosphoramide
Hydramethylnon

Idarubicin hydrochloride
Lead
Leuprolide acetate
Myclobutanil
Nifedipine
Nitrofurantoin
Oxydemeton methyl
Paclitaxel
Quizalofop-ethyl
Ribavirin
Sodium fluoroacetate
Streptozocin (streptozotocin)
Sulfasalazine
Thiophanate methyl
Triadimefon
Uracil mustard

APPENDIX 2-8

Regulated Pesticides in Food[4] with Residue Tolerances

Acephate
Acetochlor (1998 Drinking Water Contaminant Candidate List)
Alachlor (Regulated as a Primary Drinking Water Constituent)
Aldicarb **(also found in California drinking water)**[5]
Allethrin (allyl homolog of cinerin I)
Aluminum tris (O-ethylphosphonate)
Ametryn
Aminoethoxyvinylglycine
4-Aminopyridine
Amitraz
Ammoniates for [ethylenebis-(dithiocarbamato] zinc and ethylenebis [dithiocarbamic acid]
bimolecular and trimolecular cyclic anhydrosulfides and disulfides
Arsanilic acid (4-aminophenyl) arsonic acid
Asulam
Atrazine (Regulated as a Primary Drinking Water Constituent)
Avermectin B1 and its delta-8,9-isomer
Azoxystrobin
Barban
Benomyl
Benoxacor
Bentazon **(also found in California drinking water)**
Beta-([1,1-biphenyl]-4-yloxy)-alpha-(1,1-dimethylethyl)-1H-1,2,4-triazole-1-ethanol
Beta-(4-Chlorophenoxy)-alpha-(1,1-dimethylethyl)-1H-1,2,4-triazole-1-ethanol
Bifenthrin
1,1-Bis(p-chlorophenyl)-2,2,2-trichloroethanol

---

[4]CFR Title 40, Part 180, Tolerances and Exemptions From Tolerances for Pesticide Chemicals in Food, 2000.
[5]CPRIG, 1999; unregulated pesticides found in groundwater.

Bromide (inorganic)
Bromacil **(also found in California drinking water)**
Bromoxynil
Buprofezin
Butylate **(also found in California drinking water)**
n-Butyl-N-ethyl-α.α.α-trifluoro-2,6-dinitro-p-toluidine
Cacodylic acid
Captafol
Captin
Carbaryl **(also found in California drinking water)**
Carbofuran (Regulated as a Primary Drinking Water Constituent)
Carbon disulfide **(also found in California drinking water)**
Carboxin
Carfentrazone-ethyl
Chlorfenapyr
Chlorimuron ethyl
2-[[4-chloro-6-(ethylamino)-s-triazin-2-yl]amino]-2-methylpropionitrile
Chloroneb
2-Chloro-N-isopropylacetanilide
p-Chlorophenoxyacetic acid
2-(m-Chlorophenoxy) propionic acid
Chloropyrifos **(also found in California drinking water)**
Chlorothalonil **(also found in California drinking water)**
Chlorpyrifos-methyl
Chlorsulfuron
CICP
Clethodim ((E)-2-[1-[[(3-chloro-2-propenyl)oxy]imino]propyl]-5-[2-
    (ethylthio)propyl]-3-hydroxy-2-cyclohexen-1-one
Clodinafop-propargyl
Clofencet
Clofentezine
Clomazone
Clopyralid
Cloquintocet-mexyl
Cloransulam-methyl
Copper
Copper carbonate
Coumaphos
Cyano(3-phenoxyphenyl)methyl-4-chloro-α-(1-methylethyl) benzeneacetate
Cyclanilide
Cyfluthrin
Cymoxanil
Cypermethrin and an isomer zeta-cypermethrin
Cyproconazole
Cyprodinil
Cyromazine
2,4-D (Regulated as a Primary Drinking Water Constituent)
Deltamethrin
Desmedipham
N,N-diallyl dichloroacetamide

Diazinon (1998 Drinking Water Contaminant Candidate List, and also found in California drinking water)
Dicamba **(also found in California drinking water)**
Dichlobenil
4-(Dichloroacetyl)-1-oxa-4-azaspiro[4,5]decane
4-(2,4-Dichlorophenoxy) butyric acid
1-[[2-(2,4-dichlorophenyl)-4-propyl-1,3-dioxolan-2-yl]methyl]-1H-1,2,4-triazole
2-(3,5-Dichlorophenyl)-2-(2,2,2-trichloroethyl)-oxirane
Dichlorvos
Diclofop-methyl
Diclosulam
O,O-Diethyl S-[2-(ethylthio)ethyl] phosphorodithioate
N,N-Diethyl-2-(4-methylbenzyloxy)ethylamine hydrochloride
N,N-Diethyl-2-(1-naphthalenyloxy)propionamide
Difenoconazole
Difenzoquat
Diflubenzuron
Diflufenzopyr
Dihydro-5-heptyl-2(3H)-furanone
2-[4,5-Dihydro-4-methyl-4-(1-methylethyl)-5-oxo-1H-imidazol-2-yl]-3-quinoline carboxylic acid
Dihydro-5-pentyl-2(3H)-furanone
S-(O,O-Diisopropyl phosphorodithioate) of N-(2-mercaptoethyl) benzenesulfonamide
Dimethenamid, 2-chloro-N-[(1-methyl-2methoxy)ethyl]-N-(2,4-dimethylthien-3-yl)-acetamide
Dimethipin
Dimethoate **(also found in California drinking water)**
Dimethomorph
2,2-Dimethyl-1,3-benzodioxol-4-ol methylcarbamate
O-[2-(1,1-Dimethylethyl)-5-pyrimidinyl]O-ethyl-O-(1-methylethyl) phosphorothioate
O,O-Dimethyl S-[(4-oxo-1,2,3-benzotriazin-3(4H)-yl)methyl]phosphorodithioate
O,O-Dimethyl S-[(4-oxo-1,2,3-benzotriazin-3(4H)-yl)methyl]phosphorodithioate
Dimethyl phosphate of 3-hydroxy-N-methyl-cis-crotonamide
Dimethyl phosphate of 3-hydroxy-N-N-dimethyl-cis-crotonamide
Dimethyl tetrachloroterephthalate
2,6-dimethyl-4tridecylmorpholine
4,6-Dinitro-o-cresol and its sodium salt
2,4-Dinitro-6-octylphenyl crotonate and 2,6-dinitro-4octylphenyl crotonate
Diphenamid
Diphenylamine
Dipropyl isocinchomeronate
Diquat **(also found in California drinking water)**
Diuron (1998 Drinking Water Contaminant Candidate List, and also found in California drinking water)
Dodine
Emamectin benzoate
Endosulfan **(also found in California drinking water)**
Endothall (Regulated as a Primary Drinking Water Constituent)
Esfenvalerate

Ethalfluralin
Ethephon
Ethion
Ethofumesate
Ethoprop **(also found in California drinking water)**
Ethoxyquin
5-Ethoxy-3-(trichloromethyl)-1,2,4-thiadiazole
O-Ethyl S-phenyl ethylphosphonodithioate
S-Ethyl cyclohexylethylthiocarbamate
S-Ethyl dipropylthiocarbamate
Ethylene oxide
S-Ethyl hexahydro-1H-azepine-1-carbothioate
S-[2-(Ethylsulfinyl)ethyl]O,O-dimethyl phospharothioate
Fenamiphos
Fenarimol
Fenbuconazole
Fenhexamid
Fenitrothion
Fenoxaprop-ethyl
Fenoxycarb
Fenpropathrin
Fenridazon
Fenthion
Ferbam
Fipronil
Fluazifop-butyl
Fludioxonil
Flumetsulam
Flumiclorac pentyl
Fluorine compounds
N-(4-fluorophenyl)-N-(1-methylethyl)-2-[[5-(trifluoromethyl)-1,3,4-thiadiazol-
    2-yl]oxy]
Acetamide
Fluridone
Fluroxypyr 1-methylheptyl ester
Fluthiacet-methyl
Flutolanil(N-(3-(1-methylethoxy)phenyl)-2-(trifluoromethyl)benzamide)
Fluvalinate
Folpet
Formetanate hydrochloride
Furilazole
Glufosinate ammonium
Glyphosate (Regulated as a Primary Drinking Water Constituent)
Halosulfuron
Hexaconazole
Hexakis (2-methyl-2-phenylpropyl)distannoxane
Hexazinone **(also found in California drinking water)**
Hexythiazox
HOE-107892 (mefenpyr-diethyl)
Hydramethylnon
Hydrogen cyanide

Hydroprene
Imazalil
Imazamox
Imazapic-ammonium
Imazapyr
Imazethapyr, ammonium salt
Imidacloprid
Iprodione
Isoxaflutole
Kresoxim-methyl
Lactofen
Lambda-cyhalothrin
d-Limonene
Lindane (Regulated as a Primary Drinking Water Constituent)
Linuron (1998 Drinking Water Contaminant Candidate List)
Malathion
Maleic hydrazide
Mancozeb
Maneb
Mepiquat chloride
N-(Mercaptomethyl) phthalimide S-(O,O-dimethyl phosphorodithioate)
Metalaxyl
Metaldehyde
Methamidophos
Methanearsonic acid
Methidathion
Methomyl
Methoprene
Methoxychlor (Regulated as a Primary Drinking Water Constituent)
2-methyl-4chlorophenoxyacetic acid
4-(2-Methyl-4-chlorophenoxy) butyric acid
6-methyl-1,3-dithiolo [4,5-b] quinoxalin-2-one
Methyl 3-[(dimethoxyphosphinyl)oxy]butenoate
Methyl2-(4-isopropyl-4-methyl-5oxo-2-imidazolin-2yl)-p-toluate and methyl
　6-(4-isopropyl-4- methyl-5-oxo-2-imidazolin-2-yl)-m-toluate
Metolachlor (1998 Drinking Water Contaminant Candidate List, and also found
　in California drinking water)
Metsulfuron methyl
Mineral oil
Myclobutanil
Naled
α-Naphthaleneacetamide
1-Naphthaleneacetic acid
N-1-Naphthyl phthalamicacid
Nicosulfuron, [3-pyridinecarboxamide, 2-((((4,6-dimethoxypyrimidin-
　2-yl)aminocarbonyl) aminosulfonyl))-N,N-dimethyl]
Nicotine and nicotine containing compounds
Nitrapyrin
Norflurazon **(also found in California drinking water)**
n-Octyl bicycloheptenedicarboximide
Orthoarsenic acid

Oryzalin
Oxadiazon
Oxadixyl
Oxamyl (Regulated as a Primary Drinking Water Constituent)
Oxyfluorfen
Oxytetracycline
Paraquat
Parathion
Pendimethalin
Pentachloronitrobenzene
Permethrin
Picloram (Regulated as a Primary Drinking Water Constituent)
Pirimiphos-methyl
Phenmedipham
o-Phenylphenol
Phorate
Phosphamidon
Phosphine
Phosphorothioic acid, 0,0-diethyl 0-(1,2,2,2-tetrachloroethyl) ester
Piperonyl butoxide
Prallethrin(RS)-2-methyl-4-oxo-3-(2-propynyl)cyclopent-2-enyl(IRS)-cis,trans-
  chrysanthemate
Primisulfuron-methyl
Procymidone
Profenofos
Prohexadione calcium
Propamocarb hydrochloride
Propanil
Propargite **(also found in California drinking water)**
Propazine
Propetamphos
S-Propyl butylethylthiocarbamate
S-Propyl dipropylthiocarbamate
Propylene oxide
Prosulfuron
Pymetrozine
Pyrazon
Pyrethrins
Pyridaben
Pyridate
Pyrimethanil
Pyriproxyfen
Pyrithiobac sodium
Quinclorac
Quizalofop ethyl
Resmethrin
Rimsulfuron
Sethoxydim
Simazine (Regulated as a Primary Drinking Water Constituent)
Sodium dimethyldithiocarbamate
Sodium salt of acifluorfen

Sodium salt of fomesafen
Spinosad
Streptomycin
Sulfentrazone
Sulfosate (Sulfonium, trimethyl-salt with N-(phosphonomethyl)glycine (1:1))
Sulfosulfuron
Sulfur dioxide
Sulprofos
Synthetic isoparaffinic petroleum hydrocarbons
Tarta emetic
Tebuconazole
Tebufenozide
Tebuthiuron **(also found in California drinking water)**
Tefluthrin
Terbacil (1998 Drinking Water Contaminant Candidate List)
Terbufos (1998 Drinking Water Contaminant Candidate List)
Tetradifon
Tetrachlorvinphos
Tetraconazole
Thiabendazole
Thiazopyr
Thidiazuron
Thifensulfuron methyl (methy-3-[[[[(4-methoxy-6-methyl-1,3,5-triazin-2-
    yl)amino]carbonyl] amino]sulfonyl]-2-thiophene carboxylate
Thiobencarb **(also found in California drinking water)**
2-(Thiocyano-methylthio)benzothiazole
Thiodicarb
Thiophanate-methyl
Tralkoxydim
Triadimefon
Thiram
Tralomethrin
Triasulfuron
Triazamate
Tribenuron
Tribuphos
Trichlorfon
S-2,3,3-Trichloroallyl diisopropylthiocarbamate
Triclopyr
Trifloxystrobin
Triflumizole
Trifluralin **(also found in California drinking water)**
Triflusulfuron methyl
Triforine
3,4,5-Trimethylphenyl methylcarbanate and 2,3,5-trimethylphenyl
    methylcarbamate
Triphenyltin hydroxide
Vinclozolin
Zinc phosphide
Ziram

APPENDIX 2-9

Comparison of Chemicals Required to Be Monitored in Groundwater

Groundwater Appendix IX Chemicals	Drinking Water	Priority Pollutant	Ambient Criteria
Acenapthene	**yes**	**yes**	**yes**
Acenapthlyene	no	yes	no
Acetone	no	no	no
Acetophenone	no	no	no
Acetonitrile	no	no	no
2-Acetylaminoflourene	no	no	no
Acrolein	no	yes	no
Acrylonitrile	no	yes	yes
Aldrin	no	yes	yes
Allyl chloride	no	no	no
4-Aminobiphenyl	no	no	no
Aniline	no	no	no
Anthracene	no	yes	yes
Aramite	no	no	no
Arsenic	**yes**	**yes**	**yes**
Barium	yes	no	no
Benzene	**yes**	**yes**	**yes**
Benzanthracene	no	yes	yes
Benzo[b]flourantheneno	yes	yes	
Benzo[k]flourantheneno	yes	yes	
Benzo[ghi]peryleneno	yes	no	

(continued)

Full title: Comparison of Chemical Required to Be Monitored in Groundwater by RCRA. Under the Resource Conservation and Recovery Act, CFR 40, Part 264, Appendix IX with USEPA Regulate Chemicals under Drinking Water Standards, Priority Pollutant List, or Ambient Water Quality Criteria.

Groundwater Appendix IX Chemicals	Drinking Water	Priority Pollutant	Ambient Criteria
Benzo[a]pyrene	yes	yes	yes
Benzyl alcohol	no	no	no
Beryllium	yes	yes	no
alpha BHC	no	yes	yes
beta BHC	no	yes	yes
delta BHC	no	yes	no
Lindane	yes	yes	yes
Bis(2-chloroethoxy)methane	no	yes	no
Bis(2-chloroethyl)ether	no	yes	yes
Bis(2-chloro-1-methylethyl)ether	no	no	no
Bis(2-ethylhexyl)phthalate	no	yes	yes
Bromodichloromethaneno	yes	no	
Bromoform	no	yes	yes
4-Bromophenyl phenyl ether	no	yes	no
Butyl benzyl phthalate	no	yes	yes
Cadmium	no	yes	yes
Carbon disulfide	no	no	no
Carbon tetrachloride	yes	yes	yes
Chlordane	yes	yes	yes
p-Chloroaniline	no	no	no
Chlorobenzene	yes	yes	yes
Chlorobenzilate	no	no	no
p-Chloro-m-cresol	no	yes	no
Chloroethane	no	no	no
Chloroform	no	yes	yes
2-Chloronaphthalene	no	yes	yes
2-Chlorophenol	no	yes	no
4-Chlorophenyl phenyl ether	no	yes	no
Chlorprene	no	no	no
Chromium	yes	yes	yes
Chrysene	no	yes	yes
Cobalt	no	no	no
Copper	yes	yes	yes
m-Cresol	no	no	no
o-Cresol	no	no	no
p-Cresol	no	no	no
Cyanide	yes	yes	yes
2,4-D	no	no	yes
4,4'-DDD	no	yes	yes
4,4'-DDE	no	yes	yes
4,4'-DDT	no	yes	yes
Diallate	no	no	no
Dibenz[a,h]anthracene	no	yes	yes
Dibenzofuran	no	no	no
Dibromochloromethane	no	yes	no
1,2-Dibromo-3-chloropropane	yes	no	no
1,2-Dibromoethane	no	no	no

Groundwater Appendix IX Chemicals	Drinking Water	Priority Pollutant	Ambient Criteria
Di-n-butyl phthalate	no	no	yes
1,2-Dichlorobenzene	**yes**	**yes**	**yes**
1,3-Dichlorobenzene	no	yes	yes
1,4-Dichlorobenzene	**yes**	**yes**	**yes**
3,3'-Dichlorobenzidine	no	yes	yes
trans-1,4-Dichloro-2-butene	no	no	no
Dichlorodifluoromethane	no	yes	no
1,1-Dichloroethane	no	yes	yes
1,2-Dichloroethane	**yes**	**yes**	**yes**
1,1-Dichloroethylene	**yes**	**yes**	**yes**
trans-1,2-Dichloroethylene	**yes**	**yes**	**yes**
2,4-Dichlorophenol	no	yes	yes
2,6-Dichlorophenol	no	no	no
1,2-Dichloropropane	yes	yes	no
cis-1,3-Dichloropropene	no	yes	no
trans-1,3-Dichloropropene	no	yes	no
Dieldrin	no	yes	no
Diethyl phthalate	no	yes	yes
Thionazin	no	no	no
Dimethoate	no	no	no
Benzenamine	no	no	no
Benz[a]anthracene	no	yes	yes
Benzeneethanamine	no	no	no
2,4-Dimethylphenol	no	yes	no
Dimethyl phthalate	no	yes	yes
m-Dinitrobenzene	no	no	no
2,4-Dinitrophenol	no	yes	no
2,4-Dinitrotoluene	no	yes	no
2,6-Dinitrotoluene	no	yes	no
Dinoseb	yes	no	no
Di-n-octyl phthalate	no	yes	no
1,4-Dioxane	no	no	no
Diphenylamine	no	no	no
Disulfoton	no	no	no
Endosulfan I	no	yes	yes
Endosulfan II	no	yes	yes
Endosulfan sulfate	no	yes	yes
Endrin	**yes**	**yes**	**yes**
Endrin aldehyde	no	yes	yes
Ethylbenzene	yes	yes	no
Ethyl methacrylate	no	no	no
Ethyl methanesulfonate	no	no	no
Famphur	no	no	no
Fluoranthene	no	yes	yes
Fluorene	no	no	no
Heptachlor	yes	yes	no
Heptachlor epoxide	yes	yes	no

Groundwater Appendix IX Chemicals	Drinking Water	Priority Pollutant	Ambient Criteria
Hexachlorobenzene	yes	yes	yes
Hexachlorobutadiene	no	yes	yes
Hexachlorocyclopentadiene	yes	yes	yes
Hexachloroethane	no	yes	yes
Hexachlorophene	no	no	no
Hexachloropropene	no	no	no
2-Hexanone	no	no	no
Indeno(1,2,3-cd)pyrene	no	yes	yes
Isobutyl alcohol	no	no	no
Isodrin	no	no	no
Isophorone	no	yes	yes
Isosafrole	no	no	no
Kepone	no	no	no
Lead	yes	no	yes
Mercury	yes	yes	yes
Methacrylonitrile	no	no	no
Methapyrilene	no	no	no
Methoxychlor	no	no	yes
Methyl bromide	no	yes	no
Chloromethane	no	yes	yes
3-Methylcholanthrene	no	no	no
Dibromomethane	no	no	no
Methylene chloride	no	yes	yes
Methyl ethyl ketone	no	no	no
Methyl iodide	no	no	no
Methyl methacrylate	no	no	no
Methyl methanesulfonate	no	no	no
Methyl parathion	no	no	no
4-Methyl-2-pentanone	no	no	no
Naphthalene	no	yes	no
1,4-Naphthoquinone	no	no	no
1-Naphthylamine	no	no	no
2-Naphthylamine	no	no	no
Nickel	no	yes	yes
o-Nitroaniline	no	no	no
m-Nitroaniline	no	no	no
p-Nitroaniline	no	no	no
Nitrobenzene	no	yes	yes
o-Nitrophenol	no	yes	no
p-Nitrophenol	no	yes	no
4-Nitroquinoline 1-oxide	no	no	no
N-Nitrosodi-n-butylamine	no	no	no
N-Nitrosodiethylamine	no	no	no
N-Nitrosodimethylamine	no	yes	yes
N-Nitrosodiphenylamine	no	yes	yes
N-Nitrosodipropylamine	no	yes	yes
N-Nitrosomethylethylamine	no	no	no

Groundwater Appendix IX Chemicals	Drinking Water	Priority Pollutant	Ambient Criteria
N-Nitrosomorpholine	no	no	no
N-Nitrosopiperidine	no	no	no
N-Nitrosopyrrolidine	no	no	no
5-Nitro-o-toludine	no	no	no
Parathion	no	no	yes
PCB's	**yes**	**yes**	**yes**
PCDD's	no	no	no
PCDF's	no	no	no
Pentachlorobenzene	no	no	no
Pentachloroethane	no	no	no
Pentachloronitrobenzene	no	no	no
Pentachlorophenol	**yes**	**yes**	**yes**
Phenacetin	no	no	no
Phenanthrene	no	no	no
Phenol	no	yes	no
p-Phenylenediamine	no	no	no
Phorate	no	no	no
2-Picoline	no	no	no
Pronamide	no	no	no
Ethyl cyanide	no	no	no
Pyrene	no	yes	yes
Pyridine	no	no	no
Safrole	no	no	no
Selenium	**yes**	**yes**	**yes**
Silver	**yes**	**yes**	**yes**
2,4,5-TP	yes	no	no
Styrene	yes	no	no
Sulfide	no	no	no
2,4,5-T	no	no	no
2,3,7,8-TCDD	yes	no	no
1,2,4,5-Tetrachlorobenzene	no	no	yes
1,1,1,2-Tetrachloroethane	no	yes	no
1,1,2,2-Tetrachloroethane	no	yes	yes
Tetrachloroethylene	yes	no	yes
2,3,4,6-Tetrachlorophenol	no	no	no
Sulfotepp	no	no	no
Thallium	**yes**	**yes**	**yes**
Tin	no	no	no
Toluene	**yes**	**yes**	**yes**
o-Toluidine	no	no	no
Toxaphene	**yes**	**yes**	**yes**
1,2,4-Trichlorobenzene	**yes**	**yes**	**yes**
1,1,1-Trichloroethane	yes	yes	no
1,1,2-Trichloroethane	yes	no	yes
Trichloroethylene	**yes**	**yes**	**yes**
Trichlorofluoromethane	no	yes	no
2,4,5-Trichlorophenol	no	no	yes

Groundwater Appendix IX Chemicals	Drinking Water	Priority Pollutant	Ambient Criteria
2,4,6-Trichlorophenol	no	yes	no
1,2,3-Trichloropropane	no	no	no
o,o,o-Triethyl phosphorothioate	no	no	no
syn-Trinitrobenzene	no	no	no
Vanadium	no	no	no
Vinyl acetate	no	no	no

Bold-faced compounds occur in all three regulatory categories.

APPENDIX 3-1

General Drinking Water Monitoring and Warning Requirements (as of 2002)

Lead and Copper

Monitoring for lead and copper is required for both Community Water Systems (CWS) and Noncommunity Water Systems (NCWS). Tap water samples must be collected every 6 months. However, if the system meets the action levels (i.e., Maximum Contaminant Levels, see Table 1-9) and maintains optimal corrosion control treatment for two consecutive six-month periods, tap water sampling can be reduced to once a year. For those systems that exceed the action levels, samples of source water must also be collected.

All systems that exceed the lead action level must provide EPA-developed educational materials to their customers within 60 days. A warning must also be added to each water bill and repeated every 12 months as long as the lead exceeds the action level. This warning notice must be given to all daily and weekly newspapers and brochures delivered to specified locations in the service area. In addition, public service announcements must be given every 6 months to at least five radio and television stations in the service area.

Total Trihalomethane Compounds

CWSs that serve a population of more than 10,000 and add chlorine or bromine disinfectant to the water must monitor for total trihalomethane compounds.[6] If the water source used by the CWS is a surface body of water, the CWS must monitor on a quarterly basis, if it is a groundwater

---

[6]The Stage 2 Disinfection By-Product Rule (May, 2002) will add monitoring for the Haloacetic acids, bromate and chlorite.

source only annual monitoring is needed.[7] If a groundwater source exceeds the action level, however, monitoring must be increased to a quarterly sampling. If the average of four consecutive quarterly samples exceeds the total trihalomethane action level, public notification is required.

Inorganic Chemicals

The Group I inorganic chemicals are barium, cadmium, chromium, mercury, and selenium while the Group II inorganic chemicals are antimony, berylium, cyanide, nickel, sulfate, and thallium. The Group I and II inorganic chemicals for both CWS and Nontransient Noncommunity Water System (NTNCWS) must be monitored once every 3 years for groundwater systems and annually for surface water systems. Both systems having at least three previous Group I sampling results below the Maximum Contaminant Level will be granted a waiver and can then be monitored once every 9 years. Any supply exceeding the maximum contaminant level must be sampled quarterly for at least 1 year until the four consecutive samples are below the maximum contaminant level. Public notification is required if the average of the past four quarters exceeds the maximum contaminant level.

All public water systems must monitor for nitrate. CWS and NTNCWS systems that use surface water as their source must monitor quarterly. All public water systems that use groundwater as their source must monitor annually, whereas NCWS that use surface water as their source must also monitor annually. If the maximum contaminant level is exceeded, the public water supply must notify the public.

All public water systems must monitor for nitrite. Any CWS or NTNCWS that exceed the maximum contaminant level must be sampled quarterly for at least 1 year. When at least four consecutive samples are less than the maximum contaminant level, monitoring can be reduced to once a year. If the maximum contaminant level is exceeded, the public water supply must notify the public.

CWS and NTNCWS must also monitor for arsenic. Any supply exceeding the maximum contaminant level will be sampled quarterly for at least one year until the average of four consecutive samples are below the maximum contaminant level. Sampling will then resume with one sample every 3 years for public water systems that use groundwater for their source and annually for surface water sources.

---

[7]Because the formation of trihalomethane compounds results from the reaction of the disinfectant with organic matter in water and groundwater usually has less organic matter than surface water, the frequency of monitoring is reduced for groundwater sources.

CWS are required to monitor for fluoride. When the average of four original and three repeat samples exceeds the maximum contaminant level, public notification is required.

Volatile Organic Chemicals

CWS and NTNCWS must be monitored for benzene, carbon tetrachloride, o-dichlorobenzene, p-dichlorobenzene, 1,2-dichlorethane, 1,1-dichloroethylene, cis-1,2-dichloroethylene, trans-1,2-dichloroethylene, dichloromethane, 1,2-dichloropropane, ethylbenzene, monochlorobenzene, styrene, toluene, trichloroethylene, 1,1,2-trichloroethane, 1,1,1-trichloroethane, 1,2,4-trichlorobenzene, vinyl chloride, and xylenes. If any chemical is detected above the maximum contaminant level, the system must be sampled quarterly. If the average of four consecutive samples is below the maximum contaminant level, sampling can be reduced to once a year for the next three years. If the average of four consecutive quarterly samples exceeds the maximum contaminant level, the system is required to notify the public.

Synthetic Organic Chemicals

CWS and NTNCWS must be monitored for alachlor, aldicarb, aldicarb sulfone, aldicarb sulfoxide, atrazine, carbofuran, chlordane, dalapon, dibromochloropropane, dinoseb, diquat, 2,4-D, endothall, endrin, ethylene dibromide, glyphosate, heptachlor, heptachlor epoxide, lindane, methoxychlor, oxamyl, pentachlorophenol, picloram, PCB, simazine, toxaphene, 2,4,5-TP, benzo(a)pyrene, di-(ethylhexyl)adipate, di-(ethylhexyl)phthalate, hexachlorobenzene, hexachlorocyclopentadiene, and dioxin. If any of these chemicals is detected, the system must be sampled quarterly for at least a year. After 1 year of quarterly monitoring, the sampling may be reduced to once a year if the results are below the maximum contaminant level. After 3 years of annual monitoring below the maximum contaminant level, the system must return to four quarterly samples every 3 years. If the average of four consecutive quarterly samples exceeds the maximum contaminant level, the system is required to notify the public.

Treatment Chemicals

Each public water supply must certify annually that acrylamide and epichlorohydrin do not exceed the maximum contaminant level. No notification is required.

APPENDIX 3-2

National Drinking Water Contaminant Occurrence Database Data on Primary Water Quality Standards (May 18, 2001)

Chemical	Resource (*) (μg/L)	Average (μg/L)	Range (μg/L)	MCL
Inorganic Compounds				
Antimony	Surface Water(50)	3.059	0.03-50.0	6.0
	Groundwater (133)	3.0266	0.2-50.0	
Arsenic	Surface Water(98)	165.4088	0.28-1000	50.0
	Groundwater (677)	65.4899	0.2-1000	
Barium	Surface Water(259)	64.0223	0.2-2000	2000
	Groundwater (1196)	145.5923	0.19-9200	
Beryllium	Surface Water(42)	36.7646	0.01-50	4.0
	Groundwater (123)	27.9408	0.2-50	
Cadmium	Surface Water(67)	32.9276	0.01-50	5.0
	Groundwater (213)	64.147	0.6-2000	
Chromium	Surface Water(106)	242.2496	0.05-2000	100.0
	Groundwater (544)	146.7289	0.07-14000	

(continued)

Chemical	Resource (*) (μg/L)	Average (μg/L)	Range (μg/L)	MCL
Copper	Surface Water(60)	269.1479	0.5-2580	1300.0
	Groundwater (746)	194.6182	0.2-23000	
Cyanide	Surface Water(22)	2844.3333	1-10000	200.0
	Groundwater (53)	2194.19	2-32700	
Fluoride	Surface Water(316)	609.14	0.77-4060	4000
	Groundwater (1627)	720.5208	300-1500	
Lead	Surface Water(90)	7.5624	0.10-110	15.0
	Groundwater (572)	7.4019	0.1-1900	
Mercury	Surface Water(87)	12.8257	0.1-25	2.0
	Groundwater (222)	20.7166	0.1-3900	
Nitrate	Surface Water(313)	1065.15	10-12000	10000
	Groundwater (1943)	2142.55	0.6-558000	
Nitrite	Surface Water(47)	67.60	1-1770	1000
	Groundwater (231)	62.43	1-1800	
Selenium	Surface Water(85)	553.09	0.41-14000	50
	Groundwater (320)	143.2883	0.3-14000	
Thallium	Surface Water(42)	192.26	0.04-10000	2.0
	Groundwater (160)	576.63	0.039-43100	
Organic Compounds Alachlor (Lasso)				
	Surface Water(37)	0.7338	0.03-2.2	2.0
	Groundwater (9)	1.3458	0.08-4.4	
Atrazine	Surface Water(106)	1.8995	0.08-42	3.0
	Groundwater (82)	1.1293	0.06-12	

Chemical	Resource (*) (µg/L)	Average (µg/L)	Range (µg/L)	MCL
Benzene	Surface Water(42)	3.1768	0.13-128	5.0
	Groundwater (157)	104.209	0.098-43213	
Benzo(a)pyrene	Surface Water(8)	0.7167	0.02-1.0	0.2
	Groundwater (12)	0.3981	0.02-0.84	
Carbofuran	Surface Water(6)	0.6417	0.2-4.1	40.0
	Groundwater (2)	0.4252	0.2-2	
Carbon tetrachloride	Surface Water(92)	1.408	0.1-12	5.0
	Groundwater (179)	71.8988	0.13-37738	
Chlordane	Surface Water(0)			
	Groundwater (3)	0.43	0.28-0.57	2.0
Chlorobenzene	Surface Water(84)	3.8011	0.1-38	100.0
	Groundwater (72)	159.5636	0.0009-22220	
2,4-D	Surface Water(60)	1.1787	0.1-58	70.0
	Groundwater (52)	0.8706	0.083-8	
Dalapon	Surface Water(35)	12.1473	0.6-68	200.0
	Groundwater (21)	3.9365	0.37-25	
DBCP	Surface Water(87)	0.6599	0.00001-85	0.2
	Groundwater (146)	0.4261	0.0008-100	
O-Dichlorobenzene	Surface Water(41)	2229.81	0.1-100100	600.0
	Groundwater (70)	1.1767	0.11-21	
P-Dichlorobenzene	Surface Water(50)	5.1068	0.1–75	75.0
	Groundwater (94)	0.9326	0.03-11	

(continued)

Chemical	Resource (*) (µg/L)	Average (µg/L)	Range (µg/L)	MCL
1,2-Dichloroethane	Surface Water(44)	6.8174	0.1-500	5.0
	Groundwater (125)	2.0671	0.17-15	
Cis-1,2-Dichloro-ethylene	Surface Water(53)	408.3739	0.0012-100100	70.0
	Groundwater (177)	3.7655	0.0008-94.8	
1,1-Dichloroethylene	Surface Water(42)	7.9406	0.1-2300	7.0
	Groundwater (130)	2.457	0.11-35.2	
Trans-1,2-Dichloroethylene	Surface Water(31)	2277.12	0.1-100100	100.0
	Groundwater (92)	3.7371	0.11-61	
Dichloromethane	Surface Water(297)	166.1186	0.0005-100100	5.0
	Groundwater (1079)	2.576	0.0004	620
1,2-Dichloropropane	Surface Water(43)	1301.6	0.0005-100100	5.0
	Groundwater (76)	577.2092	0.0013-100000	
Di(2-Ethylhexyl) Adipate	Surface Water(32)	10.2818	0.02-180	400.0
	Groundwater (158)	1.1036	0.01	132
Di(2-Ethylehexyl) Phthalate	Surface Water(15)	2.1824	0.11-36	6.0
	Groundwater (181)	0.8267	0.01-65	
Dinoseb	Surface Water(16)	0.5273	0.1-4	7.0
	Groundwater (11)	0.4934	0.1-2	
2,3,7,8-TCDD (Dioxin)	Surface Water(0)	0.0		
	Groundwater (0)	0.0		
Diquat	Surface Water(12)	1.8857	0.3-24	20.0
	Groundwater (15)	2.7111	0.3-41.8	
Endothall	Surface Water(1)	9.0	9.0-9.0	100.0
	Groundwater (4)	1142.2	1.9-4548	

Chemical	Resource (*) (μg/L)	Average (μg/L)	Range (μg/L)	MCL
Endrin	Surface Water(11)	0.0338	0.008-0.1	2.0
	Groundwater (10)	0.1487	0.008-1.1	
Ethylbenzene	Surface Water(80)	2.6042	0.001-100	700.0
	Groundwater (331)	524.6229	0.0008-274750	
Ethylene dibromide	Surface Water(88)	1.3187	0.00001-28	.05
	Groundwater (118)	0.2316	0.0001-7.14	
Glyphosate	Surface Water(0)	700		
	Groundwater (2)		97-11	
Heptachlor	Surface Water(6)	0.107	0.043-0.388	0.4
	Groundwater (9)	0.1764	0.003-1.34	
Heptachlor exoxide	Surface Water(6)	0.0627	0.02-0.07	0.2
	Groundwater (12)	0.1095	0.005-0.99	
Hexachlorobenzene	Surface Water(5)	0.9859	0.018-1.2	1.0
	Groundwater (2)	1.2	1.2-1.2	
Hexachloro-cyclopentadiene	Surface Water(38)	0.4873	0.051-2	50.0
	Groundwater (8)	2.6335	0.184-4.4	
Lindane	Surface Water(10)	0.4949	0.005-0.8	0.2
	Groundwater (13)	0.3417	0.008-0.3417	
Methoxychlor	Surface Water(9)	0.3738	0.06-1	40.0
	Groundwater (6)	0.2729	0.05-1	
Oxamyl (Vydate)	Surface Water(3)	1.3	1.3-1.3	200
	Groundwater (2)	1.3	1.3-1.3	

(continued)

Chemical	Resource (*) (µg/L)	Average (µg/L)	Range (µg/L)	MCL
Polychlorinated biphenyls	Surface Water(1)	2.9		0.5
	Groundwater (6)	0.8485	0.1-5.7	
Pentachlorophenol	Surface Water(18)	0.4052	0.04-1	1.0
	Groundwater (23)	0.495	0.04-1.64	
Picloram	Surface Water(19)	0.7185	0.108-2	500
	Groundwater (16)	0.5188	0.11-1.1	
Simazine	Surface Water(75)	0.6383	0.07-4.89	4.0
	Groundwater (38)	0.4684	0.02-2.5	
Styrene	Surface Water(36)	18.4037	0.044-660	100
	Groundwater (90)	1.5169	0.03-10	
Total trihalomethanes	Surface Water(196)	40.2774	0.19-3295	100
	Groundwater (207)	16.0413	0.08-18400	
2,4,5-TP (Silvex)	Surface Water(16)	0.148	0.04-0.574	50.0
	Groundwater (26)	2.3825	0.01-10.5	
1,2,4-Trichloro-benzene	Surface Water(19)	2.1989	0.5-10	70.0
	Groundwater (68)	1.3614	0.15-21	
1,1,1-Trichloro-ethane	Surface Water(100)	2.8379	0.1-87	200.0
	Groundwater (355)	8.1193	0.31-2200	
1,1,2-Trichloro-ethane	Surface Water(55)	1.9938	0.0005-85	5.0
	Groundwater (70)	1.3514	15-11.7	
Tetrachloroethylene	Surface Water(114)	5.6896	0.0065-2900	5
	Groundwater (544)	5.6103	0.0004-1400	
Toluene	Surface Water(153)	2.957	0.0005-31.3	1000
	Groundwater (473)	818.33	0.0005-574750	

Chemical	Resource (*) (μg/L)	Average (μg/L)	Range (μg/L)	MCL
Trichloroethylene	Surface Water(93)	9.3806	0.1-590	5.0
	Groundwater (313)	7.4014	0.01-3668.9	
Vinyl chloride	Surface Water(26)	1.0993	0.07-7.3	2.0
	Groundwater (57)	1.3955	0.13-12.7	
Xylenes	Surface Water(103)	5.0308	0.001-110	10000
	Groundwater (405)	1639.84	0.053-1334750	

* = Number of community water systems that detected a specific chemical in the drinking water distributed to their consumers.

APPENDIX 3-3

National Drinking Water Contaminant Occurrence Database: Data on Unregulated Compounds

Chemical Data Summary - Group I

Chemical	Resource (*) (µg/Ll)	Average (µg/L)	Range
Cyanazine (Bladex)	Surface Water(72)	1.8968	0.31-12
	Groundwater(9)	2.9477	0.449-12
Methyl-tert-butyl-ether	Surface Water(19)	2.0455	0.5-8
	Groundwater(103)	16.1573	0.4-87
Toxaphene	Surface Water(3)	0.7	0.7-0.7
	Groundwater(2)	0.625	0.1-0.7
Trifluralin	Surface Water(7)	0.0771	0.032-0.2
	Groundwater(1)	0.056	0.056-0.056

Chemical Data Summary - Group II

Aldicarb	Surface Water(29)	3917.8898	0.5-231000
	Groundwater(8)	848.9462	0.5-10000
Aldicarb Sulfone	Surface Water(50)	2016.1246	0.3-231000
	Groundwater(16)	505.0142	0.0001-10000
Aldicarb Sulfoxide	Surface Water(52)	2164.2112	0.0088-231000
	Groundwater(13)	552.2183	0.0019-10000
Aldrin	Surface Water(16)	1.2834	0.07-4.7
	Groundwater(19)	0.1976	0.07-0.84
Butachlor	Surface Water(8)	11.3093	0.04-24
	Groundwater(6)	5.2226	0.04-17
Carbaryl	Surface Water(32)	1.8448	0.18-4
	Groundwater(8)	734.5587	0.18-10000

(continued)

293

Dicamba	Surface Water(71)	5.0339	0.0007-500
	Groundwater(39)	29.546	0.026-1600
1,3-Dichloropropene	Surface Water(35)	2328.6805	0.0005-100100
	Groundwater(86)	0.827	0.08-39
Trans-1,	Surface Water(6)	0.5182	0.2-0.7
3-Dichloropropene	Groundwater(7)	0.675	0.3-1.7
Dieldrin	Surface Water(12)	2.1903	0.004-4.4
	Groundwater(12)	0.2749	0.0001-1.65
3-Hydroxycarbofuran	Surface Water(44)	2.7073	0.0015-66.3
	Groundwater(13)	577.8382	0.0015-10000
Methomyl	Surface Water(29)	1.4984	0.29-3
	Groundwater(14)	525.0304	0.0001-10000
Metolachlor	Surface Water(234)	1.5316	0.00001-130
	Groundwater(53)	82.9208	0.0007-10000
Metribuzin	Surface Water(36)	4.7731	0.1-230
	Groundwater(11)	358.7057	0.0001-10000
Propachlor	Surface Water(40)	1.1259	0.0001-3
	Groundwater(5)	910.4548	0.0025-10000

Chemical Data Summary - Group III

Chemical	Resource (*) (µg/L)	Average (µg/L)	Range
Bromobenzene	Surface Water(31)	6.1237	0.0034-43.2
	Groundwater(80)	256.1072	0.0005-22220
Bromochloromethane	Surface Water(64)	4.4931	0.0005-42.1
	Groundwater(131)	3.6694	0.0001-210
Bromomethane	Surface Water(74)	4.9135	0.0005-49.9
	Groundwater(246)	73.1414	0.0001-22220
N-Butylbenzene	Surface Water(26)	1.8158	0.0001-10
	Groundwater(76)	1.8354	0.0005-84
Sec-Butylbenzene	Surface Water(16)	4.4421	0.0005-22
	Groundwater(59)	223.2537	0.0005-15590
Tert-Butylbenzene	Surface Water(15)	2.4667	0.0002-10
	Groundwater(66)	0.8206	0.0005-9
O-Chlorotoluene	Surface Water(46)	15.5815	0.0005-239
	Groundwater(80)	240.439	0.0005-22220
P-Chlorotoluene	Surface Water(43)	16.1985	0.0005-239
	Groundwater(65)	314.4116	0.0002-22220
Dibromomethane	Surface Water(99)	714.805	0.0005-100100
	Groundwater(141)	1.595	0.0005-21
M-Dichlorobenzene	Surface Water(41)	2090.1363	0.0005-100100
	Groundwater(88)	0.7825	0.0005-21
1,1-Dichloroethane	Surface Water(59)	722.7888	0.0004-100100
	Groundwater(285)	2.5805	0.0001-500

Chemical	Resource (*) (µg/L)	Average (µg/L)	Range
Dichlorodifluoromethane	Surface Water(50)	3.6793	0.0005-88
	Groundwater(211)	8.0765	0.0001-556.9
1,3-Dichloropropane	Surface Water(20)	4767.693	0.0005-100100
	Groundwater(65)	0.4645	0.0005-3.7
2,2-Dichloropropane	Surface Water(21)	1.2459	0.0005-6
	Groundwater(63)	1.4564	0.0005-61
1,1-Dichloropropene	Surface Water(25)	3851.4901	0.0001-100100
	Groundwater(59)	0.4812	0.0005-5
Hexachlorobutadiene	Surface Water(23)	1.7616	0.0005-10
	Groundwater(14)	0.4929	0.0005-8
Isopropylbenzene	Surface Water(18)	0.5178	0.0005-2.6
	Groundwater(102)	197.448	0.0005-28463
P-Isopropyltoluene	Surface Water(17)	0.5496	0.0005-5
	Groundwater(68)	466.3746	0.0005-35325
Naphthalene	Surface Water(65)	2.1892	0.0001-55
	Groundwater(217)	774.3238	0.0001-243700
N-Propylbenzene	Surface Water(22)	1.9701	0.0001-21
	Groundwater(83)	1952.4286	0.0005-206763
1,1,1,2-Tetrachloro-ethane	Surface Water(31)	0.6672	0.0005-6.1
	Groundwater(101)	61.6757	0.0005-2700
1,1,2,2-Tetrachloro-ethane	Surface Water(30)	0.5651	0.0003-2
	Groundwater(96)	0.8696	0.0005-11
1,2,3-Trichlorobenzene	Surface Water(35)	1.4514	0.0005-10
	Groundwater(102)	0.5478	0.0004-7
Trichlorofluoromethane	Surface Water(107)	1.8953	0.0001-45
	Groundwater(291)	8.5342	0.0001-1444
1,2,3-Trichloropropane	Surface Water(31)	2.1565	0.0005-21
	Groundwater(84)	31.1628	0.0001-3000
1,2,4-Trimethylbenzene	Surface Water(61)	1.3424	0.0005-16.8
	Groundwater(202)	3390.2383	0.0001-978250
1,3,5-Trimethylbenzene	Surface Water(35)	1.5561	0.0002-10.9
	Groundwater(137)	1523.8481	0.0005-288750

* = Number of community water systems that detected a specific chemical in the drinking water distributed to their consumers.

Group I: Approximately 2,500 Community Water Systems conducted analyses for the listed chemicals.

Group II: Approximately 10,000 Community Water Systems conducted analyses for the listed chemicals.

Group III: Approximately 20,000 Community Water Systems conducted analyses for the listed chemicals.

Data collected as of May 18, 2001

{Appendix 3-3}

APPENDIX 3-4

Examples of Bottled Mineral Water Chemistry

Bottled Mineral Water	Location	TDS	Total Ions	pH	Na	K	Ca	Mg	Cl	HCO3	SO4	SiO2	NO3	F	pCi/L	Bottle	Remarks
Abatilles	Arcachon, France	259	317	NR	74.5	2.8	16.4	8	95	112	7.8	NR	NR	NR	15.6	NR	No information on source
Abbey Well	Morpeth, United Kingdom	390	256	NR	62	4.8	60	28	75	NR	25	NR	0.3	0.6	10.6	GLASS	Well, white sandstone aquifer, travel time 3,000 years
Aberfoyle	Ontario, Canada	380	541	NR	33	2	100	37	68	240	61	NR	0.2	NR	11.1	NR	Spring, sand aquifer, pasted through a 0.2 micron filter, Zn = 0.11 ppm
Acetosella	Napoli, Italy	1120	1633	6.3	50.7	9.9	276	58	106	1086	20	26.5	NR	0.3	12.9	GLASS	No information on source
Acqua Amerino	Terni, Italy	470	672	7.1	11	1.8	139	8.2	21	396	77	13	4	1	2.6	GLASS	Spring, no geologic information, Sr = 1.8 ppm; also bottled as Amerino
Acqua della Madonna	Naples, Italy	1311	1831	6	73	15	304	62	170	1135	28	37	6.2	0.3	21.3	GLASS	Spring, no geologic information, Sr = 1.8 ppm
Acqua Fucoli	Siena, Italy	2680	2717	6.7	23.2	2.9	615	89	30.2	492	1450	12.5	1.5	1.1	3.5	GLASS	No information on source, Chianciano spa

Acqua Santa di Chianciano	Siena, Italy	3398	3644	6.8	40.5	6.1	714	172	19	830	1810	48	2.2	2	No Value	GLASS	Spring, no geologic information
Acquarossa	Catania, Italy	1260	1622	6.5	158.5	13.5	120.2	150.9	70.8	1108	NR	NR	NR	NR	4	GLASS	No information on source
Adelbodner	Switzerland	2040	2022	NR	5.5	1.4	520	35	5.6	291	1160	NR	3	0.2	No Value	NR	Li .01 mg/l Adelboden, 1400 m altitude
Adobe Springs	California, USA	364	672	8.4	6.2	ND	4.4	110	6	529	13	NR	3.1	ND	No Value	PLASTIC	Spring, geology unknown, trucked to botteling plant
Aemilia	Parma, Italy	630	915	7.2	48.3	3.3	76.1	71.4	23.7	543	102.6	17	29.8	NR	2.1	GLASS	No information on source, Sr = 0.8 ppm
Aix les Bains	Savole, France	312	481	7.3	2	1	84	23	3	341	27	NR	0.2	0.2	No Value	NR	No information on source
Aksu	Russia	1000	2200	NR	NR	300	NR	100	300	500	1000	NR	NR	NR	No Value	NR	No source information
Alborz	Iran	330	574	7	23	2.5	90	26	22	320	90	NR	NR	NR	2.2	NR	Alborz Mountain, no source information
Alpine	Ohio, USA	400	283	NR	48	2.7	42	ND	NR	NR	190	ND	NR	0.2	No Value	PET	Source unknown, bromate = ND
Amata	Bari, Italy	444	712	6.8	20	2.3	99	42	29	496	5.7	18	NR	NR	3.7	NR	No information on source
Ambo	Ethiopia	NR	1518	NR	252	39	22	46	32.5	1126	0.77	NR	NR	NR	No Value	NR	No source information

(continued)

Bottled Mineral Water	Location	TDS	Total Ions	pH	Na	K	Ca	Mg	Cl	HCO$_3$	SO$_4$	SiO$_2$	NO$_3$	F	pCi/L	Bottle	Remarks
Apemin Tusnad	Tusnad, Romania	1525	301	NR	90.4	10	129	71.7	NR	NR	NR	NR	NR	NR	No Value	GLASS	Volcanic region, mineral water
Apollinaris	Eifel region, Germany	2650	2650	NR	410	20	100	130	100	1810	80	NR	NR	NR	5.1	GLASS	Spring, no geologic information
Aqua	El Sadat City, Egypt	253	347	NR	38	3.5	30.4	13	24	195.2	15	28	NR	NR	2	NR	No information on sources
Aqua di Nepi	Nepi, Italy	568	797	5.7	26	44	84	27	18	459	36	94	8	1.4	No Value	GLASS	Spring, no geologic information
Aqua Luna	Bad Leonhard spfunzen, Germany	620	627	7.4	7.3	1.6	109	27	18.1	435	14.8	14.2	0.1	NR	2.2	GLASS	Spring, no geologic information
Aqua Viva	Serbia and Montenegro	329	449	NR	9.17	2.01	88.09	12.88	13.51	305	17.77	NR	NR	0.2	1	NR	No source information
Arvie	Ardes, France	2520	3656	NR	650	130	170	92	387	2195	31	NR	NR	0.9	No Value	GLASS	No information on source, Zn = 10 ppm
Avra	Greece	263	340	7.4	9.7	1.2	68.1	6.7	10.7	223.3	16.1	NR	4.4	NR	1.1	NR	No information on source
Azzurra	Vicenza, Italy	428	524	7.7	1.1	0.9	88	34	1.3	227	167	NR	5	NR	No Value	NR	No information on source
Bad Driburger	Eggegebirge, Germany	NR	712	NR	7.1	3.1	122	32.6	6.8	361	179	NR	0.3	NR	No Value	NR	No information on source

Name	Location																Notes
Badoit	St. Galmier, France	1200	1851	6	150	10	190	85	40	1300	40	35	NR	1	0.6	GLASS	Spring, granite aquifer
Balda	Venezia, Italy	298	443	7.6	9.6	2.4	56	29	4	323	19	NR	NR	0.1	No Value	NR	No information on source
Baldovska	Slovak Republic	2479	2414	NR	90	0.6	378.9	93.7	78	1557	215	NR	0.1	0.2	13.9	NR	No source information
Baraka	Baraka, Egypt	440	593	NR	62	5	60	26	50	325	45	20	NR	NR	5.8	NR	No information on source
Beckerich	Luxemburg	360	421	7.4	2.8	0.6	96.5	5	6.3	274	29.6	NR	6.5	NR	0.6	NR	No source information
Bella	Malta	326	341	7.4	64	3	53	11	80	130	NR	NR	NR	NR	12.7	NR	No source information
Bianca Neve	Bergamo, Italy	261	394	NR	0.6	0.2	59.6	26.5	0.5	271.5	28.5	3.5	3.3	NR	No Value	GLASS	No information on source
Biborteni	Biborteni, Romania	1640	464	6	102	7.2	270.7	84	NR	NR	NR	NR	NR	NR	No Value	NR	No information on source
Birute	Lithuania	4900	4490	NR	1300		160	2100	310	620			NR	NR	57.2	NR	No source information
Blue Keld	Throstle Nest, United Kingdom	340	147	7.6	8	3.5	88	3	17	NR	6.8	NR	20.2	0.1	0.7	GLASS	Artesian well, aquifer 1000 feet deep, no geologic information
Boario	Brescia, Italy	639	741	7.2	5	2	130	40	5	305	239	10	5	0.3	No Value	GLASS	No information on source, Sr = 5 ppm

(continued)

Bottled Mineral Water	Location	TDS	Total Ions	pH	Na	K	Ca	Mg	Cl	HCO_3	SO_4	SiO_2	NO_3	F	pCi/L	Bottle	Remarks
Borjomi	Georgia	NR	7500	7.5	2000		ND	ND	500	5000	ND	NR	NR	NR	92.3	NR	Borjomi Valley at 950 m, no source information
Borsec	Borsec Spa, Romania	1402	2325	6.5	53	12	310	97	29	1800	24	NR	NR	NR	No Value	GLASS	No information on source
Bracca	Bergamo, Italy	681	769	7.3	19.8	1.5	132	41.3	32.2	256.3	277.9	6.6	0.9	0.4	4.8	NR	No information on source, Sr = 1.6 ppm, Li = 0.07
Bulgaria	Bulgaria	255	179	9.4	65.5	1.6	1.6	NR	NR	85.4	20.6	NR	NR	4	No Value	NR	Zn - 0.05 mg/l As - 0.001 mg/l Li - 0.046 mg/l
Buxton	Buxton, United Kingdom	280	397	7.4	24	1	55	19	37	248	13	NR	0.1	NR	6.1	NR	Spring, limestone aquifer, travel time of 5,000 years
Calistoga, Mineral	California, USA	571	692	7.7	183.7	14.5	1.7	1.7	224.9	36.3	83.5	145	NR	0.9	31.8	GLASS	Spring, geology unknown
Capannelle Rosablu	Roma, Italy	682	1024	6.2	39	61	131	27.7	31.9	604	20	68	40	1	No Value	GLASS	No information on source, Sr = 2.5, Br = 0.1
Carola	Ribeauville, France	NR	836	NR	114	7	83	24	57	414	136	NR	1	NR	5.8	NR	No information on source

Castello	Brescia, Italy	281	455	7.5	0.9	0.1	62.3	33.9	0.9	339	10.5	2.4	5	NR	No Value	NR	No information on source
Catskill Mountains	New York, USA	530	287	6.7	39	2.2	110	22	110	NR	NR	NR	3.1	0.2	18.7	NR	Spring, geology unknown
Cesar	St Alban les Eaux, France	1750	2713	6.5	350	46	220	70	21	2000	6	NR	ND	NR	No Value	GLASS	No information on source
Chateauneuf	Chateauneuf les Bains, France	2151	3092	NR	651	40	152	36	215	1799	195	NR	1	3	13.1	GLASS	No information on source
Chateldon	Chateldon, France	1882	3031	6.2	240	35	383	49	7	2075	240	NR	NR	2	No Value	GLASS	No information on source
Chaud-fontaine	Thermale Bron, Belgium	385	510	7	44	2.5	65	18	35	305	40	NR	0.1	0.4	4.7	PLASTIC	Spring, no geologic information
Ciego Montero	Cuba	550	619	NR	NR	NR	98	38.9	61.7	300.5	120.2	NR	NR	NR	No Value	NR	No information on source
Cinar	Turkey	2714	728	6.3	385	NR	200	85.1	NR	NR	12	45.5	NR	NR	No Value	NR	Mt. Uludag, Zn .12, As .01, Cd .003, Cu .02 (all ppm)
Clever	Czech Republic	2995	2908	NR	412	NR	185.2	107.1	144.2	2058	0.185	NR	NR	0.6	No Value	NR	No source information
Comprad	Macedonia	823	279	NR	NR**	NR	82	91.2	NR	NR	98.8	NR	6.8	0.3	No Value	NR	Spring, geology unknown
Contrex	Contrexville, France	2125	2184	NR	9.1	3.2	486	84	8.6	403	1187	NR	2.7	NR	No Value	GLASS	No information on source
Cottorella	Lazio, Italy	282	427	7.4	3.9	0.4	96	2.1	9	305	6	4	0.4	NR	1.3	GLASS	No information on source

(continued)

Bottled Mineral Water	Location	TDS	Total Ions	pH	Na	K	Ca	Mg	Cl	HCO₃	SO₄	SiO₂	NO₃	F	pCi/L	Bottle	Remarks
Courmayeur Fonte Youla	Aosta, Italy	2264	2131	7.4	1	2	517	67	NR	168	1371	5	NR	NR	No Value	GLASS	No information on source
Crystal Geyser	Calistoga, California, USA	590	416	NR	130	8.7	12	3.1	260	NR	2.6	NR	NR	NR	42.2	GLASS	Spring, geology unknown
Cutolo Rionero	Potenza, Italy	505	656	6.1	72.4	29	50	13.1	34	298	54	78	26.4	0.9	No Value	GLASS	No information on source
Dad Vilbel Urquelle	Gad Vilbel, Germany	NR	1135	NR	90	13.6	174	25.8	93.6	702	36	NR	NR	NR	8.1	PLASTIC	Spring, no geologic information
DEA	Poland	340	346	7.9	6	1.5	65.7	9.7	12	226	25	NR	NR	0.3	1.1	NR	No source information
Don Carlo	Salerno, Italy	620	966	6.7	15.2	3.7	189	25.5	23.5	683	14.4	8.4	3.1	NR	1.7	GLASS	No information on source
Dorna	Karpates, Romania	NR	3907	NR	22	2.8	360	11.8	10.2	3500	NR	NR	NR	NR	No Value	NR	No information on source
Eden	Israel	NR	641	7.8	32	3.5	26	18	24	198	6	NR	15	NR	2	NR	No source information
Egeria	Roma, Italy	644	898	5.7	48	58	94.9	23.6	31.9	494	27	85.6	33	1.6	No Value	GLASS	Spring, no geologic information
Eureka	Lecce, Italy	381	495	7.3	33/5	3	66.5	30.1	59.6	311	NR	NR	24.3	NR	8.9	NR	No information on source

Evian	France	309	492	7.2	5.5	0.8	78	24	2.2	357	10	14	1	NR	No Value	PLASTIC	Spring, French/Swiss Alps, glacial aquifer
Fabia	Terni, Italy	427	617	7.4	14	1.2	134	5.4	25.6	381	28.4	9	18	NR	3.8	GLASS	No information on source
Faito	Napoli, Italy	317	474	7.5	12	9	73	18	16.6	311	12.6	20.1	1.6	0.2	No Value	NR	No information on source
Famous Mineral Water	Texas, USA	893	688	8.2	NR**	2	38	15	145	488	NR	NR	0.1	0.2	25.4	NR	Well, As was 2 ppb, C= 8 ppb and copper 6 ppb
Ferrarelle	Val D'Assano, Italy	1270	1958	6.1	49	43	362	18	21	1372	6	81	5	0.8	No Value	GLASS	Southern Italy Mtn, geology unknown
First	China	405	378	7.9	22.3	2	53.8	19	15.1	217.5	47.5	NR	NR	0.3	1.3	NR	No information on source
First	Phillipines	330	393	7	40	NR	48	14	15	259	17	NR	NR	NR	No Value	NR	No source information
Fonte Camorei	Cuneo, Italy	276	409	7.5	4.9	1.1	62.1	6.7	6.5	284	15.3	8.3	19.6	TRACE	0.3	GLASS	No information on source
Fonte Corte Paradiso	Udine, Italy	340	451	7.7	3.2	0.7	71	26	3.3	274	51	9.8	12	NR	TRACE	NR	No information on source
Fonte di Alice	Firenze, Italy	482	688	7.8	96.1	3	57.2	23.1	70.6	363	50.5	22.2	NR	NR	11	NR	No information on source, Sr = 2.2 ppm

(continued)

Bottled Mineral Water	Location	TDS	Total Ions	pH	Na	K	Ca	Mg	Cl	HCO$_3$	SO$_4$	SiO$_2$	NO$_3$	F	pCi/L	Bottle	Remarks
Fonte Santafiora	Arezzo, Italy	616	813	7.7	108	3.2	75.8	30	121.8	390	51	20.6	13	NR	20.3	GLASS	Spring, geology unknown, bottled at spring, Sr = 0.6 ppm, bottled as Perla
Gaia	Ancona, Italy	313	454	7.3	22	1.8	88.9	3.3	19.8	274	29.2	8.9	6.2	NR	2.3	NR	No information on source
Galvanina	La Galvanina, Italy	580	756	7.1	36	2.3	135	23	37	445	68	NR	9	0.4	5.2	GLASS	Spring, Italian Appenine Mtn., geology unknown
Garci	Tunisia	1798	2417	5.5	435	16	192	54	426	1229	65	NR	NR	NR	68.1	NR	No source information
Gaudianello	Potenza, Italy	1058	1502	5.8	130	47.2	140	51.3	36.8	877	117	100	2.4	NR	No Value	GLASS	No information on source, Sr = 1.8
Gesund-brunnen	Namibia	601	252	7.3	NR	NR	154	71	7.8	NR	7.8	NR	11.4	0.3	No Value	NR	No source information
Ghadeer	Petra, Jordan	380	405	7.3	34.5	0.7	54	24	50	220	22	NR	NR	NR	8.7	NR	No information on source
Hankavan-Lithia	Armenia	5000	6957	NR	1370	100	265	126	1765	2950	214	167	NR	NR	259	NR	Spring, no geologic information
Hartz	Tasmania, Australia	400	511	6.5	35	0.2	86	9	70	260	28	20	2.4	0.2	12.7	PLASTIC	Spring, no geologic information
Hassia	Bad Viebell, Germany	NR	1784	NR	228	26.7	186	36.1	121	1144	42	NR	NR	NR	4.5	PLASTIC	No information on source

Name	Source																
Hessen Quelle	Germany	NR	2243	NR	320	33/5	242	42.4	294	1296	49	NR	NR	NR	No Value	GLASS	No information on source
Hildon	Hampshire, United Kingdom	312	269	NR	7.7	NR	97	1.7	16	136	4	NR	6	ND	No Value	GLASS	Spring, no geologic information
Izvorul Alb	Romania	NR	764	7.2	1.5	0.5	49.2	6.1	5.8	701	NR	NR	NR	NR	0.6	NR	No information on source
Johanniter Quelle	Johanniter Quelle, Germany	NR	1931	NR	98	NR	264	94	80	1207	188	NR	NR	NR	No Value	GLASS	No information on source
Kaiser Friedrich Quelle	Roisdorfer Quelle, Germany	NR	4228	NR	1020	30	131	91	1034	1560	362	NR	NR	NR	171	GLASS	No information on source
Kaiserwasser	Bolzano, Italy	827	802	7.5	1.2	0.5	164	39.6	0.4	224	366	5.9	NR	0.8	No Value	GLASS	No information on source, Cu = 0.006 ppm, Sr = 1.9 ppm
Kekkuti	Hungary	1599	2243	NR	40	9.6	242	68	16	1804.3	50	12	0	1	No Value	NR	No source information
Kelechin	Ukraine	1700	1675	6	68	NR	275	41	13	1249	4	25	NR	0.2	No Value	NR	No source information
Kokshetau Mineral Water	Kazakhstan	360	249	NR	15	0.4	40	10	25	122	36	NR	NR	0.6	4.2	NR	Pb - 0.01 Ni - 0.002 Se - 0.0002 V - 0.04
Laurentina	Roma, Italy	1301	1861	6.4	98	142.5	230.5	36.5	35.5	1140	124.4	53.5	NR	0.1	No Value	GLASS	Source unknown
Lavaredo	Bolzano, Italy	1440	1496	7.4	2.4	0.7	320	72	0.6	231	860	7.6	NR	1.6	No Value	GLASS	No information on source
Le Choix du President	Feversham, Canada	NR	760	NR	20	1	132	41	32	256	278	NR	NR	0.4	5.1	GLASS	No information on source
Lete	Caseta, Italy	915	1438	6.1	5.1	2.2	321	17.5	7.6	1055	8.7	16.4	4.2	0.4	No Value	NR	No information on source

(continued)

Bottled Mineral Water	Location	TDS	Total Ions	pH	Na	K	Ca	Mg	Cl	HCO$_3$	SO$_4$	SiO$_2$	NO$_3$	F	pCi/L	Bottle	Remarks
L'Oiselle	St Amand, France	NR	1238	NR	86	21	164	78	144	312	430	NR	1.6	1	12.6	NR	No information on source
Luisen	Germany	NR	2352	NR	240	20.3	347	44.2	319	1336	45	NR	NR	NR	45.4	GLASS	No information on source
Luna	Orobic Alps, Italy	279	293	7.7	4.4	1.9	44.2	15.7	5.4	159.9	40.5	8.5	12.3	NR	No Value	PLASTIC	Spring, no geologic information
Madonna della Mercede	Parma, Italy	618	803	7.2	48.4	3.6	106.1	44/2	39.3	475	86.4	15.3	28.6	NR	4.8	GLASS	No information on source
Mangali	Latvia	800	800	NR	95	15	75	40	245	210	120	NR	NR	NR	35.2	NR	No source information
Manitou Mineral Water	Colorado, USA	1600	626	NR	150	23	290	43	NR	ND	120	ND	NR	NR	No Value	GLASS	Pikes peak location, geology unknown
Miers Alvignac	Alvignac, France	NR	4107	NR	445	9.2	420	257	NR	206	2770	NR	NR	NR	No Value	NR	Spring, no geologic information, Fe=0.1 ppm
Montinverno dei Colli di Parma	Parma, Italy	605	809	7.2	17.5	2.8	132.8	38.9	7.1	401	183.5	25	NR	0.2	No Value	GLASS	Spring, no geologic information
Montrolland	Senegal	NR	511	NR	69	4.2	64	9.7	121	219	24.5	NR	0.1	NR	19.5	NR	No source information
Nash	Ireland	450	527	7.2	22	2.3	25	32	436	9.2	13.8	NR	NR	NR	79	NR	No source information
Ninfa	Potenza, Italy	352	466	5.7	41.1	25.2	32.2	8.7	23.5	219	13.8	100	2.8	NR	No Value	NR	No information on source

Noah's	California, USA	380	570	8.3	6	ND	4.4	120	9	410	16	NR	5	ND	No Value	NR	Spring, geology unknown, trucked to bottling plant, Cr = 0.011 ppm
Oulmes	Morocco	1485	1842	NR	252	27.3	171.3	52.3	294.2	994	16	34.8	NR	NR	36.1	NR	No source information
Palmense del Piceno	Ascoli Piceno, Italy	665	633	7.2	35.5	2	120	21.5	81	323	39.5	9.8	NR	0.6	13.5	GLASS	Well, no information on source, Fe and Li = 0.005 ppm
Parot	St Romain le Puy, France	3100	4747	6.4	968	103	99	88.1	88	3380	18	NR	1	1.6	No Value	GLASS	No information on source
Passugger Heilquellen	Passugg, Switzerland	NR	1446	NR	46	3.4	286	24	19	1,020	48	NR	NR	NR	1.1	NR	Spring, digestive, geology unknown
Penafiel	Mexico	880	604	5.3	159	11	131	41	131	NR	130	NR	NR	0.5	16.8	NR	No source information
Penafiel	Tehuacan, Mexico	NR	604	5.3	159	11	131	41	131	130	NR	NR	NR	0.5	16.8	NR	Volcanic origin
Peria Harghitel	Sancraleni, Romania	687	1121	NR	70.6	8.4	112.2	43.8	16	854	16	NR	NR	NR	No Value	NR	Artesian, geology unknown
Perrier	Vergeze, France	478	615	5.9	9	0.4	147.3	3.4	21.5	390	33	9.9	ND	0.1	3.6	GLASS	Spring, geology unknown
Prata	Caserta, Italy	442	714	6.8	3.9	2.9	162.4	12.2	6.1	512	4.5	4.3	5.1	0.1	No Value	NR	No information on source
Prealpi	Bergamo, Italy	346	508	8	30.8	1.6	70.4	19.4	12.8	311	43.8	11.4	7	NR	1.2	NR	Spring, no geologic information

(continued)

Bottled Mineral Water	Location	TDS	Total Ions	pH	Na	K	Ca	Mg	Cl	HCO₃	SO₄	SiO₂	NO₃	F	pCi/L	Bottle	Remarks
Puits St Georges	St Romain le Puy, France	1615	2354	6.8	390	19	55	38	41	1762	9	33	4.2	3	No Value	GLASS	Spring, no geologic information
Quezac	Quezac, France	NR	2508	NR	255	49.7	241	95	38	1685	143	NR	1	NR	No Value	NR	No information on source
Ramiosa	Helsingborg, Sweden	817	271	NR	222	1.5	2.2	0.5	23	12	7.3	NR	NR	2.8	3.1	GLASS	Aquifer travel time, 70 years, geology unknown
Resan	Moldova	1288	1288	7.1	113		62.1	132.5	58.8	530.9	391.1	NR	NR	NR	10.7	NR	No source information
Rosbacher	Rosbach, Germany	NR	1623	NR	85	3.9	209	93	141	1079	12	NR	NR	NR	23.4	GLASS	No information on source
S. Andrea	Parma, Italy	606	806	7.6	73	3.9	59.8	56	17.2	457	139	NR	NR	NR	0.5	GLASS	Spring, no geologic information
S. Giorgio	Cagliari, Italy	278	343	7.1	45.5	1.9	23.3	16.7	79	138.5	9.3	28.7	0.2	0.2	13.2	GLASS	Spring, no geologic information
Saint Jean	France	905	1517	6	228	36	76	25	16	908	228	NR	NR	NR	No Value	NR	No information on source
Saint Justine	Canada	NR	1341	NR	415	3	7	6	350	560	NR	NR	NR	NR	62.6	NR	No information on source
Saint Martial	Martin Le Redon, France	331	496	7	4.8	0.4	114	3	11	356	6	NR	1	NR	1.6	NR	No information on source
Saint Nicholas	Cyprus	290	417	NR	18	NR	49	28	19	292	11	NR	ND	NR	No Value	NR	No information on source

San Pellegrino	North of Milan, Italy	1074	1143	5.2	43.8	2.9	208.6	58	66.2	222.7	540	NR	NR	0.7	10.2	GLASS	Spring, Mountains north of Milan, limestone and volcanic aquifers
Sanfaustino	Villa Sanfaustino, Italy	1207	2025	6.1	17.4	2.5	414	17.2	221	1244	90	15.2	3.8	0.2	39.1	GLASS	Spring, no geologic information, Sr = 0.9 also contains lithium
Schlossquelle Friedrichsroda	Naturpark Thuringer Wald, Germany	NR	1738	NR	110	3	296	106	105	171	945	NR	NR	1.5	17.3	GLASS	No information on source
Sergente Angelica	Umbria Region, Italy	300	447	7.2	4.3	0.7	104	1.4	7.2	317	4.4	7.7	NR	0.1	0.7	GLASS	No information on source
Sidi el Kebir	Algeria	297	384	7	34	NR	55	11	22	230	21	10.9	NR	NR	No Value	NR	No information on source
Socosani	Peru	NR	955	NR	134.3	10.1	91	69		651		NR	NR	NR	No Value	NR	No source information
Sole	Piedmont Region, Italy	412	659	NR	2.6	43	108	31.1	2.9	439	19.3	6	6.6	0.1	No Value	GLASS	No information on source
Solfurea	Pesaro, Italy	3260	3101	6.7	130	13	320	152	125	587	1750	23	NR	1	14.4	GLASS	No information on source
Sorgente del Bucaneve	Veneto Region, Italy	312	155	7.6	4	1	71	27	6	NR	46	NR	NR	NR	0.3	NR	No information on source
Sorgente dell'Amore	Calabria Region, Italy	332	479	7.4	10.3	1.6	89.6	15.5	8.5	320	22	8.7	2.5	NR	0.4	NR	No information on source

(continued)

Bottled Mineral Water	Location	TDS	Total Ions	pH	Na	K	Ca	Mg	Cl	HCO₃	SO₄	SiO₂	NO₃	F	pCi/L	Bottle	Remarks
Sorgente Pozzillo	Siclia Region, Italy	1280	1031	6.8	252	36.8	82.1	75	294	NR	220	27.7	43	0.5	29.9	GLASS	No information on source
Source Blanche	Fontaine Bonneleau, France	NR	501	NR	10	1	110	5	14	325	10	NR	26	NR	1.8	NR	Spring, no geologic information, also bottled as Valoise
Source Louise	Merignies, France	456	632	7.3	37	12	91	20	25	395	50	NR	2	NR	No Value	NR	No information on source
Source Pioule	Luc en Provence, France	647	731	7.3	33.5	1.1	140.8	11.4	38.2	351	117.7	37	NR	NR	6.2	NR	No information on source
St. Leonhards-quelle	Bavaria, Germany	485	554	NR	5.4	1	93	24	13.1	393	9.4	15	0.1	0.1	1.6	GLASS	Spring, geology unknown, As=0.008, Cu=0.082, Zn=0.01, Sr=0.14, Fe=3.7
Tesalia	Ecuador	NR	1492	NR	147	8	46	148	NR	1120	23	NR	NR	NR	No Value	NR	No source information
Tipperary	Tipperary, Ireland	272	410	7.7	25	17	37	23	15	282	10	NR	0.5	NR	No Value	GLASS	Devil's Bit Mtn., geology unknown
Vals	Vals les Bains, France	NR	2022	6.2	453	32.8	45.2	21.2	27.2	1403	38.9	NR	1	NR	No Value	NR	Spring, no geologic information
Valser St. Peters-quelle	Swiss Alps, Switzerland	NR	1883	NR	10.1	NR	436	54.7	2.4	390.5	988	NR	NR	1.1	No Value	NR	Artesian, thermal, diretic and purgative, geology unknown

Name	Location																Geology
Vernet	Lalevade, France	675	1029	NR	192	28.7	33.5	17.6	6.4	734	14	NR	1	1.3	No Value	NR	Spring, geology unknown
Verniere	Commune des Aires, France	NR	1654	NR	154	49	190	72	18	1170	NR	NR	NR	1	No Value	NR	No information on source
Vichy Celestins	Vichy, France	3325	4719	6.8	1172	66	103	10	235	2989	138	NR	NR	6	No Value	GLASS	Spring, no geologic information
Vila	Bosnia - Herzegovina	843	819	NR	34.5	7.8	154	22.8	42.2	488	70	NR	NR	NR	2.5	NR	Sr - 220 mg/l, no source information
Viscachani	Bolivia	1473	1705	6.9	280	20.5	19.6	20.6	64	1192	60.2	45	3	NR	No Value	NR	No source information
Vittel	Vosge region, France	403	497	7.5	7.3	4.9	91	19.9	NR	258	106	9	0.6	NR	No Value	NR	Spring, mountain region, sandstone aquifer
Wattwiller	du Haut-Rhin, France	1092	1136	NR	3	1.4	288	20.1	3.9	142	678	NR	ND	NR	No Value	NR	Spring, no geologic information
Wihelmsthaler Brunnen	Kassel, Germany	NR	2188	NR	91.4	NR	310	111	92.7	1347	236	NR	NR	NR	No Value	GLASS	No information on source
Yukon Spring	Whitehorse, Canada	342	498	7.3	6.1	1.7	84.7	13.9	29.1	305	46.9	9.9	0.3	0.1	4.1	NR	Spring, no geologic information, As=0.001, Fe=0.13, Zn=0.009, Mn = 0.001

** = Based on the ion balance this water may contain Na
pCi = calculated using equation 3-1
No Value = No data for calculation or a non-positive number
TRACE = Less than 0.1 ppm
NR = Not reported
ND = Not detected
Note: All elemental concentrations are in mg/L (ppm)

APPENDIX 3-5

Examples of Bottled Water Chemistry

Bottled Water	Location	TDS	Total Ions	pH	Na	K	Ca	Mg	Cl	HCO₃	SO₄	SiO₂	NO₃	F	pCi/L	Bottle	Remarks
Abita Springs	Louisiana, USA	160	157	8.2	47	NR	ND	ND	3.3	97	9	NR	NR	0.2	No Value	PLASTIC	Well, 1800 feet deep aquifer
Acqua Panna	Tuscany, Italy	188	187	8.2	6.5	0.9	30.2	6.9	7.1	100	21.4	8.2	5.7	0.1	0.6	GLASS	Spring, Apennines Mtn., Sr = 0.2 ppm
Acquafine	Brazil	157	497	4.9	14	4.4	6.6	2.3	0.7	459	0.9	NR	8.8	0.1	No Value	NR	Sr - 0.04 ppm Ba - 0.035 ppm, no information on source
Acqual del Cardinale	L'Aquila, Italy	110	168	7.7	0.8	0.3	39.3	0.5	3.7	122	0.6	NR	0.8	NR	0.4	NR	No information on source
Adalsan	North Korea	NR	9	7	0.9	0.5	4	1.2	NR	2.5	TRACE	NR	NR	TRACE	OR	NR	No source information
Adelholzener	Bad Adelholzen, Germany	NR	488	NR	10	1.1	69.8	31	22.5	330	23.1	NR	0.1	0.2	3.3	PLASTIC	No information on source
Agua Salus	Uruguay	124	66	NR	7.2	0.71	36	8.9	7.1		5.7	NR	0.7	NR	0.7	NR	Sierra de Minas, no source information
Agua Sana	Spain	28	23	NR	5.8	NR	1.1	0.6	8.7	6.7	NR	NR	NR	0.4	OR	NR	No source information
Aiqua d'Andorra	Andorra	NR	128	7.6	3.7	0.5	26.9	1.2	5.6	54.3	26.9	6.5	2.5	0.2	0.6	NR	No information on source
Al Nabek	Qasr al Mushatta, Jordan	246	295	7.3	23	5.1	44	12	43	149	19	NR	NR	NR	4.4	NR	No information on source
Alardo	Portugal	25	27	NR	3.4	NR	0.8	NR	2	6.7	NR	12	2	NR	OR	NR	No source information
Alba Sergente	Vicenza, Italy	44	61	7.3	1.8	0.2	3	3	1	30.5	7.7	7	2.5	NR	OR	GLASS	No information on source

Alhambra	California, USA	NR	12	7.2	3.8	NR	0.7	0.8	3.5	NR	3.4	NR	NR	NR	OR	PET	Well water, no geologic information
Almaha	Lebanon	250	230	7.5	13.1	3.2	65.2	6.1	22.5	90	22	NR	7.4	NR	1.9	NR	Zn 0.016 mg/l Ba 0.0049 mg/l Be 0.0035 mg/l
Alpi Bianche	Cuneo, Italy	49	65.7	7.5	1.2	NR	14	NR	NR	39	9	2.5	NR	NR	OR	NR	No information on source
Alpia	Verbania, Italy	55	78	8.1	3.5	1.2	5.7	3.4	5.7	33.7	4.7	16.7	3.4	0.2	OR	NR	No information on source
Alpine	Malaysia	75	393	7.2	5	4.2	20	16	12	335.6	ND	NR	NR	NR	No Value	NR	No source information
Alpine	Rwanda	NR	116	7.4	27.8	3.8	18.9	7.3	43		11	NR	4.3	NR	5.3	NR	No source information
Al-Quassim	Saudi Arabia	160	127	7.5	31.2	0.54	7.6	1.2	34.8	18.2	19.3	NR	13.3	0.8	6	NR	No source information
Amaro	St-Cutbert, Canada	NR	140	NR	12	1	17	4	2	98	6	NR	NR	NR	No Value	NR	Spring, no geologic information
Amorosa	Carerra, Italy	20	28	5.7	4	0.1	6.7	0.6	6.7	4.3	0.9	4.2	0.4	TRACE	OR	NR	No information on source
Antrim Hills	N. Ireland, United Kingdom	208	98	6.9	12	NR	30	13	20	NR	23	NR	NR	NR	No Value	PLASTIC	Spring, basalt aquifer
Appalachian Springs	Georgia, USA	24	0	6	NR	NR	NR	NR	NR	NR	<1	NR	NR	<0.1	OR	PLASTIC	Spring, Appalachian Mountains, wilderness area
Aqua Africa	South Africa	95	35	7.6	17	1	7	2	5	NR	ND	NR	2	0.5	0.1	NR	No source information
Aqua Fiji	Viti Levu, Fiji	250	288	7.5	10.2	1.3	33	16	6.8	160	NR	61	NR	NR	0.3	PLASTIC	Artesian, volcanic aquifer
Aqua Silva	Emillian Apennines, Italy	159	202	7.8	4.9	0.7	31.5	9.5	4.7	107	9.5	33.3	0.8	ND	0.3	NR	Spring, no geologic information

(continued)

Bottled Water	Location	TDS	Total Ions	pH	Na	K	Ca	Mg	Cl	HCO$_3$	SO$_4$	SiO$_2$	NO$_3$	F	pCi/L	Bottle	Remarks
Aqua Siwa	Egypt	170	206	NR	26	19.5	6.4	8.3	24.5	86.6	22	12	NR	0.6	No Value	NR	No information on source
Aqua-Pura	Armathwaite, United Kingdom	125	43	6.4	8	2	7	2.3	11	NR	8	NR	5	NR	0.6	PLASTIC	Well, 100 meters deep, sand stone aquifer, travel time is decades
Aquarel	Belgium	208	282	NR	2	NR	70	NR	NR	210	NR	NR	NR	NR	No Value	NR	No information on source
aQuelle	Ekhamanzi Spring, South Africa	52	35	6.4	8	4	3	2	11	NR	5	BR	2.3	0.1	OR	PLASTIC	No information on source
Arrowhead	California, USA	125	159	8.1	11.1	1.8	19.7	3.8	10.8	80.2	3.4	28.1	NR	NR	0.7	PLASTIC	Spring, San Bernardino Mountains
Aura	Thailand	242	375	7.1	4.8	1.3	69	8.1	0.6	237	4.6	47	1.9	0.3	No Value	NR	Mae Rim, Chian Mai Zn .08 mg/l As <.005 mg/l
Aurele	Aurele, France	NR	426	NR	4.1	1.9	98	4	3.6	269	43	NR	2	NR	No Value	NR	No information on source
Auzat	Auzat, France	28	29	6.8	1.5	0.4	3	0.6	0.6	5.2	8.7	7.5	1	NR	OR	NR	No information on source
Avita	Michigan, USA	160	215	NR	1	0.3	45.9	10.3	NR	151	6	NR	NR	0.1	No Value	GLASS	Artesian, Ausable State Forest
Awa	Ivory Coast	NR	319	NR	21.5	4.3	56.7	2.7	8.7	216	8.6	NR	NR	NR	No Value	NR	No source information
Beber	Vicenza, Italy	128	219	NR	0.7	0.3	28	16	1	168	4.5	NR	NR	NR	No Value	NR	No information on source

Name	Location																Source description
Bernina	Sondrio, Italy	39	53	7.2	1	0.9	8.6	0.8	0.8	22.6	8	6.5	3.5	NR	OR	NR	No information on source
Biovive	Dax, France	193	256	7.7	18.6	2.4	42	3.8	17.1	157.3	14.9	NR	ND	NR	1.4	NR	Spring, no geologic information
Bistra	Slovenia	225	359	7.5	9.5	0.9	58.7	11.9	1.5	252	2.3	19.5	2.2	NR	No Value	NR	No source information
Bistra	Croatia	200	100	NR	1.8	0.8	81.8	10.4	1.6	NR	2.5	NR	NR	1	No Value	NR	No source information
Black Mountain	California, USA	44	45	7.2	8.3	0.7	25	0.7	10	NR	NR	NR	NR	NR	OR	PLASTIC	Spring, San Francisco Peninsula, no geologic information, carbon filtered
Brecon Carreg	Wales, United Kingdom	198	95	7	5.7	0.4	47.5	16.5	9	NR	9	5.1	2.2	NR	1.3	PLASTIC	Spring, 240 feet deep sandstone and basalt aquifer (National Park)
Bru	Bru, Belgium	160	290	NR	8	1.5	23	22	4	209	5	17	0.7	NR	No Value	GLASS	Spring, geology unknown
Cachantun	Chile	NR	41	NR	2.1	0.4	7.6	1	2.8	10.3	8.5	8.5	NR	NR	OR	NR	No information on source
Café De Rome	Signes, France	249	416	NR	3	1	55	23	55	271	7	NR	1	NR	9.4	GLASS	Spring, no geologic information
Calabria	Catanzaro, Italy	92	97	7.4	10	1.5	9.2	4.1	18.7	33.9	4.9	13.5	0.8	NR	2.2	NR	No information on source
Calistoga	California, USA	130	71	7.3	10.2	5.4	4.4	1.9	NR	47.6	1.5	NR	NR	ND	No Value	PLASTIC	Spring, from Mayacmas, Sierra Nevada and Palomar Mtn.

(continued)

Bottled Water	Location	TDS	Total Ions	pH	Na	K	Ca	Mg	Cl	HCO$_3$	SO$_4$	SiO$_2$	NO$_3$	F	pCi/L	Bottle	Remarks
Canaqua	Chilliwack, Canada	200	109	7.8	5	NR	86	16	1.3	NR	NR	NR	NR	0.3	No Value	PLASTIC	Spring, faulted bedrock
Capes	Guadeloupe	NR	363	6.9	22	5.5	58	24	53	140	59	NR	1.3	0.1	6	NR	Eau de source de Dole', Gourbeyre
Carlsberg Kurvand	Denmark	220	309	NR	30	2	50	5	40	147	35	NR	0.1	NR	5.9	NR	No information on source
Carolina Mountain	North Carolina, USA	60	21	6.9	3.5	1.5	6.7	0.3	NR	NR	9.3	NR	NR	TRACE	OR	PLASTIC	Spring, National Forest
Castellina	Isernia, Italy	160	244	7.4	2.1	0.3	54	2.5	2.8	175	2.2	3.5	1	0.2	0.2	GLASS	No information on source
Cavagrande	Catania, Italy	150	308	7.7	34.4	7.8	10.4	7	24.8	NR	223	NR	NR	0.4	No Value	NR	No information on source, Sr = 0.044
Cedar Springs	Oro Mountain, Canada	236	58	8.1	3	NR	41.5	NR	1.1	NR	12.7	NR	NR	NR	No Value	PLASTIC	Spring, sand aquifer, transported to bottleing plant
Cerist	Llawr Cae, United Kingdom	19	24	6.3	3	0.4	3.2	1.5	8.1	NR	6.6	NR	0.8	NR	OR	NR	No information on source
Chamonix	Cape Chamonix, South Africa	NR	33	NR	11.8	1	2.2	1.2	15.4	NR	0.3	NR	0.6	0.1	OR	PLASTIC	Spring, no geologic information
Chispal	Argentina	120	88	NR	31.2	NR	10.3	1.4	NR	30	5	NR	10	0.1	No Value	NR	No information on source
Cloud Juice	Tasmania, Australia	NR	32	NR	9.4	0.9	0.5	NR	19	NR	2	NR	0.5	NR	OR	GLASS	Rain water
Club	Indonesia	215	209	7.2	8.7	3.4	20.4	13.3	4.8	152.6	5.1	NR	0.1	0.2	No Value	NR	Desa Lemahbang, Pandaan, Sukorejo

																Source
Colorado Crystal	Colorado, USA	95	4	7	1	18	3.2	25.2	0.3	12.7	NR	NR	NR	No Value	GLASS	Spring, 10,000 ft. above sea level, volcanic aquifer
Cool Blue	Putaruru, New Zealand	123	86	6.9	NR	NR	NR	NR	NR	NR	NR	NR	NR	No Value	PLASTIC	Spring, no geologic information
Cooroy Mountain	Queensland, Australia	NR	91	NR	19.2	0.7	NR	NR	1.2	NR	2	74.6	NR	OR	PLASTIC	Spring, igneous aquifer
Coraiba	Fonti San Damiano, Italy	203	319	7.8	0.4	0.5	2.4	1.4	5.4	NR	1.7	NR	NR	No Value	NR	Spring, Maritime Alps, no geologic information
Corona	Pisa, Italy	175	223	6.7	31.8	0.9	8.3	2.3	NR	43.5	16.5	NR	1.8	No Value	NR	No information on source
Cristal	Costa Rica	183	31	7.7	5.8	2.8	49	19.2	1.4	231	9.2	2.5	0.1	No Value	NR	No information on source
Cristal Roc	Ardenay sur Merize, France	NR	325	NR	10.5	1.8	14.8	12.1	0.4	33.7	125	10	NR	2.1	NR	No information on source
Crystal	Mauritius	NR	84	7	17	1	71.6	7.9	NR	NR	NR	NR		OR	NR	No source information
Crystal	Djibouti	NR	60	NR	15	3.4	18	2.2	18.5	198.5	19.5	2	NR	OR	NR	No information on source
Crystal Geyser	USA	120	20	NR	ND	1.9	0.3	14	21.6	NR	10	2.5	NR	No Value	PLASTIC	Springs, Multiple mountain sources, ND pesticides, trihalomethanes

(continued)

Bottled Water	Location	TDS	Total Ions	pH	Na	K	Ca	Mg	Cl	HCO$_3$	SO$_4$	SiO$_2$	NO$_3$	F	pCi/L	Bottle	Remarks
Crystal Spring	Adelaide Hills, Australia	NR	89	NR	24	2	2	4	45	8	4	NR	NR	NR	OR	PLASTIC	Spring, Mount Lofty Ranges, no geoloic information
Deep Rock	Colorado, USA	180	85	NR	15	63	5.2	NR	ND	NR	NR	NR	NR	1.4	No Value	NR	Artesian well
Deer Park	Maryland, USA	90	122	8.1	1.1	0.4	26.5	2.6	0.7	88.8	1.8	NR	NR	NR	No Value	PET	Allegheny Mt., woodland, carbon filtered, reverse osmosis & distillation
Delta	Egypt	240	331	NR	36	3.5	25.5	14.5	12	205	12	22	NR	0.2	No Value	NR	No information on source
Diamond	Arkansas, USA	170	105	7.6	NR**	1.6	74	3	7.2	NR	19	NR	NR	NR	0.1	GLASS	Limestone and shale aquifer, carbon filtration and ozonation for iron removal
Diamond	Tasmania, Australia	NR	49	NR	3.3	0.4	35	3.2	3.7	NR	3.3	NR	NR	NR	OR	PLASTIC	Spring, no geologic information
Ducale	Monte Zuccone, Italy	56	75	8.3	3	0.4	12.5	1.3	3.8	39.7	6.9	5.4	1.9	NR	OR	GLASS	Spring, bottled at source, no geologic information
Earth2O	Oregon, USA	94	0.3	NR	NR	NR	NR	NR	ND	NR	ND	NR	0.2	0.1	No Value	PLASTIC	Spring, contains 3 ppm arsenic

Name	Location																Source
Ech Chaafi	Mauritania	NR	165	7.8	34.8	3.1	16.3	5.1	NR	106	NR	NR	NR	NR	No Value	NR	No source information
Edge	New Zealand	98	47	7.9	6.9	0.7	12	1.9	3.9	NR	3.6	18	NR	NR	0.1	NR	Canterbury Plains, South Island
Edins X.O.	Sweden	NR	238	8	7.9	20	49	6.1	TRACE	140	15	NR	NR	TRACE	No Value	NR	Well, 150ppm dissolved oxygen
El Castano	Venezuela	74	36	7.6	NR	NR	NR	1.8	4.2	28	1	NR	NR	0.06	No Value	NR	No source information
Eldorado	Colorado, USA	70	49	NR	NR	2.6	8.7	2.3	3	31	NR	NR	0.7	0.3	No Value	PLASTIC	Spring, surrounded by federal/state parks, sandstone aquifers: As was ND
English Mountain	Tennessee, USA	68	94	7.7	0.5	NR	22	4.3	NR	67	NR	NR	NR	NR	OR	PLASTIC	Spring, Great Smoky Mountains
Esker	Quebec, Canada	NR	131	NR	2	1	21	4	NR	96	7	NR	NR	NR	No Value	PLASTIC	No information on source
Ethernal Water	Otakiri Valley New Zealand	135	152	7	NR	NR	23	10	NR	34	NR	85	NR	NR	No Value	PLASTIC	Artesian, 1000 feet deep aquifer, no geologic information
Falcon	Pakistan	242	306	7.8	28	2.4	50	46	6	152	20	NR	0.6	0.8	No Value	NR	No source information
Famous Drinking Water	Texas, USA	9	14	6.3	3	1	ND	1	2	6	1	NR	TRACE	TRACE	OR	NR	Well water
Fiji Water	Figi's main island	160	253	7.5	NR	NR	17	13	NR	140	NR	83	NR	NR	No Value	PET	Artesian, island volcanic aquifer
Fine	Shuzenji, Japan	178	261	7.6	9.6	1.7	9.7	4.7	5.3	63.7	5.7	76.2	NR	NR	No Value	GLASS	Well, 2100 feet deep volcanic aquifer

(continued)

Bottled Water	Location	TDS	Total Ions	pH	Na	K	Ca	Mg	Cl	HCO₃	SO₄	SiO₂	NO₃	F	pCi/L	Bottle	Remarks
Finn Spring	Finland	NR	50	6.5	20	NR	<3.9	1.1	26	3	NR	NR	<0.6	NR	OR	NR	Cd - <0.002 mg/l Pb - <0.01 mg/l Hg 0.0001 mg/l
Fiuggi	Fiuggi, Italy	122	148	6.8	6.4	4.4	15.9	6.3	13.9	81.7	NR	12.8	7	NR	No Value	GLASS	Spring, volcanic aquifer, Sr = 0.8 ppm
Font Selva	Catallunya, Spain	NR	350	NR	49.7	NR	33.9	5	12.4	215.8	10.6	22.5	ND	NR	No Value	PLASTIC	Spring, Les Guilleries Mtn., granite aquifer
Fonte Abrau	Cuneo, Italy	135	205	NR	0.8	0.2	26	12.5	NR	140	NR	25.4	NR	NR	No Value	NR	No information on source
Fonte Annia	Udine, Italy	190	339	7.8	NR	NR	53	23.7	NR	183	75	NR	4	NR	No Value	NR	No information on source
Fonte Azzurrina	Lucca, Italy	57	77	8.1	4	1.7	12.2	1	4.7	42.7	2.6	7	1.3	NR	OR	NR	No information on source
Fonte Caudana	Biella, Italy	48	23	7.6	2.6	0.7	7.6	3.3	1.4	NR	4.2	NR	3.5	NR	OR	NR	No information on source
Fonte della Buvera	Verbania, Italy	54	69	7.7	0.9	0.3	13	0.8	0.9	36.6	6.9	8	1.6	NR	OR	NR	No information on source
Fonte Laura	Como, Italy	211	332	7.8	1.7	0.7	46.6	22.4	1.4	237.9	10.1	5.5	5.9	NR	No Value	NR	No information on source
Fonteviva	Massa, Italy	31	35	5.7	4.8	0.1	2.7	1.2	8.8	8.2	2.2	6.3	0.5	NR	OR	GLASS	No information on source
Fonti Bauda	Savona, Italy	47	61	7.1	3.4	NR	6.4	1.1	3.7	27.5	6.1	10.3	2.7	NR	OR	NR	No information on source

Brand	Location														Source
Fountainhead	South Carolina, USA	55	10	7.5	ND	NR	9	0.3	NR	NR	NR	NR	0.4	OR	NR	Artesian, Sumter National Forest, granite aquifer, ozonated
Four Sixty Four	Twyford, United Kingdom	NR	251	NR	11	1.4	106	2.1	24	106	NR	NR	NR	3.4	NR	Well, no geologic information
Gibraltar	Ontario, Canada	NR	412	7.2	3	2	80	25	5	290	6	1.4	NR	No Value	PET	Well, dolomite aquifer, undeveloped land
Glacier	Victoria, Australia	53	53	NR	8	1	4	3.5	8.5	28	NR	NR	NR	OR	GLASS	Spring, no geologic information
Goccia di Carnia	Forni Avoltri, Italy	69	107	8.3	1.2	0.2	17.6	4	0.3	79	2.8	1.6	NR	OR	NR	No information on source
Good Hydration	Maryland, USA	NR	312	NR	5.1	1.1	75	19	9.4	202	NR	NR	NR	0.9	PLASTIC	No information on source
Great Bear	New York, USA	24	17	6.6	1.7	0.7	1.3	1	1.4	5.5	5.3	NR	NR	OR	POLYC	Spring, Woodland area, gray and white sand aquifer, carbon filtered, UV
Gulfa	United Arab Emirates	120	115	7.7	14	1.1	4.8	6.7	52	21	13	1.7	0.5	8.7	NR	No source information
Hawaiian Springs	Hawaii, USA	64	67	7.7	6	2.2	7.2	3.3	NR	44	4	NR	NR	OR	PLASTIC	Spring, volcanic aquifer
Hello	India	175	161	7.6	38	1.8	18	12	25	50	16	NR	0.5	3.3	NR	No information on source
Highland	Uganda	97	73	7.1	19.5	3.4	14	10.8	20	NR	NR	5	NR	1.3	NR	No source information

(continued)

Bottled Water	Location	TDS	Total Ions	pH	Na	K	Ca	Mg	Cl	HCO$_3$	SO$_4$	SiO$_2$	NO$_3$	F	pCi/L	Bottle	Remarks
Highland Spring	Perthshire, Scotland	136	200	7.8	6	0.6	35	8.5	7.5	136	6	NR	ND	ND	0.9	GLASS	Well, Ochill Hills, red sandstone and basalts
Hillcrest	Wyoming, USA	210	66	NR	NR**	2.4	47.2	16.4	NR	NR	NR	NR	NR	0.2	No Value	NR	Spring, Bigh Horm Mt. Range
Himalaya	Langtang, Himalaya Mountains	14	32	NR	0.8	0.7	1.3	0.4	NR	25.8	0.2	3	NR	ND	OR	NR	Spring, from snow melt through bedrock above 4,500 ft.
Ice Age	British Columbia, Canada	4	0	NR	ND	NR	NR	NR	NR	NR	NR	NR	NR	NR	OR	GLASS	Alpine creek from glacial melt at Toba Inlet
Ice Blue	Iceland	57	38	7.9	11.6	0.7	5.5	2.5	13.9		3.9	NR	0.1		OR	NR	No information on source
Ice Mist	Sweden	NR	166	8.7	115	8	33	9.5	NR	NR	NR	NR	NR	0.6	No Value	PLASTIC	Well, carbonaceous clay and slate aquifer
Ice Mountain	Maine, USA	19	56	7	1.1	0.8	1.3	1.3	1.1	46.1	4.7	NR	NR	NR	OR	PLASTIC	Spring, Mt. Zircon woodland, sand and gravel aquifer
Isbre	Hardanger Fjord, Norway	19	5	6.7	1	ND	2.6	ND	1.2	ND	ND	NR	0.2	ND	OR	PLASTIC	No information on source
Izvorul Minunilor	Stana de Vale, Romania	115	100	7.3	1.2	0.7	17.7	3.8	0.6	70	2.3	NR	3.7	0.1	No Value	PLASTIC	No information of origin, As = 0.001ppm, Co = 0.002, Br = 0.05

Name	Location																
Julia Sorgente Geu	Udine, Italy	112	187	8.1	0.3	0.2	28.5	7.9	0.3	132	1.8	0.6	15.1	NR	No Value	NR	No information on source
Keringet	Kenya	143	143	NR	34	NR	1.9	0.02	12.4	12.4	80	NR	0.3	1.7	No Value	NR	No source information
Kilimanjaro	Tanzania	NR	56	NR	15.5	5.2	NR	NR	35.5	NR	NR	NR	NR	0.1	OR	NR	No source information
Kingshill Forest Glade	Scotland, United Kingdom	NR	61	4.6	9.9	NR	9.9	3.8	35.5	NR	NR	NR	0.1	1.5	OR	NR	Spring, no geologic information
La Cascade	Togo	131	102	6.1	5.1	10	8.8	5	69.5	1	NR	2.4	NR	NR	6.1	NR	No source information
La Vie	Vietnam	232	346	7.6	60	4	23	8	NR	251		NR	ND	0.2	No Value	NR	No source information
Label'O	St Martin Le Redon	NR	127	NR	4.7	0.5	98.6	1.4	7.5	NR	14	NR	NR	NR	0.9	NR	No information on source
Laure Pristine	Tennessee, USA	55	24	7.6	ND	1.3	12	10	NR	NR	ND	NR	0.1	0.1	OR	PLASTIC	No information on source
L'eau de Source Alizee	Chambon la Foret, France	NR	444	NR	8.8	2.6	93	8.1	18	306	5.2	NR	2	NR	1.5	NR	No source information
Levico Casara	Trento, Italy	42	53	6.4	1.3	0.5	6.6	1.8	0.4	25.2	12	4.9	0.7	NR	OR	NR	No information on source
Llanllyr Source	Wales, United Kingdom	84	86	5.8	14	2	12	6	30	NR	17	NR	5	NR	4.1	GLASS	Spring, no geologic information
Loon Country	Maine, USA	120	52	8.1	ND	0.9	26	15	0.8	NR	9.1	NR	NR	ND	No Value	PET	Spring, Allagash Wilderness, naturally sodium free
Loubiere	Dominica	NR	54	NR	13	2.6	12	4.5	8.5	7.6	3	NR	2.3	NR	OR	NR	Loubiere Springs
Lynx	Fonti di San Fermo, Italy	165	245	7.5	2.4	NR	51.4	4.8	4	165	12.4	4.5	NR	NR	No Value	GLASS	Spring, no geologic information
Majan	Oman	120	132	NR	12	1.4	15.6	5.7	23	42	32	NR	NR	0.6	3.2	NR	No source information

(continued)

Bottled Water	Location	TDS	Total Ions	pH	Na	K	Ca	Mg	Cl	HCO₃	SO₄	SiO₂	NO₃	F	pCi/L	Bottle	Remarks
Mangiatorella	Reggio Calabria, Italy	67	77	6	9.8	0.8	5.8	1.4	12	26.5	4.5	15.8	NR	NR	OR	NR	No information on source
Mareb	Yemen	NR	460	NR	80.3	NR	47.7	22.1	NR	215	95	NR	NR	NR	No Value	NR	No source information
Maxim's	Arezzo, Italy	82	103	7.1	3.9	0.5	20.2	1.6	7	53.1	7.7	3.7	5	NR	0.8	NR	Spring, Maritimes Alps
Mayo	Republic of Congo	214	295	7.8	1.4	5.1	33	22	3	202	8	20	ND	0.2	No Value	NR	196 meter well Zn 0.02 mg/. Al <.01mg/l
Metzeral	Metzeral, France	NR	47	NR	2.7	0.9	7.1	2	2	24	6.6	NR	2	NR	OR	NR	No information on source
Monashee	Valemount, Canada	81	134	7.6	3	10.9	23.4	6.4	0.4	89.8	NR	NR	NR	NR	No Value	NR	Spring, Monashee Mountains, no geologic information
Mont Roucous	Mont Roucous, France	19	25	6	2.8	0.4	1.2	0.2	3.2	4.9	3.3	6.9	2.3	NR	OR	NR	No information on source
Montagne d'Arree	Commana, France	NR	30	NR	6.7	0.2	0.8	1	14	3.6	2	NR	1.6	NR	OR	NR	No information on source
Montclair	Cedar Valley Springs, Canada	NR	117	NR	1	0.2	66	17	5.7	NR	25	NR	1.6	NR	0.8	NR	Spring, no geologic information, Zn = 0.01ppm
Moschetta	Catanzaro, Italy	113	120	7	9.9	1.1	9.5	5.5	18.4	56.3	6.2	13	NR	0.1	2.5	NR	No information on source
Mount Franklin	Australia	NR	45	NR	14	1	3	5	20	2	NR	NR	NR	NR	OR	PLASTIC	No information on source

Mount Olympus	Utah, USA	56	29	7.2	3.4	0.5	7.9	2.4	5.9	NR	8.8	NR	NR	OR	POLYC	Spring, distillation, ozonation, reverse osmosis, carbon filter, deionization
Mountain Lite	Hinton, Canada	NR	72	8.1	1	1	58.7	11.5	ND	NR	NR	NR	NR	OR	NR	Spring, Wilmore Wilderness Park, no geologic information
Mountain Valley	Arkansas, USA	221	272	7.7	2.7	1	71	7.5	NR	190	NR	NR	0.2	No Value	PET	Spring, limestone/sandstone aquifer, wilderness area
Nanton Water	Nanton, Canada	NR	97	NR	NR	7.5	36.7	48.6	4	NR	NR	NR	0.5	OR	NR	Well, no source information
Naturally Boulder	Colorado, USA	160	68	8.2	61	1	3.9	0.2	1.9	NR	NR	0.1	NR	No Value	NR	Well, Arapahoe aquifer
NAYA	St. Andre d'Argenteuil, Canada	250	358	7.7	20	3	42	21	13	234	24	NR	0.2	0.3	PLASTIC	Spring, glacial aquifer, wilderness area; As, Cu and Zn not detected
NAYA	Mirabel, Canada	220	354	7.5	7	3	45	25	1.5	245	27	NR	0.5	0.2	No Value	PLASTIC	Spring, glacial aquifer, wilderness area; As, Cu and Zn not detected

(continued)

Bottled Water	Location	TDS	Total Ions	pH	Na	K	Ca	Mg	Cl	HCO$_3$	SO$_4$	SiO$_2$	NO$_3$	F	pCi/L	Bottle	Remarks
Nerea	Macerata, Italy	163	257	7.5	1.9	0.4	52.8	0.6	3.5	189	2	5.8	1.2	NR	0.3	NR	No information on source
Northern Crystal	Grafton, Canada	160	145	7.3	13	1	20	7	44	46	13	NR	1	0.4	7.3	NR	No information on source
Nubia	Egypt	NR	329	NR	40	3.5	26	12	20	191	12	24	NR	0.2	1.2	NR	No information on source
Ozarka	Texas, USA	32	16	6.6	2.3	NR	2	1.2	3.5	4.9	2.1	NR	NR	NR	OR	NR	Spring, carbon filtered, dionized, ozonated
Palomar Mountain	California, USA	95	118	7.2	9.3	2.4	12	5.2	NR	80	9	NR	NR	NR	No Value	POLYV	Spring, mountain region, 1 micron filter, UV and ozonated
Pampara	Adour, France	NR	345	7.6	19.4	2.4	63.8	5	17.9	211.1	25.8	NR	ND	NR	1.6	NR	No information on source
Panther Creek	Florida, USA	<5	12	NR	0.5	1	10.5	0.1	NR	NR	ND	ND	NR	0.2	OR	NR	Cypress spring, white sand aquifer
Para Springs	Suriname	NR	96	7.5	NR	0.7	16	0.9	8	70	ND	NR	NR	NR	OR	NR	No source information
Paul Gaugin	French Polynesia	NR	214	NR	48.7	4.5	13	5.8	67	69.5	5	NR	0.9	NR	9.3	NR	Te Vai Arii Source Royale
Pian della Mussa	Torino, Italy	40	43	6.9	0.8	0.3	6.1	3	0.2	24.1	6.1	NR	2.2	TRACE	OR	NR	No information on source
Pioda	Bergamo, Italy	153	214	7.9	0.9	0.4	36	13.8	NR	152	3.7	NR	6.8	NR	No Value	NR	No information on source

Poland Spring	Maine, USA	37	44	6.4	2.9	0.5	8.3	0.8	6.1	20	5	NR	ND	OR	PET	Spring, pine forest, rain/snow infiltration, micron filtered and UV
Pradis	Pordenone, Italy	154	59	8.2	0.6	0.1	30.8	16.7	1.2	NR	5	5	NR	TRACE	NR	No information on source
Purely Scottish	Scotland, United Kingdom	NR	103	NR	16	2	30	NR	29	NR	15	11	NR	OR	NR	Spring, igneous and sedimentary rocks
Radnor Hills	Wales, United Kingdom	NR	67	NR	28.3	1.6	6.2	6.8	1.4	NR	10.9	12	0.1	OR	PLASTIC	Spring, no geologic information
Rayyan	Qatar	NR	152	7.8	9	6.2	10	12	20	55	39	NR	0.7	No Value	NR	No source information
Riwa	Syria	NR	222	NR	2.3	0.3	33	16	7	150	12	1.5	0.1	1	NR	No source information
Roc Vert	Mali	NR	337	NR	2.3	6.2	17.6	46.5	3	260		1		No Value	NR	No source information
Royal Mountains	Appalachian Rock, Canada	58	0	6.9	NR	NR	NR	NR	NR	NR	NR	NR	ND	OR	NR	Spring, no source information, Fe = 0.01, Mn = ND
S.Antonio	Como, Italy	133	200	7.7	3.6	0.7	33.5	5.4	1.3	133	1.6	4.4	NR	No Value	GLASS	Spring, no geologic information
Saint-Elle	Saint-Elle de Caxton, Canada	180	129	8.2	NR**	3.9	13	5	34	60	13	NR	NR	3.6	NR	Spring, igneous and metamorphic rock aquifer
Santa Clara Oligomineral	Dominican Republic	145	247	6.7	7.3	0.3	20	15	NR	158	5.5	1.2	NR	No Value	NR	No information on source

(continued)

Bottled Water	Location	TDS	Total Ions	pH	Na	K	Ca	Mg	Cl	HCO₃	SO₄	SiO₂	NO₃	F	pCi/L	Bottle	Remarks
Saratoga	New York, USA	17	57	7	7.2	0.7	11	2.2	2.4	26	7.1	NR	NR	ND	OR	GLASS	Spring, Adirondack mountains, natural carbonation
Scotch Mist	Scotland, United Kingdom	NR	110	NR	12.1	0.9	55	20	21	NR	0.6	NR	NR	NR	OR	GLASS	Well, fault zone flow, no specific geology
Snow Valley	Pennsylvania, USA	29	9	5.4	2.7	NR	NR	ND	4.2	NR	1	NR	1.3	ND	OR	HDPE	Spring, Appalachian mountains, National Forest, ozonated
Sollar	Azerbijan	NR	204	7.6	NR	41	65.8	34.5	17.5	NR	42	NR	3	NR	No Value	NR	No information on source
Sorgente Dolomiti	Veneto Region, Italy	114	163	8.2	1.3	0.7	23.8	8.7	1.1	94.6	22	7.3	3.4	NR	No Value	NR	No information on source
Soria	Sore, France	175	265	NR	54	2.3	12	6	18	152	19.5	NR	0.5	0.7	1.7	NR	No information on source
Source d'Or	Seychelles	63	81	6.8	2.4	5.7	26.85	14.25	22	NR	10	NR	NR	NR	OR	NR	No source information
Sourcy 1	Netherlands	170	245	NR	10	0.9	50	3.5	12	150	3.7	15	ND	NR	1.5	NR	No source information
Spa	Belgium	33	42	NR	3	0.5	4.5	1.3	5	15	4	7	1.9	NR	OR	PLASTIC	Spring, Ardennes Mountains
Spa Reine	Ardennes, Belgium	33	40	NR	3	0.5	3.5	1.3	5	11	6.5	7	1.9	NR	OR	PET	Spring, snow melt through bedrock
Sparkletts	California, USA	NR	12	7.2	3.8	NR	0.7	0.8	3.5	NR	3.4	NR	NR	NR	OR	PLASTIC	Well water, no geologic information

Spirit	British Columbia, Canada	3	1.8	NR	0.3	0.2	0.5	TRACE	0.3	NR	0.4	NR	0.1	NR	OR	NR	Glacial water from remote coastal range mountains
Spring Water	Ontario, Canada	NR	270	NR	4	1	46	16	1	190	12	NR	NR	NR	No Value	No Value	No information on source
Spritzer Bhd	Malasia	135	69	NR	6	3.5	12	1.6	<1	NR	<3	46	NR	NR	No Value	PET	No information on origin
St. Antonio	Oregon, USA	79	20	NR	4	1	NR	2.6	6	NR	6	NR	NR	ND	0.3	PET	Spring, Cascade Mountains, granite aquifer
St-Georges	Serra-Cimaggia, France	NR	85	NR	14.1	1.2	5.2	2.4	25	30.5	6	NR	ND	0.1	OR	NR	Mountain spring, no geologic information
Stretton Hills	Shropshire, United Kingdom	226	130	7.7	15.2	1.2	44	8.5	17	NR	31	NR	13	0.1	2.2	GLASS	Well, no geologic information
Supermont	Cameroon	128	179	NR	12.7	5.2	16	10.2	2.1	128	1	NR	4.1	NR	No Value	NR	Mount Cameroon, no information on source
Thunder Mtn.	North Carolina, USA	8	26	NR	NR	NR	3	1	1	20	1	ND	NR	NR	OR	PET	Spring, Blue Ridge Mtn., igneous rock aquifer
Thundr Rock	Alabama, USA	160	49	NR	1	1.1	45	1.5	NR	NR	NR	NR	NR	0.2	No Value	NR	Spring, geology unknown, Cr = 0.001 ppm, As = ND

(continued)

Bottled Water	Location	TDS	Total Ions	pH	Na	K	Ca	Mg	Cl	HCO$_3$	SO$_4$	SiO$_2$	NO$_3$	F	pCi/L	Bottle	Remarks
Trinity	Idaho, USA	195	146	9.6	10	1.2	1.4	NR	8	47.6	NR	74.6	NR	3.6	0.6	PET	Spring, granite batholith aquifer in wilderness area
TyNant	Wales, United Kingdom	165	191	6.8	22	1	22	11.5	14	116	4	NR	ND	0.1	1.8	GLASS	Spring, igneous aquifer
Valvita	South Africa	225	285	7.8	5	NR	42	24	5	202	6	NR	1.4	NR	No Value	PLASTIC	No information on source
Virgin Kiwi	Canterbury Plains, New Zealand	97	21	7.9	6.9	NR	12	1.9	NR	NR	NR	NR	NR	NR	No Value	PLASTIC	Well, Mountain origin to aquifer in plain, travel time to well is 70 years
Volvic	Auvergne region, France	109	142	7	9	6	10	6	8	65	7	30	1	NR	No Value	NR	Spring, Volcanic region, national park
Voss	Norway	22	19	6.4	1	ND	3.5	0.8	9	NR	5	NR	NR	0.1	OR	GLASS	Artesian, wilderness area, under ice and rock for centuries
Whistler Water	British Columbia, Canada	48	47	7.3	2	1	12	1	3	20	8	NR	NR	<0.1	OR	GLASS	Spring, Garibaldi National Park, granite aquifer

Winifred Springs	Rylstone, Australia	NR	21	5.4	6.7	0.6	0.1	0.6	10	1	1	1	NR	1	NR	OR	PLASTIC	Spring, Wollemi National Park, no geologic information
Zephyrhills	Florida, USA	185	226	7.7	5.1	0.2	58	3.9	11	140	8	NR	NR	NR	0.2	1.8	PET	Spring, limestone aquifer, carbon filtered, UV and ozonated

** = Based on the ion balance this water may be expected to contain Na
pCi = calculated using equation 3-1
No Value = No data for calculation or a non-positive number
TRACE = Less than 0.1 ppm
NR = Not reported
ND = Not detected
OR = Outside Range of regression equation
Note: All elemental concentraions are mg/L (ppm)

APPENDIX 3-6

Trace Element Analysis of Mineral Waters (ppb) that Appear in Either Appendix 3-4 or Appendix 3-5

Name	As	Cd	Cr	Cu	Ni	Pb	Zn	U
Aqua Panna	0.38	0.09	1.92	8.0	10.5	0.66	4.41	0.61
A Quelle	0.44	0.09	0.56	1.24	2.89	0.73	7.25	0.07
Abbey Well	0.49	0.03	0.73	3.0	12.3	0.47	2.58	NR
Appollinaris	2.85	0.15	4.04	3.21	5.93	0.74	14.8	0.19
Aqua pura	0.28	0.07	1.03	7.56	11.2	0.41	3.32	0.03
Badoit	7.26	0.09	3.53	2.16	1.96	0.57	13.6	0.81
Baraka	0.68	0.09	2.84	1.24	0.95	0.41	2.91	NR
Birute	2.91	0.29	1.4	33.7	7.2	2.22	37.9	0.28
Blue Keld	0.81	0.06	2.02	10.8	7.78	0.66	6.18	0.18
Contrex	1.31	0.33	1.61	6.7	14.5	1.18	14.9	1.39
Crystal	4.31	0.05	1.75	3.28	14.0	1.1	3.9	NR
Evian	1.24	0.18	0.84	4.79	10.8	1.05	5.23	1.91
Hildon	NR	1.1	0.07	2.82	3.07	1.94	0.52	3.56
Perrier	0.37	0.22	4.66	4.41	4.87	0.99	12.8	2.41
Ramlosa	0.38	0.11	3.48	1.41	1.54	0.61	2.94	NR
San Pellegrino	NR	1.95	0.23	3.41	4.58	2.92	1.05	7.87
SPA	0.34	0.2	6.32	1.31	3.49	0.44	10.9	NR
Ty Nant	NR	NR	0.38	4.43	0.56	0.97	7.78	NR
Vittel	13.2	0.23	0.76	3.63	2.41	1.03	20.4	0.73
Volvic	6.55	0.24	2.11	9.96	9.26	1.14	6.83	0.4

NR = Not reported.

APPENDIX 4-1

Glossary of Terms Adapted from the International Union of Pure and Applied Chemistry (IUPAC) (1993)

This brief appendix is adapted from the IUPAC Clinical Chemistry Division Commission on Toxicology *Glossary for Chemists of Terms Used in Toxicology* (IUPAC, 1993). The following is a summary of relevant toxicologic terms from the IUPAC. A full version has been reorganized into relevant topics including toxicology, dose-response, biological modeling, genetics, epidemiology, cancer, risk assessment, and regulatory jargon. To view the full version go to the online version of the glossary.

Toxicology

acute: 1. Short-term, in relation to exposure or effect. In experimental toxicology, "acute" refers to studies of 2 weeks or less in duration (often less than 24 hours). 2. In clinical medicine, sudden and severe, having a rapid onset.

additive effect: Consequence that follows exposure to two or more physico-chemical agents that act jointly but do not interact. Commonly, the total effect is the simple sum of the effects of separate exposure to the agents under the same conditions. Substances of simple similar action may show dose or concentration addition.

antagonism: Combined effect of two or more factors that is smaller than the solitary effect of any one of those factors. In bioassays, the term may be used when a specified response is produced by exposure to either of two factors, but not by exposure to both together.

biotransformation: Any chemical conversion of substances that is mediated by living organisms or enzyme preparations derived therefrom. Nagel et al. (eds), 1991.

chronic exposure: Continued exposures occurring over an extended period of time, or a significant fraction of the test species' or of the group of individuals', or of the population's lifetime.

chronic toxicity: 1. Adverse effects following chronic exposure. 2. Effects that persist over a long time whether or not they occur immediately on exposure or are delayed (IRIS, 1986).

cumulative effect: Overall adverse change that occurs when repeated doses of a harmful substance or radiation have biological consequences that are mutually enhancing.

delayed effect: Consequence occurring after a latent period following the end of exposure to a toxic substance or other harmful environmental factor.

detoxification: 1. Process or processes of chemical modification that make a toxic molecule less toxic. 2. Treatment of patients suffering from poisoning in such a way as to promote physiologic processes that reduce the probability or severity of harmful effects.

dosage: Dose expressed as a function of the organism being dosed and time, for example mg/(kg body weight)/day.

exposed: Subject to a factor that is under study in the environment, for instance an environmental hazard.

first-pass effect: Biotransformation of a substance in the liver after absorption from the intestine and before it reaches the systemic circulation.

hydrophilic/ adj., **-ity** n.: Describing the character of a molecule or atomic group that has an affinity for water.

hydrophobic/ adj., **-ity** n.: Describing the character of a molecule or atomic group that is insoluble in water, or resistant to wetting or hydration.

idiosyncrasy: Genetically based unusually high sensitivity of an organism to the effect of certain substances.

in vitro: In glass, referring to a study in the laboratory usually involving isolated organ, tissue, cell, or biochemical systems.

in vivo: In the living body, referring to a study performed on a living organism.

lethal: Deadly; fatal; causing death.

lipophilic/ adj., **-ity** n.: Having an affinity for fat and high-lipid solubility. A physicochemical property that describes a partitioning equilibrium of solute molecules between water and an immiscible organic solvent, favoring the latter, and that correlates with bioaccumulation.

lipophobic/ adj., **-ity** n.: Having a low affinity for fat and a high affinity for water.

local effect: Circumscribed change occurring at the site of contact between an organism and a toxicant.

long-term exposure: Continuous or repeated exposure to a substance over a long time, usually of several years in humans, and of the greater part of the total life span in animals or plants (IRPTC, 1982).

pharmacodynamics: Process of interaction of pharmacologically active substances with target sites, and the biochemical and physiologic consequences leading to therapeutic or adverse effects.

pharmacokinetics: Process of the uptake of drugs by the body, the biotransformation they undergo, the distribution of the drugs and their metabolites in the tissues, and the elimination of the drugs and their metabolites from the body. Both the amounts and concentrations of the drugs and their metabolites are studied. The term has essentially the same meaning as toxicokinetics, but the latter term should be restricted to the study of substances other than drugs.

phase 1 reaction (of biotransformation): Enzymic modification of a substance by oxidation, reduction, hydrolysis, hydration, dehydrochlorination, or other reactions catalyzed by enzymes of the cytosol, of the endoplasmic reticulum (microsomal enzymes), or of other cell organelles.

phase 2 reaction (of biotransformation): Binding of a substance, or its metabolites from a phase 1 reaction, with endogenous molecules (conjugation), making more water-soluble derivatives that may be excreted in the urine or bile.

phase 3 reaction (of biotransformation): Further metabolism of conjugated metabolites produced by phase 2 reactions. It may result in the production of toxic derivatives.

potency: Expression of chemical or medicinal activity of a substance as compared to a given or implied standard or reference.

potentiation: Dependent action in which a substance or physical agent at a concentration or dose that does not itself have an adverse effect enhances the harm done by another substance or physical agent.

precursor: Substance from which another, usually more biologically active, substance is formed.

predictive validity: Reliability of a measurement expressed in terms of its ability to predict the criterion. An example is an academic aptitude test that was validated against subsequent academic performance (Last, 1988).

reference concentration: Term used for an estimate of air exposure concentration to the human population (including sensitive subgroups) that is likely to be without appreciable risk of deleterious effects during a lifetime (USEPA, 1989).

regulatory dose: Term used by the USEPA to describe the expected dose resulting from human exposure to a substance at the level at which it is regulated in the environment (Barnes and Dourson, 1988).

route of exposure: Means by which a toxic agent gains access to an organism by administration through the gastrointestinal tract (ingestion), lungs (inhalation), skin (topical), or by other routes such as intravenous, subcutaneous, intramuscular, or intraperitoneal routes.

slope factor: Value, in inverse concentration or dose units, derived from the slope of a dose-response curve; in practice, limited to carcinogenic effects with the curve assumed to be linear at low concentrations or doses. The product of the slope factor and the exposure is taken to reflect the probability of producing the related effect.

subacute: Term used to describe a form of repeated exposure or administration usually occurring over about 21 days, not long enough to be called *long-term* or *chronic*.

subchronic: Related to repeated dose exposure over a short period, usually about 10 percent of the life span; an imprecise term used to describe exposures of intermediate duration.

synergism: Pharmacologic or toxicologic interaction in which the combined biological effect of two or more substances is greater than expected on the basis of the simple summation of the toxicity of each of the individual substances.

systemic effect: Consequence that is of either a generalized nature or that occurs at a site distant from the point of entry of a substance. A systemic effect requires absorption and distribution of the substance in the body.

target organ(s): Organ(s) in which the toxic injury manifests itself in terms of dysfunction or overt disease (WHO, 1979).

threshold: Dose or exposure concentration below which an effect is not expected.

toxic dose: Amount of a substance that produces intoxication without lethal outcome.

toxicity: Capacity to cause injury to a living organism defined with reference to the quantity of substance administered or absorbed, the way in which the substance is administered (inhalation, ingestion, topical application, injection) and distributed in time (single or repeated doses), the type and severity of injury, the time needed to produce the injury, the nature of the organism(s) affected, and other relevant conditions. Adverse effects of a substance on a living organism defined with reference to the quantity of substance administered or absorbed, the way in which the substance is administered (inhalation, ingestion, topical application, injection) and distributed in time (single or repeated doses), the type and severity of injury, the time needed to produce the injury, the nature of the organism(s) affected, and other relevant conditions. Measure of incompatibility of a substance with life. This quantity may be expressed as the reciprocal of the absolute value of median lethal dose $(1/LD_{50})$ or concentration $(1/LC_{50})$.

toxicodynamics: Process of interaction of potentially toxic substances with target sites, and the biochemical and physiologic consequences leading to adverse effects.

toxicokinetics: Process of the uptake of potentially toxic substances by the body, the biotransformation they undergo, the distribution of the substances and their metabolites in the tissues, and the elimination of the substances and their metabolites from the body. Both the amounts and the concentrations of the substances and their metabolites are studied. The term has essentially the same meaning as pharmacokinetics, but the latter term should be restricted to the study of pharmaceutical substances (WHO, 1979).

toxicology: Scientific discipline involving the study of the actual or potential danger presented by the harmful effects of substances (poisons) on living organisms and ecosystems, of the relationship of such harmful effects to exposure, and of the mechanisms of action, diagnosis, prevention, and treatment of intoxications.

toxin: Poisonous substance produced by a biological organism such as a microbe, animal, or plant.

xenobiotic: Strictly, any substance interacting with an organism that is not a natural component of that organism. Manufactured compounds with chemical structures foreign to a given organism (Nagel et al., 1991).

References

Barnes, D.G. and Dourson, M.L., 1988, *Regl Toxicol Pharmacol*, Vol. 8, pp. 471–486.

IRIS, 1986, "Integrated Risk Information System of the United States," EPA/600/6-86/032a, U.S. Environmental Protection Agency, Washington, D.C.

IRPTC, 1982, "International Register of Potentially Toxic Substances, English-Russian Glossary of Selective Terms in Preventive Toxicology," Interim Document, United Nations Environment Programme, Moscow.

Last, J.M. (ed.), 1988, *A Dictionary of Epidemiology*, 2nd ed., Oxford Medical Publications, Oxford University Press, New York. Nagel, B., Dellweg, H., and Gierasch, L.M. (eds.), 1992, Pure Appl. Chem., 64, 143-168.

USEPA, 1989, "Glossary of Terms Related to Health, Exposure, and Risk Assessment,", EPA/450/3-88/016, U.S. Environmental Protection Agency, Washington, D.C.

WHO, 1979, "Agreed Terms on Health Effects Evaluation and Risk and Hazard Assessment of Environmental Agents," Internal Report of a Working Group, (EHE/EHC/79.19), Word Health Organization, Geneva.

APPENDIX 4-2

Chemical Examples on the Toxicology of Drinking Water Standards

The following information regarding drinking water standards and the health effects associated with regulated compounds have been adapted from the Natural Resource Defense Council's 2003 report on drinking water quality across the United States entitled "What's on Tap? Grading Drinking Water in U.S. Cities." Information regarding standards for California and the basis of the standards are also provided. The chemicals selected for this appendix (arsenic, chromium, lead, perchlorate, atrazine, DBCP, Di-(2-Ethylhexyl)phthalate, PCE, and TCE) represent compounds that are frequently detected in drinking water supplies, have clear adverse effects associated with the consumption of the compound, and have a significant economic impact on drinking water supplies. Standards for three of the compounds, arsenic, perchlorate, and TCE, have been subjected to rereviews by the National Academy of Science of the existing toxicologic summaries to reevaluate proposed (e.g., restrictive or health protective) standards. The reevaluations have lead to a spirited discussion regarding the objectivity of those setting drinking water standards.

Inorganic Contaminants

Arsenic

National Standard (MCL)
50 ppb (average) through 2005
10 ppb (average) effective in 2006

National Health Goal (MCLG)
0—no known fully safe level
50 ppb (average) through 2005
10 ppb (average) effective in 2006

California Health Goal (PHG)
0.004 ppb (based on the mortality of arsenic-induced lung and urinary bladder cancers observed in epidemiological studies of populations in Taiwan, Chile, and Argentina)

Health Effects
Arsenic is toxic to humans and causes cancer. For this reason, no amount of arsenic is considered fully safe. Many scientific studies, including no fewer than seven reviews of the problem by the National Academy of Sciences (NAS), have determined that arsenic in drinking water is known to cause cancer of the bladder, skin, and lungs; likely causes other cancers; and is responsible for a variety of other serious health ailments. The NAS (2002) reviews culminated in the important reports Arsenic in Drinking Water (issued in 1999) and Arsenic in Drinking Water: 2001 Update, which counter the long-standing water utility and industry arguments that arsenic in tap water poses no significant threat (NRC, 1999, 2001). The NRC found in its 2001 report that a person who drinks 2 liters of water a day containing 10 ppb arsenic—the new EPA standard—has a lifetime total fatal cancer risk greater than 1 in 333 (that is, about 1 in 333 people who drink water containing this level of arsenic will die of arsenic-caused cancer) (NRC, 2001). The EPA traditionally has allowed no greater than a 1 in 10,000 lifetime fatal cancer risk for any drinking water contaminants.

Chromium

National Standard (MCL)
100 ppb (average)
100 ppb

California Standard (MCL)
50 ppb (total chromium) through 2005

California Health Goal (PHG)
10 ppb (for total chromium): withdrawn in November 2001
2.5 ppb (hexavalent chromium): withdrawn in November 2001. OEHHA assumed chromium-6 to be carcinogenic when ingested, and calculated *de minimis* cancer risk corresponding to a drinking water concentration of 0.2 µg/L, based on a mouse study (Borneff et al., 1968).

Health Effects
Health effects from human exposure to chromium range from skin irritation to damage to kidney, liver, and nerve tissues. A heated debate has taken shape recently over whether states and the EPA should adopt a separate standard for chromium VI (hexavalent chromium), a form of chromium known to cause cancer when inhaled. The EPA has refused so

far to consider it a carcinogen when it is consumed in tap water (USEPA, 2002a).

Lead

National Standard (TT)
15 ppb (action level, at 90th percentile)[8]

National Health Goal (MCLG)
0 ppb—no known fully safe level

California Standard (MCL)
15 ppb

California Health Goal (PHG)
2 ppb–The PHG for lead is based on the neurobehavioral effects of lead in children and the hypertensive effects of lead in adults.

Source
Lead is a heavy metal that generally enters drinking water supplies from the corrosion of pipes, plumbing, or faucets.

Health Effects
Lead is a major environmental threat and is often referred to as the number one environmental health threat to children in the United States. No amount of it is considered safe (USEPA, 2002b). Infants, young children, and pregnant women's fetuses are particularly susceptible to the adverse health effects of lead. Lead poisoning can cause permanent brain damage in serious cases; in less severe cases children suffer from decreased intelligence and problems with growth, development, and behavior. Lead can also increase blood pressure, harm kidney function, adversely affect the nervous system, and damage red blood cells (USEPA, 2002b).

Perchlorate

National Standard (MCL)
None established

---

[8]The action level standard for lead is different from the standard for most other contaminants. Water utilities are required to take many samples of lead in the tap water at homes they serve, including some high-risk homes judged likely to have lead in their plumbing or fixtures. If the amount of lead detected in the samples is more than 15 ppb at the 90th percentile (which means that 90 percent of the samples have 15 ppb or less), then the amount is said to exceed the action level. Under the complex USEPA lead rule, a water system that exceeds the action level is not necessarily in violation. If a system exceeds the action level, additional measures such as chemical treatment to reduce the water's corrosivity (ability to corrode pipes and thus its ability to leach lead from pipes) must be taken. If this chemical treatment does not work, the water system may have to replace lead portions of its distribution system if they are still contributing to the lead problem.

National Draft Safe Level ("Drinking Water Equivalent Level" or DWEL)70[9]
1 ppb

California Standard (MCL)
None established

California Health Goal (PHG)
6 ppb–The PHG is based on its ability to interfere with iodide uptake into the thyroid gland.

Source
Perchlorate is an inorganic contaminant that usually comes from rocket fuel spills or leaks at military facilities. Perchlorate contaminates the tap water of much of southern California via the Metropolitan Water District's Colorado River Aqueduct. It also is in the water of Phoenix, Las Vegas, and many other cities and towns that rely on the Colorado River for their water. The source of the Colorado's contamination is reportedly a Kerr-McGee site in Henderson, Nevada, where perchlorate was manufactured and whose waste leaks into the Colorado River.[10] Perchlorate also contaminated water sources for many other towns and cities across the nation, where it has been manufactured or used at military bases or in commercial applications. In addition to its heavy use in rocket fuel, perchlorate is also used, in far lower quantities, in a variety of products and applications, including electronic tubes, automobile air bags, leather tanning, and fireworks.[11]

Health Effects
Perchlorate harms the thyroid and may cause cancer.[12] According to the EPA, perchlorate disrupts how the thyroid functions. In adults, the thyroid helps to regulate metabolism. In children, the thyroid plays a major

---

[9]A drinking water equivalent level is the presumed level of perchlorate that one would need to consume in tap water to reach the *reference dose*—the maximum safe level. See EPA, "Perchlorate," fact sheet available online at www.epa.gov/safewater/ccl/perchlor/perchlo.html.

[10]MWD is well aware that this Henderson facility is the source of this perchlorate. See MWD, "In the News: Perchlorate," available online at www.mwdh2o.com/mwdh2o/pages/yourwater/ccr02/ccr03.html; MWD press release, "Water Officials Report Significant Progress in Perchlorate Removal," April 17, 2002. This release puts an unduly optimistic face on the problem, as MWD is well aware that the cleanup of the facility remains problematic and partially unsuccessful. See also Environmental Working Group, Rocket Science (2001), available online at www.ewg.org/reports/rocketscience.

[11]See California Department of Health Services, "Perchlorate in California Drinking Water," available online at www.dhs.cahwnet.gov/ps/ddwem/chemicals/perchl/perchlindex.htm; see also EWG, Rocket Science (2001), available online at www.ewg.org/reports/rocketscience/.

[12]Ibid.; see also California Office of Environmental Health Assessment, Draft Public Health Goal for Perchlorate in Drinking Water (March 2002), available online at www.oehha.org/water/phg/pdf/PHGperchlorate372002.pdf.

role in proper development in addition to metabolism. Impairment of thyroid function in expectant mothers may affect the fetus and newborn and result in effects including changes in behavior, delayed development, and decreased learning capability. Changes in thyroid hormone levels may also result in thyroid gland tumors. [The EPA finds that] perchlorate's disruption of iodide uptake is the key event leading to changes in development or tumor formation.[13]

Organic Contaminants

Atrazine

National Standard (MCL)
3 ppb (average)

National Health Goal (MCLG)
3 ppb

California Standard (MCL)
1 ppb

California Health Goal (PHG)
0.15 ppb–The PHG is based on mammary tumors (adenocarcinoma and fibroadenoma) observed in females in a carcinogenicity study in Sprague-Dawley rats (70/sex/dose) fed atrazine at dietary concentrations of 0, 10, 70, 500, or 1,000 ppm for 24 months and using a linear dose response approach with a carcinogenic slope factor (CSF) of 0.23 (mg/kg-day)$^{-1}$.

Source
Atrazine is among the most widely used pesticides in this country, applied to corn and other crops to protect from broad-leaved and grassy weeds. Atrazine enters source waters through agricultural runoff, and also volatilizes, or evaporates, and is then redeposited with rain. It is among the most commonly detected pesticide in drinking water, particularly during spring runoff season throughout most of the Mississippi River basin and virtually anywhere else that corn is grown.

Health Effects
Atrazine is an animal carcinogen. According to the EPA, short-term human exposure to atrazine may cause prostate cancer; congestion of the heart, lungs, and kidneys; low blood pressure; muscle spasms; weight loss; and damage to the adrenal glands. Over the long term, the USEPA reports, atrazine may cause weight loss, cardiovascular damage, retinal

---

[13]EPA, "Perchlorate," fact sheet available online at www.epa.gov/safewater/ccl/perchlor/perchlo.html.

and some muscle degeneration, and possibly cancer. In addition, as noted previously, atrazine is a known endocrine disrupter, meaning that it interferes with the body's hormonal development and may cause cancer of the mammary gland (NRDC, 2002).

Dibromochloropropane (DBCP)[14]

National Standard (MCL)
200 ppt (average)

National Health Goal (MCLG)
0—no known fully safe level

California Standard (MCL)
200 ppt

California Health Goal (PHG)
1.7 ppt—The PHG is based on carcinogenic effects observed in experimental animals and reported by Hazleton (1977, 1978).

Health Effects
DBCP has been shown to cause cancer, kidney and liver damage, and atrophy of the testes leading to sterility.[15]

Di-(2-Ethylhexyl)phthalate (DEHP or Phthalate)

National Standard (MCL)
6 ppb (average)

National Health Goal (MCLG)
0—no known fully safe level

California Standard (MCL)
4 ppb

California Health Goal (PHG)
12 ppb—The PHG is based on chronic studies of DEHP, which showed a dose-dependent increase in hepatocellular carcinomas in a National Toxicology Program (NTP) bioassay using B6C3F1 mice and Fischer 344 rats, as well as in industry-sponsored repeat studies conducted with additional doses and biochemical analyses for peroxisome proliferation.

---

[14]DBCP is a banned pesticide still detected in some cities' tap water.
[15]Health effects and other general information on DBCP derived from EPA, "Consumer Fact Sheet on Dibromochloropropane," available online at www.epa.gov/safewater/dwh/c-soc/dibromoc.html.

Source

Di-(2-Ethylhexyl)Phthalate (DEHP) is a plasticizing agent used widely in the chemical and rubber industries. It is also contained in many plastics.[16]

Health Effects

The EPA has listed DEHP as a probable human carcinogen, but it also causes damage to the liver and testes. As a result, the agency set a health goal of 0 for DEHP.

Tetrachloroethylene (Also Called Perchloroethylene, PCE, or PERC)

National Standard (MCL)

5 ppb (average)

National Health Goal (MCLG)

0—no known fully safe level

California Standard (MCL)

5 ppb

California Health Goal (PHG)

0.06 ppb–The PHG is based on carcinogenic effects observed in experimental animals.

Source

Tetrachloroethylene is used in dry cleaning and industrial metal cleaning or finishing. It enters the water system via spills or releases from dry cleaners or industrial users, waste dumps, leaching from vinyl liners in some types of pipelines used for water distribution, and in some cases during chlorination water treatment.[17]

Health Effects

Prolonged consumption of water contaminated by PERC can cause liver problems and may cause cancer.

Trichloroethylene (TCE)

National Standard (MCL)

5 ppb (average)

National Health Goal (MCLG)

0—no known fully safe level

---

[16]USEPA, "Consumer Fact Sheet on DEHP," www.epa.gov/safewater/dwh/c-soc/phthalat.html.
[17]USEPA, "Consumer Fact Sheet on Tetrachloroethylene," available online at www.epa.gov/safewater/dwh/cvoc/tetrachl.html.

California Standard (MCL)
5 ppb

California Health Goal (PHG)
0.8 ppb–The PHG is based on the occurrence of hepatocellular carcinomas and adenocarcinomas in mice in two studies, in both sexes, by inhalation and oral routes of administration, and a linear dose-response approach.

Source
TCE is a colorless liquid used as a solvent to remove grease from metal parts. It is present in many underground water sources and surface waters as a result of the manufacture, use, and disposal of TCE at industrial facilities across the nation.

Health Effects
TCE is a likely carcinogen, and people exposed to high levels of it in their drinking water may experience harmful effects to their nervous system, liver and lung damage, abnormal heartbeat, coma, and possibly death.[18]

References

Borneff, I., K. Engelhardt, W. Griem, H. Kunte, and J. Reichert, 1968, "Kankzerogene Substanzen in Wasser und Boden XXII. Mäusentränkversuch mit 3,4-Benzpyren und Kaliumchromat," *Archiv für Hygiene und Bakteriologie*, Vol. 152, pp. 45–53.

Hazleton Laboratories, 1977, Final Report of 104 week dietary study in rats, 1,2-dibromo-3-chlorpropane (DBCP). Submitted to the Dow Chemical Company, Midland, Michigan, October, 1977. Hazleton Laboratories America, Inc., Vienna, VA, Project No. 174–122.

Hazleton Laboratories, 1978, Final Report of 78 week dietary study in mice, 1,2-dibromo-3-chlorpropane (DBCP). Submitted to the Dow Chemical Company, Midland, Michigan, November, 1978. Hazleton Laboratories America, Inc., Vienna, VA, Project No. 174–125.

NAS, 2002, September 11, 2002 press release, available online at www4. nationalacademies.org/news.nsf/isbn/0309076293?OpenDocument.

NRC, 1999, *Arsenic in Drinking Water Subcommittee on Arsenic in Drinking Water Committee on Toxicology Board on Environmental Studies and Toxicology Commission on Life Sciences National Research Council*, National Academy Press, Washington, D.C., available online at www.nap.edu/catalog/6444.html.

NRC, 2001, *Arsenic in Drinking Water: 2001 Update*, National Academy Press, 2001, available online at www.nap.edu/catalog/10194.html.

---

[18] Agency for Toxic Substances and Disease Registry, "ToxFAQs—Trichloroethylene (TCE)"(September 1997), available online at www.atsdr.cdc.gov/tfacts19.html.

NRDC, 2002, Atrazine: An Unacceptable Risk to America's Children and Environment (June 2002), Natural Resources Defense Council, Washington, D.C.

USEPA, 2001, National Primary Drinking Water Regulation for Arsenic, 66 Fed. Reg. 6976, at 6981 (January 22, 2001).

USEPA, 2002a, "National Primary Drinking Water Regulations; Announcement of the Results of EPA's Review of Existing Drinking Water Standards and Request for Public Comment; Proposed Rule," 67 Fed. Reg. 19030, at 19057-58 (April 17, 2002).

USEPA, 2002b, "Consumer Fact Sheets on Lead," www.epa.gov/safewater/Pubs/lead1.html and www.epa.gov/safewater/standard/lead&col.html, and IRIS summary for lead available online at www.epa.gov/iris/subst/0277.htm.

APPENDIX 4-3

Suspected Endocrine-Disrupting Chemicals

Acetochlor (Contaminant Candidate List)*
Alachlor (Regulated)*
Aldicarb
Aldrin (Contaminant Candidate List)
Amitrole
Atrazine (Regulated)
Benomyl
Bifenthrin
Bromacil
Bromaxynil
Carbaryl
Carbofuran (Regulated)
Chlordane (Regulated)
Chlordecone
Chlorfentezine
Cyanazine
8-Cyhalothrin
2,4-D (Regulated)
DBCP (Regulated)
DCPA
DDT DDE (Contaminant Candidate List)
DDD
Deltamethrin
Dicofol
Dieldrin (Contaminant Candidate List)
Dimethoate
Dinitrophenol (Contaminant Candidate List)
Dioxin (Regulated)
Endosulfan
Endrin
Ethiozin

Ethofenprox
Etridiazole
Fenarimol
Fenbuconazole
Fenitrothion
Fenvalerate
Fipronil
n-2-Fluorenylacetamide
Glufosinate-ammonium
alpha-HCH
beta-HCH
Heptachlor (Regulated)
Heptachlor epoxide (Regulated)
Hexachlorobenzene (Regulated)
Ioxynil
Lindane (gamma-HCH) (Regulated)
Linuron (Contaminant Candidate List)
Malathion
Mancozeb
Maneb
MethomylMethoxychlor (Regulated)
MetiramMetribuzin
Mirex
Molinate (Contaminant Candidate List)
Nabam
Nitrofen
Oryzalin
Oxyacetamide/fluthamide
Oxychlordane
Paraquat
Parathion
Penachloronitrobenzene
Pendimethalin
Pentachlorophenol (Regulated)
Penta- to nonylphenols
Perchlorate (Contaminant Candidate List)
Phthalates
Photomirex
Picloram (Regulated)
Polychlorinated biphenyls (Regulated)
Prodiamine
Pronamide
Pyrethrins
Pyrethroids
Ronnel (fenchlorfos)
Simazine (Regulated)
Styrenes (Regulated)
2,4,5-T
Terbutryn
Thiazopyr
Toxaphene (Regulated)

Transnonachlor
Triadimefon
Tributyltin
Trichlorobenzene (Regulated)
Trifluralin
Vinclozolin
Zineb
Ziram

---

*Regulated chemicals are listed in Table 1-9 and the Contaminant Candidate List chemicals are in Table 3-1.

Source: *Pesticides News*, No. 46 (December, 1999); Colborn, T. et al., 1997, "Our Stolen Future: Are We Threatening Our Fertility, Intelligence and Survival," Penguin, East Rutherford, N.J. (March, 1997); and USEPA, 2001, "Perchlorate," Office of Water, Groundwater and Drinking Water, April 26.

APPENDIX 4-4

U.S. Geological Survey Target Compounds, National Reconnaissance of Emerging Contaminants in U.S. Streams (2000)

Human Drugs

Prescription

Metformin (antidiabetic agent)
Cimetidine (antacid)
Ranitidine (antacid)
Enalaprilat (antihypertensive)
Digoxin
Diltiazem (antihypertensive)
Fluoxetine (antidepressant)
Paroxetine (antidepressant)
Warfarin (anticoagulant)
Salbutamol (antiasthmatic)
Gemfibrozil (antihyperlipidemic)
Dehydronifedipine (antianginal metabolite)
Digoxigenin (digoxin metabolite)

Nonprescription

Acetaminophen (analgesic)
Ibuprofen (anti-inflammatory, analgesic)
Codeine (analgesic)
Caffeine (stimulant)
1,7-Dimethylxanthine (caffeine metabolite)
Cotinine (nicotine metabolite)

Veterinary and Human Antibiotics

Carbadox
Chlortetracycline
Ciprofloxacin
Doxycycline
Enrofloxacin
Erythromycin-H_2O (metabolite)
Lincomycin
Norfloxacin
Oxytetracycline
Roxithromycin
Sarafloxacin
Sulfachlorpyridazine
Sulfamerazine
Sulfamethazine
Sulfathiazole
Sulfadimethoxine
Sulfamethiazole
Sulfamethoxazole
Tetracycline
Trimethoprim
Tylosin
Virginiamycin

Sex and Steroidal Hormones

Biogenetics

17b-Estradiol
17a-Estradiol
Estrone
Estriol
Testosterone
Progesterone
cis-Androsterone

Pharmaceuticals

17a-Ethynylestradiol (ovulation inhibitor)
Mestranol (ovulation inhibitor)
19-Norethisterone (ovulation inhibitor)
Equilenin (hormone replacement therapy)
Equilin (hormone replacement therapy)

Sterols

Cholesterol (fecal indicator)
3b-Coprostanol (carnivore fecal indicator)
Sigmastanol (plant sterol)

Industrial and Household Wastewater Products

Antioxidants

2,6-di-ter-Butylphenol
5-Methyl-1H-benzotriazole
Butylatehydroxyanisole (BHA)
Butylatedhydroxytoluene (BHT)
2,6-di-tert-Butyl-p-benzoquinone

Detergent metabolites

p-Nonylphenol
Nonylphenol monoethoxylate (NPEO1)
Nonylphenol diethoxylate (NPEO2)
Octylphenol monoethoxylate (OPEO1)
Octylphenol diethoxylate (OPEO2)

Fire retardants

Tri(2-chloroethyl)phosphate
Tri(dichlorisopropyl) phosphate

Insecticides

Diazion
Carbaryl
Chlorpyrifos
cis-Chlordane
N,N-diethyltoluamide (DEET)
Lindane
Methyl parathion
Dieldrin
Plasticizers
bis(2-Ethylhexyl)adipate
Ethanol-2-butoxy-phosphate
bis(2-Ethylhexyl)phthalate
DiethylphthalateTriphenyl phosphate
Polycyclic aromatic hydrocarbons
Anthracene
Benzo(a)pyrene
Fluoranthene
Naphthalene
Phenanthrene
Pyrene
Others
Tetrachloroethylene (solvent)
Phenol (disinfectant)
1,4-Dichlorobenzene (fumigant)
Acetophenone (fragrance)

p-Cresol (wood preservative)
Phthalic andydride (used in plastic)
Bisphenol A (used in polymers)
Triclosan (antimicrobial disinfectant)

Glossary

Advanced treatment This is an additional treatment step following primary and secondary treatment. Technologies employed might include activated carbon, ultraviolet (UV) light, reverse osmosis, nano-filtration, or any combination of these technologies. The purpose of advanced treatment is to remove those chemical and biological constituents not removed (in whole or in part) by primary and secondary treatment.

Aquifer An aquifer is a permeable formation (e.g., sand and gravel or fractured rock) that is saturated with water so that a well or spring in that formation will yield a significant quantity of water.

Assimilative capacity Assimilative capacity is a term used to describe the ability of a body of water, usually surface water, to "accept" without causing harm, the discharge of certain chemical and biological constituents of a waste stream.

Best available technology The technology that is currently available and will, on implementation, meet specific effluent discharge standards.

Community water system A public water system that supplies water to the same population year-round is defined as a community water system. A small system serves less than 3,300 people, a medium system serves up to 10,000 people, and large systems serve over 10,000 people.

Confined aquifer Artesian or confined aquifers occur where groundwater is confined under pressure by overlying relatively impermeable formations or strata.

Contamination An impairment of the quality of water by chemical, biological, or radiological substances to a degree that creates an actual hazard to public health through poisoning or through the spread of disease.

Disinfection by-products   During water treatment, chlorine or bromine may be added as a disinfectant. Dissolved organic compounds in the treated water may react with the chlorine or bromine and form new chlorinated or bromated organic compounds. The most common compounds formed are the trihalomethanes and haloacetatic acids.

Distribution system   When water leaves the treatment plant, the water is distributed throughout the consumer community by means of a series of pipes collectively referred to as the distribution system.

Dual distribution system   Dual distribution usually means two separate piping systems, each carrying water of different quality, with the higher quality directed to drinking water while the lower quality might be directed for use as irrigation water or for firefighting.

Effluent limitations   Limitations placed on selected chemicals and biologicals which cannot be exceeded in the treatment plant effluent.

Groundwater   The saturated portion of an aquifer is usually referred to as groundwater. Groundwater can be seen to escape at the earth's surface through a spring, be extracted through a water well (or groundwater well), or may seep into surface water.

Legalized pollution   Federal and state regulatory programs that allow chemicals to be released or discharged into surface water or groundwater as long as the chemical concentration of each compound is below set governmental standards.

Manufactured chemical   A man-made or synthetic organic compound that does not occur in nature.

Membrane separation technology   Artificially manufactured materials (commonly referred to as membranes) that allow the passage of water molecules but prevent the passage of many chemical molecules as well as bacteria and viruses. Examples of this technology include reverse osmosis, micro-, and nano-filtration.

Organic compound   A molecule that is composed of either two carbon atoms bonded to each other; or one carbon atom bonded to at least one hydrogen atom or halogen atom (e.g., chlorine, bromine, fluorine); or one carbon atom bonded to at least one nitrogen atom.

Pollution   An impairment of the quality of water by chemical, biological or radiological substances but may not create an actual hazard to public health but which does adversely and unreasonably affect such waters for domestic, industrial, agricultural, navigational, recreational, or other beneficial uses.

Potable water   Potable water is generally understood to mean "drinkable" water—for example, water containing no biological hazard.

Purified water Purified water is water treated so as to remove contaminants and thus considered to be essentially free of contaminants.

Recycled water Water that is reused or used again, usually by the original user. For example, treated wastewater might be recycled (reused) for irrigation purposes.

Regulated chemical Regulated chemicals have set standards. Examples are the chemical compounds listed in the Primary Drinking Water Standards or in the National Ambient Water Quality Criteria.

Risk assessment In the context of this text, a risk assessment identifies potential human health effects that may result from exposure to various chemicals in drinking water.

Sewage treatment plant A facility where municipal (community) wastewater is collected and treated (to some specified level of cleanliness) prior to discharge back to the environment. These facilities are also known as Publicly Owned Treatment Works (POTW).

Standard or criteria A numerical value determined by a federal or state governmental agency that establishes the concentration at which a chemical compound can occur in water.

State-of-art As used in this text, state-of-art generally applies to a technology or technologies that are currently available and represent the best available technology to address a contaminant removal problem. An example might be the use of membrane separation technology to remove impurities from drinking water.

Surface water Water that accumulates as a result of rain and runoff into bodies of water that include ponds, lagoons, lakes, creeks, streams, or rivers. Surface water can then infiltrate into groundwater.

Tertiary treatment Tertiary treatment, often referred to as advanced treatment, is an additional (sometimes considered the third step) measure applied in a sequential treatment process (see Advanced treatment).

Unconfined aquifer An unconfined aquifer has a water table that is not confined, tends to follow the slope of the land, and may rise and fall depending on how much water is recharged into or removed from the aquifer (i.e., pumped from a groundwater well). Unconfined aquifers usually do not have any impermeable formation or strata between the earth's surface and the water table.

Unregulated chemical Any chemical compound that does not have a defined standard at which it may occur in any drinking water source, surface water, or groundwater resource. This is in direct contrast to regulated chemicals, which would be listed, for example, in the Primary Drinking Water Standards or in the National Ambient Water Quality Criteria.

Wastewater   Water discharged to domestic sewers or water discharged by commercial and industrial facilities to sewers are generally referred to as wastewater. Wastewater is assumed to contain chemical or biological pollutants.

Water infrastructure   Includes sources of water (surface water and groundwater resources), storage facilities (reservoirs), treatment works (waterworks), posttreatment storage of water (reservoirs), distribution systems (pumps and piping), as well as collection, treatment (POTW), and disposal of collected wastewater.

Water table   A water table forms the upper surface of the zone of saturation in an aquifer.

Water treatment   Includes all of the elements of treating water, including the removal of suspended matter, bacteria, viruses, and chemicals. The degree of treatment is determined by the specific unit operations (e.g., filtration, disinfection, etc.) incorporated into the water treatment process.

Index